Algal Cultivation for Obtaining High-Value Products

Algal Cultivation for Obtaining High-Value Products

Guest Editors

Cecilia Faraloni
Eleftherios Touloupakis

Basel • Beijing • Wuhan • Barcelona • Belgrade • Novi Sad • Cluj • Manchester

Guest Editors

Cecilia Faraloni
Institute of BioEconomy
National Research Council
Sesto Fiorentino
Italy

Eleftherios Touloupakis
Research Institute on
Terrestrial Ecosystems
National Research Council
Sesto Fiorentino
Italy

Editorial Office
MDPI AG
Grosspeteranlage 5
4052 Basel, Switzerland

This is a reprint of the Special Issue, published open access by the journal *Marine Drugs* (ISSN 1660-3397), freely accessible at: https://www.mdpi.com/journal/marinedrugs/special_issues/W1D46N1945.

For citation purposes, cite each article independently as indicated on the article page online and as indicated below:

Lastname, A.A.; Lastname, B.B. Article Title. *Journal Name* **Year**, *Volume Number*, Page Range.

ISBN 978-3-7258-3553-9 (Hbk)
ISBN 978-3-7258-3554-6 (PDF)
https://doi.org/10.3390/books978-3-7258-3554-6

Contents

About the Editors

Cecilia Faraloni

Cecilia Faraloni is a permanent researcher at the Institute of BioEconomy of the National Research Council, Florence, Italy. She has a PhD on "Science of Drugs and Bioactive Substances", "Physiology of photosynthetic microorganisms, analysis and extraction of carotenoids with antioxidant activity", completed at Dipartimento di Chimica Bioorganica e Biofarmacia, University of Pisa, Italy.

Her main research areas concern the cultivation of microalgae and cyanobacteria for biotechnological applications and the study of physiology under stress conditions for the production of high-added-value biomass and hydrogen. This activity includes the scaling-up of cultivation at the laboratory scale in massive culture in open-air photobioreactors, focusing on the use of biomass as a bioresource (with applications in the food, nutraceutical, agronomic, cosmetic, and pharmacological sectors). She has authored 64 scientific publication and 13 book chapters.

Eleftherios Touloupakis

Eleftherios Touloupakis (BSc., PhD) is a permanent researcher at the Research Institute on Terrestrial Ecosystems of the National Research Council, Florence, Italy. He holds a PhD and a Master's degree in chemistry from the University of Crete, Greece. He has 25 years of laboratory experience in the fields of biochemistry and plant biotechnology. From 2011 to 2013, he worked as an assistant professor at the University of Crete, Greece. His research interests include the development of technologies for the mass cultivation of photosynthetic microorganisms for applications in the fields of renewable energy and bioplastics. He has authored 58 scientific publications, 8 book chapters, and 1 patent.

Editorial

Algal Cultivation for Obtaining High-Value Products

Cecilia Faraloni [1,*] and Eleftherios Touloupakis [2]

[1] Istituto per la Bioeconomia, Consiglio Nazionale delle Ricerche, Via Madonna del Piano 10, Sesto Fiorentino, 50019 Firenze, Italy

[2] Istituto di Ricerca sugli Ecosistemi Terrestri, Consiglio Nazionale delle Ricerche, Via Madonna del Piano 10, Sesto Fiorentino, 50019 Firenze, Italy; eleftherios.touloupakis@cnr.it

* Correspondence: cecilia.faraloni@cnr.it

Received: 27 February 2025
Accepted: 28 February 2025
Published: 28 February 2025

Citation: Faraloni, C.; Touloupakis, E. Algal Cultivation for Obtaining High-Value Products. *Mar. Drugs* 2025, 23, 107. https://doi.org/10.3390/md23030107

Interest in renewable biomass sources has increased due to global population growth, the growing need for sustainable resources, and a surge in consumer demand for natural ingredients driven by concerns regarding the harmful effects of synthetic chemicals, leading to a rise in the use of high-value products from natural sources in the fields of human health, food, cosmetics, and animal nutrition. Microalgae are considered an attractive solution to this problem because of their photosynthetic efficiency, the diversity of their metabolic pathways, and their ability to thrive in harsh conditions.

Over the last 50 years, scientists and engineers have paid great attention to microalgae due to the increasing demand for sustainable development [1]. By using sunlight for conversion into valuable bioproducts, microalgae can contribute significantly to creating links between the water, energy, food, and climate cycles [2]. Microalgae can withstand harsh conditions in natural environments and grow in a variety of water sources [3]. Several species have attracted much attention due to their high growth rate, high capacity for CO_2 sequestration, and lack of need for arable land [2,4–6].

Microalgae have evolved different strategies to survive under complex and extreme environmental conditions (high light, high salinity, extreme temperatures, nutrient deficiency, UV radiation) by adapting their metabolism. Various species of microalgae are capable of producing a large amount of secondary metabolites such as carotenoids, polyphenols, and essential oils, which have a wide range of therapeutic properties due to their antioxidant activity [7–12]. This peculiarity mainly depends on the species, strains, genetic diversity, and/or abiotic stress. Therefore, it is becoming increasingly important to understand the biology of these microorganisms in order to make informed decisions about their use in food, feeds, pharmaceuticals, and cosmetics. Numerous studies have been carried out to increase knowledge in this field and to optimize the recovery of natural antioxidant compounds under different growing conditions and with different stress factors.

The commercialization of microalgae as feedstock for natural products requires the use of efficient cultivation systems. These microorganisms can be cultivated using a variety of methods, including open and closed systems.

1. Advancements in Algal Cultivation Techniques

1.1. Design and Optimization of Photobioreactors

Photobioreactors provide controlled environments to optimize growth conditions and maximize productivity. Recent advances include the following:

- The development of novel photobioreactor geometries that improve light penetration and mixing efficiency.
- LED lighting for the precise control of light intensity and wavelength, tailored to specific microalgae species and target products.

- Automated control systems: the implementation of advanced sensors and control systems for real-time monitoring and adjustment of parameters such as temperature, pH, dissolved oxygen, and nutrient levels.

1.2. Nutrient Optimization

In order to optimize the growth of microalgae, their biomass, and the productivity of their by-products, a cultivation medium tailored to the specific needs of the desired microalgae species must be selected. At the same time, research into alternative and cost-effective cultivation media is essential to overcome economic barriers and thus facilitate the scaling up and commercialization of microalgae-based products.

1.3. Genetic Engineering and Strain Improvement

Metabolic and genetic engineering and synthetic biology are used to enhance target product synthesis, improve growth rate and stress tolerance, and reduce feedstock costs.

1.4. Downstream Processing

- Cell disruption techniques: efficient cell disruption is critical for releasing intracellular products.
- Extraction and purification: developing efficient and environmentally friendly extraction and purification techniques.

2. Challenges and Future Prospects

Despite significant progress, several challenges in the commercialization of microalgae-derived high-value products still remain, such as high production costs, challenges to scaling up, regulatory hurdles, and public perception.

The future of algal cultivation for high-value products is promising. Continued research and development efforts in the following areas are critical:

- Developing more efficient and cost-effective cultivation systems;
- Harnessing the power of synthetic biology;
- Developing sustainable and scalable downstream processing techniques;
- Addressing regulatory challenges and promoting public awareness.

This Special Issue, entitled "Algal Cultivation for Obtaining High-Value Products", focuses on the promotion of algae capable of producing high-value products, as well as cultivation technologies, strategies, and growth conditions that will lead to the proliferation of these compounds; techniques for the extraction and purification of these compounds and their potential applications are also explored. This Special Issue comprises ten original articles and a review. Below, we provide an overview of the research results and a review of the existing literature to help readers find suitable articles for their fields of interest. The contributions are listed in the List of Contributions.

Contribution 1 investigates the impact of salinity and brewery wastewater on the mixotrophic cultivation of *Arthrospira platensis*. This study aimed to clarify the specific effects of salinity on the production of phycocyanin by *A. platensis* under mixotrophic conditions in a continuous mode, and to evaluate the effects of seawater on biochemical composition in terms of lipids, proteins, and carbohydrates in view of its use in the food market.

In Contribution 2, Cao et al. used *Synechocystis* sp. PCC6803 as a microbial chassis for heme production. The objectives of their study were to construct a genetically engineered *Synechocystis* sp. PCC6803 with high heme production, to explore effective methods to increase its heterotrophic capacity, and to investigate the optimal glucose concentration that would promote its growth in order to increase heme production.

In Contribution 3, Yang et al. demonstrated an increase in lipid production in *Aurantiochytrium* sp. DECR-KO using a two-stage strategy.

Contribution 4 investigates the optimization of biomass and fucoxanthin production of *Isochrysis galbana*, isolated from the coast of Tadjoura (Djibouti), by testing various culture media.

Contribution 5 investigates the biological profiles of the *Chromochloris zofingiensis* mutant LUT-4 under different light intensities by linking the physiological properties and molecular characteristics to assess the potential of LUT-4 in producing lutein.

In Contribution 6, Yi et al. investigated the effects of calcium gluconate, magnesium gluconate, and bainengtai as additional supplements in a culture of *Porphyridium purpureum*.

Contribution 7 investigated the production of fucoxanthin and fatty acids in *Conticribra weissflogii*. The authors proved that the content of the active substance *C. weissflogii* can be increased by adjusting the iron concentration.

In Contribution 8, An et al. identified and analyzed an indigenous *Odontella aurita* strain isolated from the coastal waters of Sonyang-myeon, Republic of Korea. The study aimed to determine the optimal culture conditions, including temperature, salinity, irradiance, and nutrient concentration, that affect the growth of this species and to analyze the composition of fatty acids and carotenoid pigments to investigate its potential use in various industries.

Contribution 9 investigated scalable methods to produce a highly concentrated biomass of *Scenedesmus rubescens* under heterotrophic conditions. Their aim was to find an optimal culture medium for the cultivation of *S. rubescens* under heterotrophic conditions.

In Contribution 10, the authors review recent advances in genetic engineering and cultivation strategies to improve the production of lutein by microalgae. Techniques such as random mutagenesis; genetic engineering, including CRISPR technology; and multi-omics approaches are discussed in detail in terms of their impact on improving lutein production. Innovative cultivation strategies are compared, and their advantages and challenges are highlighted.

This Editorial's concluding remarks are as follows:

Microalgae provide a sustainable and versatile platform to produce a wide range of high-value products. Recent advances in algae cultivation techniques, including the development of photobioreactors, nutrient optimization, genetic engineering, and downstream processing, are paving the way for more efficient and cost-effective production. Addressing the remaining challenges and seizing opportunities in this area will be critical to realizing the full potential of algae as a source of sustainable and valuable resources. Continued innovation and collaboration between researchers, industry, and policy makers will be crucial to driving the commercialization of high-value algae products and contributing to a bio-based economy.

We would like to thank the Editorial Board, the Managing Editors, and the Editorial Assistant. We greatly appreciate the efforts of the authors who have contributed their results to this Special Issue. Finally, we thank the reviewers who carefully evaluated the submitted manuscripts for their support.

Conflicts of Interest: The authors declare no conflicts of interest.

List of Contributions

1. Russo, N.P.; Ballotta, M.; Usai, L.; Torre, S.; Giordano, M.; Fais, G.; Casula, M.; Dessì, D.; Nieri, P.; Damergi, E.; et al. Mixotrophic Cultivation of *Arthrospira platensis* (Spirulina) under Salt Stress: Effect on Biomass Composition, FAME Profile and Phycocyanin Content. *Mar. Drugs* **2024**, *22*, 381. https://doi.org/10.3390/md22090381.

2. Cao, K.; Sun, F.; Xin, Z.; Cao, Y.; Zhu, X.; Tian, H.; Cao, T.; Ma, J.; Mu, W.; Sun, J.; et al. Enhanced Production of High-Value Porphyrin Compound Heme by Metabolic Engineering Modification and Mixotrophic Cultivation of *Synechocystis* sp. PCC6803. *Mar. Drugs* **2024**, *22*, 378. https://doi.org/10.3390/md22090378.

3. Yang, X.; Wei, L.; Liang, S.; Wang, Z.; Li, S. Comparative Transcriptomic Analysis on the Effect of Sesamol on the Two-Stages Fermentation of *Aurantiochytrium* sp. for Enhancing DHA Accumulation. *Mar. Drugs* **2024**, *22*, 371. https://doi.org/10.3390/md22080371.

4. Mohamed Abdoul-Latif, F.; Ainane, A.; Achenani, L.; Merito Ali, A.; Mohamed, H.; Ali, A.; Jutur, P.P.; Ainane, T. Production of Fucoxanthin from Microalgae *Isochrysis galbana* of Djibouti: Optimization, Correlation with Antioxidant Potential, and Bioinformatics Approaches. *Mar. Drugs* **2024**, *22*, 358. https://doi.org/10.3390/md22080358.

5. Chen, Q.; Liu, M.; Mi, W.; Wan, D.; Song, G.; Huang, W.; Bi, Y. Light Intensity Enhances the Lutein Production in *Chromochloris zofingiensis* Mutant LUT-4. *Mar. Drugs* **2024**, *22*, 306. https://doi.org/10.3390/md22070306.

6. Yi, S.; Zhang, A.-H.; Huang, J.; Yao, T.; Feng, B.; Zhou, X.; Hu, Y.; Pan, M. Maximizing Polysaccharides and Phycoerythrin in *Porphyridium purpureum* via the Addition of Exogenous Compounds: A Response-Surface-Methodology Approach. *Mar. Drugs* **2024**, *22*, 138. https://doi.org/10.3390/md22030138.

7. Peng, K.; Amenorfenyo, D.K.; Rui, X.; Huang, X.; Li, C.; Li, F. Effect of Iron Concentration on the Co-Production of Fucoxanthin and Fatty Acids in *Conticribra weissflogii*. *Mar. Drugs* **2024**, *22*, 106. https://doi.org/10.3390/md22030106.

8. An, S.M.; Cho, K.; Kim, E.S.; Ki, H.; Choi, G.; Kang, N.S. Description and Characterization of the *Odontella aurita* OAOSH22, a Marine Diatom Rich in Eicosapentaenoic Acid and Fucoxanthin, Isolated from Osan Harbor, Korea. *Mar. Drugs* **2023**, *21*, 563. https://doi.org/10.3390/md21110563.

9. Santo, G.E.; Barros, A.; Costa, M.; Pereira, H.; Trovão, M.; Cardoso, H.; Carvalho, B.; Soares, M.; Correia, N.; Silva, J.T.; et al. *Scenedesmus rubescens* Heterotrophic Production Strategies for Added Value Biomass. *Mar. Drugs* **2023**, *21*, 411. https://doi.org/10.3390/md21070411.

10. Coleman, B.; Vereecke, E.; Van Laere, K.; Novoveska, L.; Robbens, J. Genetic Engineering and Innovative Cultivation Strategies for Enhancing the Lutein Production in Microalgae. *Mar. Drugs* **2024**, *22*, 329. https://doi.org/10.3390/md22080329.

References

1. Gallego, I.; Medic, N.; Pedersen, J.S.; Ramasamy, P.K.; Robbens, J.; Vereecke, E.; Romeis, J. The microalgal sector in Europe: Towards a sustainable bioeconomy. *New Biotechnol.* **2025**, *86*, 1–13. [CrossRef] [PubMed]

2. Wang, R.; Wang, X.; Zhu, T. Research progress and application of carbon sequestration in industrial flue gas by microalgae: A review. *J. Environ. Sci.* **2025**, *152*, 14–28. [CrossRef] [PubMed]

3. Patel, A.K.; Albarico, F.P.J.B.; Perumal, P.K.; Vadrale, A.P.; Ntan, C.T.; Chau, H.T.B.; Anwar, C.; Mud Wani, H.; Pal, A.; Saini, R.; et al. Algae as an emerging source of bioactive pigments. *Bioresour. Technol.* **2022**, *351*, 126910. [CrossRef] [PubMed]

4. Katiyar, R.; Banerjee, S.; Arora, A. Recent advances in the integrated biorefinery concept for the valorization of algal biomass through sustainable routes. *Biofuels Bioprod. Biorefining* **2021**, *15*, 879–898. [CrossRef]

5. Tounsi, L.; Ben Hlima, H.; Hentati, F.; Hentati, O.; Derbel, H.; Michaud, P.; Abdelkafi, S. Microalgae: A Promising Source of Bioactive Phycobiliproteins. *Mar. Drugs* **2023**, *21*, 440. [CrossRef] [PubMed]

6. Bhatnagar, P.; Gururani, P.; Joshi, S.; Singh, Y.P.; Vlaskin, M.S.; Kumar, V. Enhancing the bio-prospects of microalgal-derived bioactive compounds in food industry: A review. *Biomass Conv. Bioref.* **2024**, *14*, 23275–23291. [CrossRef]

7. Rea, G.; Antonacci, A.; Lambreva, M.; Pastorelli, S.; Tibuzzi, A.; Ferrari, S.; Fischer, D.; Johanningmeier, U.; Oleszek, W.; Doroszewska, T.; et al. Integrated plant biotechnologies applied to safer and healthier food production: The Nutra-Snack manufacturing chain. *Trends Food Sci. Technol.* **2011**, *22*, 353–366. [CrossRef]

8. Narayanasamy, A.; Patel, S.K.S.; Singh, N.; Rohit, M.V.; Lee, J.-K. Valorization of Algal Biomass to Produce Microbial Polyhydroxyalkanoates: Recent Updates, Challenges, and Perspectives. *Polymers* **2024**, *16*, 2227. [CrossRef] [PubMed]

9.   Kaur, M.; Bhatia, S.; Bagchi, D.; Tak, Y.; Kaur, G.; Kaur, C.; Kaur, A.; Sharma, N. Enhancing microalgal proteins for nutraceutical and functional food applications. *Future Foods* **2025**, *11*, 100564. [CrossRef]
10.   Menegol, T.; Soriano-Jerez, Y.; López-Rosales, L.; García-Camacho, F.; Contreras-Gómez, A.; Molina-Grima, E.; Rech, R.; Cerón-García, M.C. Effect of phosphorus and nitrogen levels and light emissions diodes illumination on growth and carotenoid synthesis in *Jaagichlorella luteoviridis*. Optimization of carotenoid extraction. *Algal Res.* **2025**, *86*, 103936. [CrossRef]
11.   Yu, Z.; Zhao, W.; Sun, H.; Mou, H.; Liu, J.; Yu, H.; Dai, L.; Kong, Q.; Yang, S. Phycocyanin from Microalgae: A Comprehensive Review Covering Microalgal Culture, Phycocyanin Sources and Stability. *Food Res. Int.* **2024**, *186*, 114362. [CrossRef] [PubMed]
12.   Chen, H.; Qi, H.; Xiong, P. Phycobiliproteins—A Family of Algae-Derived Biliproteins: Productions, Characterisation and Pharmaceutical Potentials. *Mar. Drugs* **2022**, *20*, 450. [CrossRef] [PubMed]

marine drugs

Article

Mixotrophic Cultivation of *Arthrospira platensis* (Spirulina) under Salt Stress: Effect on Biomass Composition, FAME Profile and Phycocyanin Content

Nicola Pio Russo [1], Marika Ballotta [1], Luca Usai [2], Serenella Torre [3], Maurizio Giordano [4], Giacomo Fais [5,6], Mattia Casula [5,6], Debora Dessì [7], Paola Nieri [3], Eya Damergi [8], Giovanni Antonio Lutzu [2,*] and Alessandro Concas [5,6,*]

[1] Department of Life Sciences, University of Modena and Reggio Emilia, Via Giuseppe Campi 287, 41123 Modena, MO, Italy; nicolarussopio01@gmail.com (N.P.R.); ballottamarika@gmail.com (M.B.)
[2] Teregroup Srl, Via David Livingstone 37, 41123 Modena, MO, Italy; luca.usai@teregrpoup.net
[3] Department of Pharmacy, University of Pisa, Via Bonanno Pisano 12, 56126 Pisa, PI, Italy; serenella.torre@phd.unipi.it (S.T.); paola.nieri@unipi.it (P.N.)
[4] Check Lab, Via Acquasanta 16, 84131 Salerno, SA, Italy; giordano.maurizio70@gmail.com
[5] Department of Mechanical, Chemical and Materials Engineering, University of Cagliari, Piazza d'Armi, 09123 Cagliari, CA, Italy; giacomo.fais@unica.it (G.F.); mattia.casula@unica.it (M.C.)
[6] Interdepartmental Center of Environmental Science and Engineering (CINSA), University of Cagliari, Via San Giorgio 12, 09124 Cagliari, CA, Italy
[7] Department of Life and Environmental Sciences, University of Cagliari, Cittadella Universitaria, Blocco A, SP8 Km 0.700, 09042 Monserrato, CA, Italy; deboradessi95@gmail.com
[8] Algaltek SARL, R&D Departments, Route de la Petite-Glane 26, 1566 Saint Aubin, FR, Switzerland; eya.damergi@algatek.com
* Correspondence: gianni.lutzu@teregroup.net (G.A.L.); alessandro.concas@unica.it (A.C.)

Abstract: *Arthrospira platensis* holds promise for biotechnological applications due to its rapid growth and ability to produce valuable bioactive compounds like phycocyanin (PC). This study explores the impact of salinity and brewery wastewater (BWW) on the mixotrophic cultivation of *A. platensis*. Utilizing BWW as an organic carbon source and seawater (SW) for salt stress, we aim to optimize PC production and biomass composition. Under mixotrophic conditions with 2% BWW and SW, *A. platensis* showed enhanced biomass productivity, reaching a maximum of 3.70 g L^{-1} and significant increases in PC concentration. This study also observed changes in biochemical composition, with elevated protein and carbohydrate levels under salt stress that mimics the use of seawater. Mixotrophic cultivation with BWW and SW also influenced the FAME profile, enhancing the content of C16:0 and C18:1 FAMES. The purity (EP of 1.15) and yield (100 mg g^{-1}) of PC were notably higher in mixotrophic cultures, indicating the potential for commercial applications in food, cosmetics, and pharmaceuticals. This research underscores the benefits of integrating the use of saline water with waste valorization in microalgae cultivation, promoting sustainability and economic efficiency in biotechnological processes.

Keywords: brewery wastewater; circular bio-economy; *Arthrospira platensis*; salt stress; Phycocianin; FAME profile

Citation: Russo, N.P.; Ballotta, M.; Usai, L.; Torre, S.; Giordano, M.; Fais, G.; Casula, M.; Dessì, D.; Nieri, P.; Damergi, E.; et al. Mixotrophic Cultivation of *Arthrospira platensis* (Spirulina) under Salt Stress: Effect on Biomass Composition, FAME Profile and Phycocyanin Content. *Mar. Drugs* **2024**, *22*, 381. https://doi.org/10.3390/md22090381

Academic Editors: Cecilia Faraloni and Eleftherios Touloupakis

Received: 30 July 2024
Revised: 20 August 2024
Accepted: 22 August 2024
Published: 24 August 2024

1. Introduction

Microalgae represent a promising frontier in biotechnology due to their rapid growth rates, high biomass productivity, and ability to produce a wide array of bioactive compounds [1,2]. These microorganisms can be exploited for various applications, such as biofuel production, wastewater (WW) treatment, CO_2 sequestration, and the generation of nutraceuticals and pharmaceuticals [3–5]. Their ability to thrive in various and harsh environmental conditions makes them versatile candidates for sustainable biotechnological processes [6,7]. Cyanobacteria can produce a large set of valuable compounds [8]. Among them, phycocyanin (PC) is a blue pigment-protein complex which has garnered significant interest due to its extensive applications in the food, cosmetic, and pharmaceutical industries [9]. Indeed, PC is of interest for its antioxidant, anti-inflammatory, and neuroprotective properties and is used for the production of a drug called phycocyanobilin [10]. The filamentous cyanobacterium *A. platensis* (universally known as *Spirulina*), beside its high protein content, productivity and resilience, is also renowned for accumulating high PC levels [11]. Among the various cultivation strategies employed to optimize biomass and metabolite production, mixotrophy stands out as an effective approach [12,13]. Mixotrophy combines autotrophic (photosynthesis) and heterotrophic (organic carbon utilization) modes, potentially enhancing the yield of PC by providing a flexible nutrient environment [14]. One of the most promising aspects of mixotrophic cultivation is the use of organic sources derived from food industry wastes to sustain growth [15]. Dairy wastewater (DWW), rich in lactose and other organic compounds, provides a viable nutrient source for *A. platensis*, promoting biomass and PC production while simultaneously addressing waste management issues [16]. Brewery wastewater (BWW), containing residual sugars and proteins, can similarly support mixotrophic growth, enhancing the economic and environmental sustainability of both the microalgae cultivation and brewing industries [17]. Molasses, a byproduct of sugar processing, offers another potent carbon source, rich in sucrose and other nutrients, that can be utilized to boost PC yield in *Spirulina* cultures [18,19]. Using these food industry wastes not only reduces the cost of cultivation by substituting expensive chemical nutrients but also mitigates environmental pollution by recycling organic waste streams. This synergy between waste valorization and microalgae cultivation promotes sustainability when producing strategic molecules such as PC [16]. Salinity is a crucial environmental factor that significantly influences the physiological and biochemical processes in *A. platensis* metabolism. It affects osmoregulation, enzyme activity, and the structural integrity of cellular components [20]. Under mixotrophic conditions, the interaction between salinity and nutrient availability can further complicate these effects. Salinity stress induces the synthesis of compatible solutes and stress proteins, which can alter the metabolic flux towards secondary metabolite production, including PC [21]. Studies have shown that moderate salinity levels can stimulate the production of PC by enhancing the expression of genes involved in its biosynthesis. However, excessive salinity may lead to osmotic stress, reducing cell viability and PC yield [22]. The balance between salinity levels and nutrient supply is thus critical in optimizing PC production under mixotrophic conditions [23]. Understanding how different waste-derived organic sources interact with salinity levels under mixotrophic conditions will be key to optimizing the sustainable production of PC. This study aims to elucidate the specific impacts of salinity levels on the production of PC by *A. platensis* under mixotrophic conditions in a continuous mode. A second aim is to assess the effects of seawater (SW) on the biochemical composition of *A. platensis* in terms of lipids, proteins, and carbohydrates in view of its use in the food market. To the best of our knowledge, the combined effect of SW and brewery wastewater (BWW) on the content and composition of fatty acids (FAs) is unknown. Therefore, BWW is used to guarantee mixotrophic conditions while cultivating *A. platensis* using SW to assure salt stress. By exploring the biochemical response of *A. platensis* to salinity under mixotrophy, this research seeks to identify optimal cultivation parameters that enhance biomass composition and PC yield.

2. Results

2.1. BWW Characterization

Table 1 presents the fundamental characteristics of the clarified BWW utilized in this work in comparison with those of other studies. It can be inferred that the loads of organic matter and N in the form of ammonium were consistent compared to other reported BWWs, with total organic carbon (TOC) > 11 g L^{-1} and NH$_4^+$ > 15 mg L^{-1}, respectively. It should be considered that, despite the fact that brewery effluents may exhibit different chemical compositions, based on the technological steps employed for manufacturing beer (mashing, boiling, fermentation and maturation, wort separation, wort clarification, and rough beer clarification), these effluents are usually characterized by the presence of easily biodegradable sugars, soluble starch, ethanol, and volatile fatty acids [24]. Their typical high organic content is confirmed by biological oxygen demand (BOD_5), chemical oxygen demand (COD), and a total suspended solids (TSS) range of 1.2–3.6 g L^{-1}, 2–6 g L^{-1}, and 0.2–1 g L^{-1}, respectively [25]. Usually, BWW is not toxic and is characterized by very low levels of heavy metal as well as N and P levels, which can vary depending on the handling of the raw material used for beer production. The BWW temperature can range from 18 to 40 °C, occasionally reaching values higher than this range [25]. The pH levels, ranging from 4.5 to 12, are highly dependent on the cleaning and sanitizing chemicals used [26]. A common procedure to allow microalgae growth inside a culture medium with a huge organic content, such as for instance a BWW, is its physical and chemical pre-treatment [24]. Due to several chemical and biochemical reactions and several solid–liquid separations performed for beer production, BWW pre-treatment is mandatory. This allows us to reduce and separate the large amounts and different varieties of wastes produced including water, spent grains and trub, hops, caustic and acid cleaners, surplus yeast, waste labels, and waste beer [27].

Table 1. Physiochemical characteristics of brewery wastewater.

Parameter	This Work	[26]	[28]	[29]	[30]	[31]
TSS	-	0.2–1	-	-	-	0.66–0.77
BOD	-	1.2–3.6	-	-	-	-
COD	-	2–6	2.1–5.8	-	2	2.1–3.54
TOC	11.2	25–80 *	-	16.5	-	-
TN	0.427	10–50 *	5 *	1.2	23 *	23–69 *
$N\text{-}NO_3^{2-}$	-	-	-	-	-	0.1–9 *
$N\text{-}NH_4^+$	26.55 *	-	0.004 *	11.3	18 *	18–48 *
TP	0.1 *	-	0.44 *	95.2 *	-	8.2–18.4 *
$P\text{-}PO_4^{3-}$	-	-	-	-	-	4–6 *
Fe	0.1 *	-	-	-	-	-
Mg	0.1 *	-	-	-	-	-
pH	6.7	4.5–1.2	6.9	6.1	-	7.66–7.89

Note: TSS = total suspended solid, BOD = biological oxygen demand, COD = chemical organic demand, TOC = total organic carbon. All the concentrations are expressed in terms of g L^{-1}, except those with * which are reported in terms of mg L^{-1}.

The choice of using very low concentrations of BWW (2% v v^{-1}) was dictated by the elevated TOC of the effluent (Table 1). Previous studies have suggested that the optimal BWW concentration for mixotrophic cultivation of microalgae is less than 3% v v^{-1}, with higher concentrations (>30% v v^{-1}) resulting in growth inhibition [32]. This was recently confirmed by Miotti et al. [33], who supplemented a mix of synthetic medium and BWW with lower organic carbon content in terms of glycerol, ranging from 0.2% to 2% v v^{-1}, to monitor the growth of *Chlorella vulgaris*.

Additionally, when assessing the N:P molar ratio for the BWW used in this study, considering N (nitrate and ammonia) and P (phosphate) values, it appears to be much lower than the N:P ratios of approximately 10:1 reported in Table 1 for other BWWs by other studies. Moreover, the N:P ratio is smaller than the Redfield ratio (N:P of 16:1), suggesting that the BWW serves as an N-limited medium for microalgae.

2.2. Growth Profile and Biomass Composition of A. platensis in BWW and SW

A. platensis was cultivated in both photoautotrophic and mixotrophic conditions utilizing brewery effluent (JB) and SW (JBS) at a concentration of 2% v v^{-1} for a duration of 15 days until the early stationary phase of growth was achieved. This investigation aimed to assess key kinetic parameters, including maximum biomass concentration (X_{max}), average biomass productivity (ΔX), doubling time (t_d), and specific growth rate (μ).

Figure 1 depicts the progression of *A. platensis* growth curves under both mixotrophic and photoautotrophic conditions. The lag phase persisted for 96 h in both JB and JBS systems while in the control there was no lag phase at all. This was primarily due to the time needed by *A. platensis* to adapt to the new growth conditions represented by the addition of the brewery and SW to the JM. Similar behavior was also reported when cultivating *A. platensis* in both freshwater and SW [34]. At the end of the batch phase, maximum biomass concentrations of 0.98 g L^{-1} and 0.90 g L^{-1} were obtained in freshwater and SW, respectively. In addition, the duration of the batch phase was affected since 9 and 12 days were required to reach the maximum concentration in the two systems.

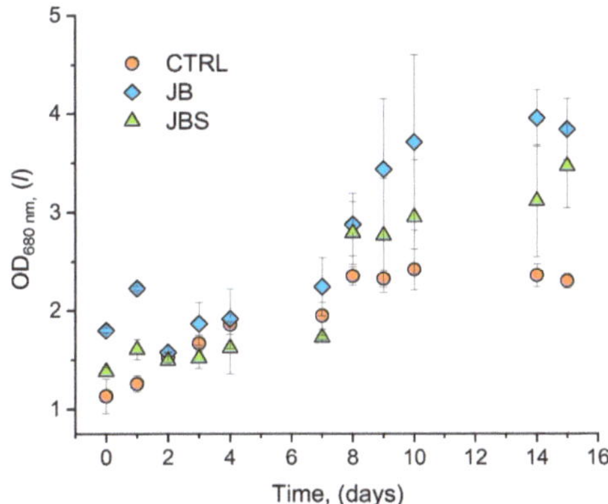

Figure 1. Kinetic evolution of optical density at 680 nm in the investigated cultures.

In our study, the longer adaptation phase was primarily due to the time needed by *A. platensis* to acclimate to the novel growth conditions represented by the addition of both BWW and SW to the JM. Acclimation is a key phase in cyanobacteria adaptation affecting the overall performance of the culture. Subsequently, the exponential growth phase in all samples persisted for up to 15 days, displaying varying patterns. Around the midpoint of the cultivation period (7 days), all systems exhibited an $OD_{680nm} > 1.5$. By the end of the cultivation period, the JB and JBS systems outperformed the control, with the JBS system still increasing OD values on the 16th day.

The maximum biomass production for JB and JBS cultures occurred at the end of the stationary growth phase, while the control was attained after 10 days. The control achieved a maximum concentration of biomass equal to 2.22 g L^{-1}, i.e., the lowest among the investigated mixotrophic conditions (Figure 2a).

Remarkably, the photoautotrophic X_{max} value was 2.5–5 times higher than the values (0.55–0.89 g L^{-1}) reported for *A. platensis* by Sassano et al. [35] for the same strain grown in continuous mode using NH_4Cl as N source. The mixotrophic cultivation of *S. platensis* in JBS exhibited the highest biomass yield of 3.70 g L^{-1}. In Figure 2b, it is apparent that the JB and JBS systems displayed a superior specific growth rate (0.083 and 0.073 day, respectively) compared to the control (0.055 day). Additionally, JB also showed the highest average biomass productivity at 302 mg L^{-1} day^{-1}. This represented about a 50% increase in biomass compared to the control (172 mg L^{-1} day^{-1}), as illustrated in Figure 2d. Mixotrophic cultures exhibit elevated growth rates compared to photoautotrophic and heterotrophic cultures. Their ability to assimilate both growth substrates, coupled with the advantage of performing photosynthesis, provides independence. This is advantageous because the acetyl-CoA pool is preserved for both CO_2 fixation through the Calvin cycle and extracellular organic carbon production [36]. The suitability of BWW to boost mixotrophic metabolism in microalgae and cyanobacteria has been addressed in many scientific reports. Miotti et al. cultivated *Chlorella vulgaris* in BWW with varying concentrations of glycerol under both autotrophic, heterotrophic, and mixotrophic conditions [33]. When *C. vulgaris* was cultivated under mixotrophic conditions, it exhibited a significantly higher biomass yield (1.33 g L^{-1}) compared to the autotrophic cultivation (1.08 g L^{-1}). The FAME profile analysis revealed that, in comparison to the autotrophic control, higher PUFA contents were obtained under mixotrophy within the range of glycerol mL tested. Similarly, our study observed that mixotrophic cultivation of *A. platensis* in JBS yielded the highest biomass concentration (3.70 g L^{-1}), which significantly surpassed the control. This outcome underscores the potential of mixotrophic systems to enhance biomass productivity, corroborating Miotti et al.'s findings. Moreover, their observation of higher PUFA content under mixotrophic conditions aligns with the increased biomass productivity and superior specific growth rates we reported in the JB and JBS systems. BWW allowed the vigorous growth of a filamentous microalga, *Tribonema aequale*, to a final biomass concentration as high as 6.45 g L^{-1}.

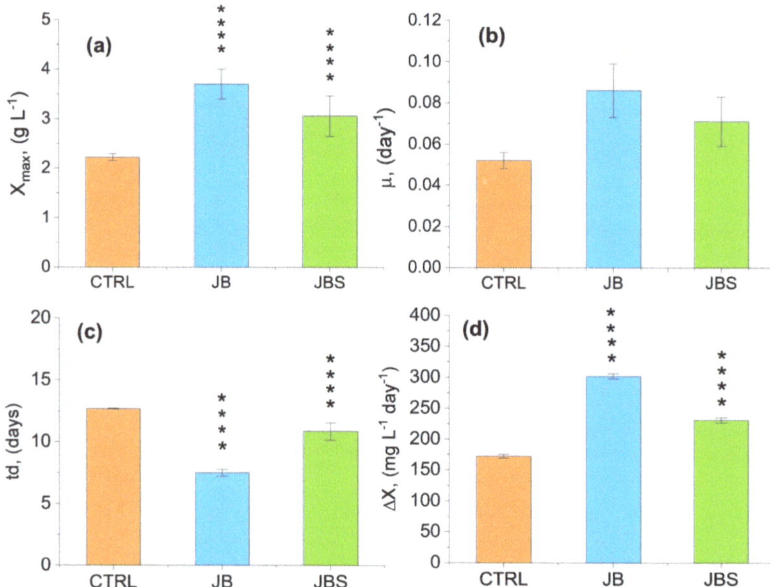

Figure 2. Comparison of growth performance indicators including maximum biomass content X_{max} (**a**), specific growth rate μ (**b**), doubling time t_d (**c**), and average biomass productivity ΔX (**d**) for the three investigated media. Mean differences were compared using 2-way ANOVA ($n = 3$, **** indicates $p < 0.0001$).

The resulting algal biomass obtained under mixotrophy contained high amounts of high-added value products such as chrysolaminarin, palmitoleic acid, and eicosapentaenoic acid (EPA) [31]. *A. platensis* was also successfully cultivated in non-diluted and non-pretreated BWW under non-sterile and alkaline growth conditions. The mixotrophic growth of this cyanobacterium was evaluated in terms of pollutant consumption, biomass productivity, and composition as well as pigment production. It was obtained that the combination of sodium chloride with sodium bicarbonate determined a maximum final pigment production. Moreover, the use of a photoperiod of 16:8 h caused an increase in the pollutant removal rate (up to 90% of initial concentrations) and biomass concentration (950 mg L^{-1}) [37]. Their findings, in terms of biomass productivity, pigment production, and pollutant removal, are indeed noteworthy. In comparison, our research, which utilized a diluted BWW blend, also demonstrated significant improvements in biomass production and μ (Figure 2b) under mixotrophic conditions. However, the use of non-diluted BWW in their study suggests the possibility of further optimizing our culture conditions to potentially achieve even higher yields. This comparison highlights the importance of exploring a range of BWW concentrations to balance the benefits of increased organic load with the risks of contamination and growth inhibition.

The effect of mixotrophy on the biomass composition of *A. platensis* in terms of macronutrients, such as total carbohydrates (TC), total proteins (TP), and total lipids (TP), is summarized in Figure 3.

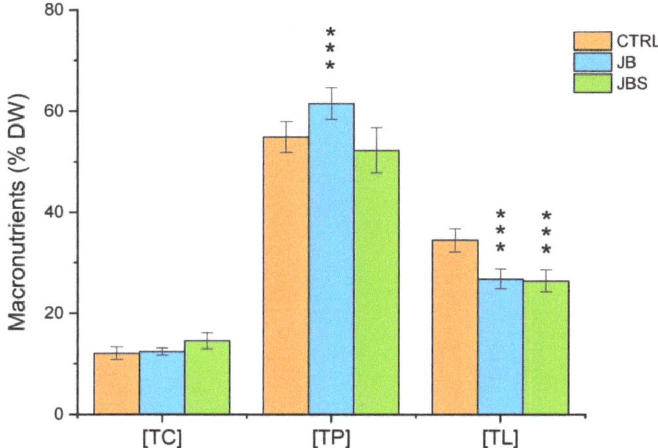

Figure 3. Total carbohydrates (TC), lipids (TL), and proteins (TP) obtained with the investigated media. Mean differences were compared using Tukey's test ($n = 3$, *** indicates $p < 0.001$).

TP represented the main fraction, followed by TL and TC, both under photoautothropy and mixotrophy. These macronutrients' distribution, characterized by a remarkable fraction of lipids, differs significantly from a typical chemical composition reported for *A. platensis* which consists of 15–25% carbohydrates, 55–70% proteins, and 4–7% lipids according to Bezerra et al. [38]. Batch or continuous growth impacts not only microalgae growth and biomass productivities but also their biochemical composition [39]. Compared to photoautothropy, there was a considerable rearrangement of proteins and lipids under mixotrophy, with an increase in TP and a decrease in TL after BWW addition. The addition of SW to the JB under mixotrophy produced a considerable decrease in TP with an enhancement of TC fraction.

The effect of salt on the biochemical composition of biomass is not well understood. Some studies have demonstrated a direct relationship between cultivating *A. platensis* in SW and a reduction in protein content and FAs [34] along with an increased accumulation of carbohydrates [39], essential amino acids including valine, leucine, and isoleucine as well as carotenoids [34]. In another study, a direct relationship between increased salinity and the enhancement of lipid fraction was demonstrated in *Spirulina* [40]. These findings are corroborated by this study in terms of TP, TL, and TC modifications. Salinity is known to influence the morphology of trichomes and the biochemical composition of biomass, making it a potential method for manipulating the morphological and biochemical properties of produced biomass. These effects are obtained after cell acclimation to increasing salinity. However, there are only a few studies on the impact of salt stress after cell acclimation on the biochemical composition and the growth rate of *A. platensis* [21]. In mixotrophic cultivation, several approaches can be applied to improve the production of high-value metabolites. Among them, salinity (along with pH and nutrient availability) as well as the selection of the operation mode (batch, fed-batch, and continuous cultivation) are regarded as valid bioengineering approaches that can be used to optimize microalgae biomass production and composition [41]. It should be noted that not all the microalgae strains adapt to the high salinity levels typical of SW. However, *A. platensis* is known to thrive at the SW salinity level (35 g L^{-1}), and the strain used in this study exhibited a tolerance of up to 41 g L^{-1}. Studies confirm that biomass production can be enhanced in salinity up to 40 g L^{-1} NaCl, while at 60 g L^{-1} NaCl biomass production decreases slightly [21]. Other authors demonstrated that it is possible to produce *A. platensis* biomass using SW after an adaptation period producing a biomass suitable for food applications [34].

2.3. Phycobiliproteins Production by A. platensis under Mixotrophy and Salt Stress

Figure 4 depicts the concentration of PC, APC, PE, and total phycobiliproteins (PBPs) in extracts from *A. platensis* biomass cultivated both under photoautotrophic and mixotrophic conditions with the addition of BWW (JB) and SW (JBS).

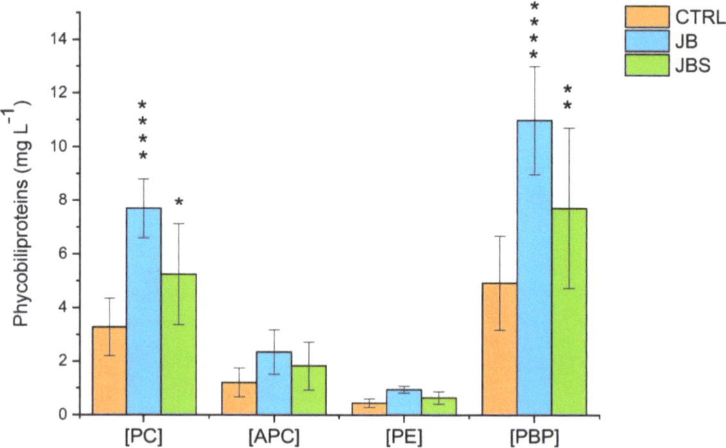

Figure 4. Phycobiliproteins concentration in extracts of *A. platensis* grown in the investigated media. Mean differences were compared using Tukey's test ($n = 3$, * $p < 0.05$; ** $p < 0.01$; **** indicates $p < 0.0001$).

The analysis revealed that pigment concentrations, especially under mixotrophic conditions with BWW, were particularly relevant. Under these conditions, *A. platensis* showed a notable increase in PC production, more than doubling the concentration in JB and nearly doubling it in JBS when compared to the control (7.70 mg L^{-1} and 5.25 mg L^{-1} vs. 3.28 mg L^{-1}). This increase highlights the beneficial effects of mixotrophy on both microalgal growth rates and pigment synthesis, with notably higher levels of PC and APC compared to the control.

Previous research suggests that the increase in PC production is influenced by a complex interaction between culture medium composition, the presence of an organic carbon source, and the physiological responses of microalgae to specific culture conditions, which create a stress environment favorable for PC synthesis.

In this study, it was evident that mixotrophic cultures had a significant increase in overall PBP concentration compared to the photoautotrophic control. Specifically, JB showed the highest total PC content at 7.70 mg L^{-1}, confirming the effectiveness of BWW supplementation in enhancing pigment synthesis. Similarly, APC and PE concentrations increased significantly, reaching 2.35 mg L^{-1} and 0.95 mg L^{-1}, respectively, compared to the control values of 1.21 mg L^{-1} and 0.45 mg L^{-1}.

These findings are consistent with the trend observed during the mixotrophic growth of *A. platensis* using cheese whey as an organic carbon source. In this case, the positive correlation between organic load and increased PBP content was demonstrated in an upscaled volume process [16]. Similarly, higher PC content compared to the conventional control Zarrouk medium was reported for *A. platensis* cultivated mixotrophically in tofu WW [42], and for the microalga *Galdieria sulphuraria* ACUF 064 grown in media containing buttermilk as a carbon source [43]. Regarding the correlation between salinity and PBP production, the addition of SW to JB resulted in a slight increase in PC, APC, and PE. These results are consistent with documented behavior in *A. platensis*, which exhibited increased protein and PC content when different NaCl concentrations were added [21,40].

The suitability of PC for various applications largely depends on its purity, which is typically assessed using an absorbance ratio. This ratio measures PC's absorbance at 620 nm (A620) against that of other proteins at 280 nm (A280). A PC with an A620/A280 ratio, defined as extraction purity (EP), of 0.7 or higher is classified as food grade, making it apt for use as a food additive or a natural blue pigment in cosmetics. When EP falls between 0.7 and 3.9, PC is considered a reagent grade, with EP values of 1.5 or higher suitable for cosmetic applications. An EP of 4 or more qualifies PC as an analytical grade, which is suitable for pharmaceutical uses [44].

In the current study, PC purity under mixotrophy conditions was found to have an EP of 1.15 with JB and 0.92 with JBS, compared to a control of 0.77 (Figure 5a). Correspondingly, the PC yields were 100, 50, and 35 (mg g^{-1}) for JB, JBS, and CTRL cultures, respectively (Figure 5b). These findings are consistent with previous research by Khandual et al. [45] on *A. platensis* and by Chini Zittelli et al. [46] on various strains, which demonstrate that an organic source (such as dairy WW or BWW) in concentrations of 0.5–2% v v^{-1} can enhance PC synthesis in mixotrophic cultures of *A. platensis*.

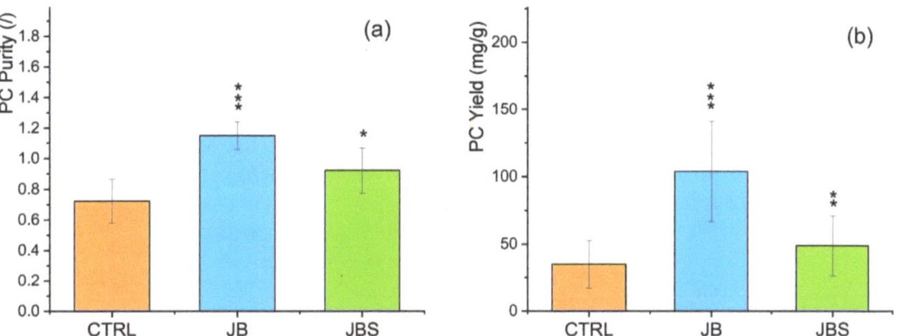

Figure 5. Purity (**a**) and yield (**b**) of PC in the investigated media. Mean differences were compared using 2-way Tukey's test ($n = 6$, * $p < 0.05$; ** $p < 0.01$; *** indicates $p < 0.001$).

PC purity is closely linked to the extraction methods used, which are influenced by physical and chemical parameters such as temperature, pH, solvent type, biomass-to-solvent ratio, and the form of biomass (dried or fresh). The commercial value of PC varies greatly depending on its purity level. According to the literature, the cost of PC increases with its purity, particularly in industries like cosmetics, agro-chemistry, and food [47]. For instance, analytical-grade PC with a purity exceeding 4 can cost up to 60 US$ g^{-1}, and for the highest purity levels, prices can soar up to 19,500 US$ g^{-1}. Recent studies indicated that PC for biocolorant use costs around 0.35 US$ g^{-1}, while analytical-grade PC can reach up to 4500 US$ g^{-1} [48]. The global PC market is projected to grow to \$245.5 million by 2027 and \$279.6 million by 2030 [49]. These high prices are due to the challenges in the extraction and purification processes, making PC an expensive protein pigment. Overall, the findings from this study on PC yield and purity offer new insights into developing algal cultivation strategies that maximize the production of bioactive components while promoting sustainable and eco-friendly practices.

The increased yield of PC observed in JB and JBS compared to the control may be due to the cellular stress from the abrupt shift from autotrophic to mixotrophic conditions and the high turbidity and salinity of BWW. BWW is known for its turbidity and salinity [50]. When microalgae are transferred to a more opaque medium, lower light intensity might inhibit photosynthesis, promoting cells to overexpress light-harvesting pigments like PC to capture light more efficiently [51].

Conversely, high light intensity can downregulate PC content to avoid excessive electron formation and oxidative damage [52]. High PC content in some strains is often observed under very low light intensity [53,54]. Using continuous cultivation modes can mitigate excessive turbidity, enhancing medium transparency and thereby potentially increasing PC content [55]. Therefore, the increased PC content observed when passing from the control to BWW and SW-supplemented media could be ascribed to this phenomenon. Additionally, osmotic stress can lead specific microalgae strains to boost PBP production [22]. High salinity may cause rapid sodium ion absorption, leading to phycobilisome separation from thylakoid membranes [56]. Cells responding to increased salinity may overregulate PC production to counteract this effect. Thus, the high salinity of BWW likely contributed to the increased PC observed in microalgae cultivated under mixotrophic conditions in JB and especially in JBS.

2.4. FAME Profile by A. platensis under Mixotrophy

The fatty acid methyl ester (FAME) profile of *A. platensis* under mixotrophic conditions with (JBS) and without (JB) salt stress, compared to the control (CTRL), is shown in Table 2. Overall, there were no significant differences in the FAME profile between the photoautotrophic and mixotrophic cultures systems, except for certain FAs. Specifically, myristic acid (C14:0), palmitic acid (C16:0), and oleic acid (C18:1) exhibited higher levels under mixotrophy, whereas stearic acid (C18:0), linoleic acid (C18:2), and y-linolenic acid *n*-6 (C18:3) had lower values. In CTRL, the most abundant FA was C16:0 (40.57%), followed by C18:3 (15.59%), C18:2 (15.18%), and stearic acid C18:0 (15.6%). In the JB group, C16:0 was also the most prevalent (43.77%), with C18:0 (12.73%), C18:2 (12.47%), and C18:3 *n*-6 (12.27%). For the JBS group, the sequence of FAs' predominance resulted in C16:0 (41.48%) > C18:2 (14.68%) > C18:3 *n*-6 (14.57%) > C18:0 (13.43).

Table 2. Effect of different growth media on FAME composition. Values are reported as mean % $\pm$ standard deviation (n = 6). The percentages refer to the total weight of FAMEs produced.

FAME	C:N $	CTRL	JB	JBS
Myristic acid	14:0	1.98 ± 0.30	3.14 ± 0.40 ****	2.70 ± 0.91 **
Palmitic acid	16:0	40.57 ± 2.97	43.77 ± 4.00	41.48 ± 4.48
Hexadecenoic acid	16:1	4.12 ± 0.62 ****	3.24 ± 0.80	3.20 ± 0.45 ****
Heptadecenoic acid	17:0	0.19 ± 0.06	0.27 ± 0.06	0.18 ± 0.04
cis-10-Heptadecanoic acid	17:1	0.56 ± 0.10	0.61 ± 0.14	0.54 ± 0.09
Stearic acid	18:0	15.16 ± 1.69	12.73 ± 6.14	13.43 ± 4.59
Elaidic acid	18:1	0.71 ± 0.09	0.80 ± 0.34	0.68 ± 0.15
Oleic acid	18:1	3.50 ± 1.65	7.85 ± 2.09	6.49 ± 2.63 *
Linoleic acid	18:2	15.18 ± 1.32	12.47 ± 1.14	14.68 ± 1.61
α-Linolenic acid	18:3 (ω-3)	1.19 ± 0.21	1.43 ± 0.43	0.75 ± 0.51
y-Linolenic acid	18:3 (ω-6)	15.59 ± 1.49	12.27 ± 2.57	14.57 ± 2.30
8,11,14-Eicosatrienoic acid	20:3	0.87 ± 0.10	1.04 ± 0.08	0.89 ± 0.14
13-Docosenoic acid	22:1	0.39 ± 0.05	0.38 ± 0.09	0.42 ± 0.08
Σ SFAs	/	57.84	59.91	57.79
Σ UFAs	/	42.11	40.09	42.22
Σ MUFAs	/	9.28	12.88	11.33
Σ PUFAs	/	32.83	27.21	30.89
PUFA:SFA	/	0.57	0.45	0.53
C16-C18	/	6.02	94.56	95.28
h/H	/	0.85	0.74	0.84

Note: $, which represents C:N, refers to number of atom carbon (C) and double bonds (N). Mean differences were tested using two-way ANOVA. No asterisk denotes p-value > 0.05; * p-value < 0.05; ** p-value < 0.01; **** p-value < 0.0001.

The C16:0 was the predominant FA across all the cultivation systems. The total percentages of C16–C18 FAs in *A. platensis* showed minor differences between photoautotrophic (96.02%) and mixotrophic conditions (94.56% and 95.28%). These findings are consistent with the study by Cavallini et al. [16], who observed a similar FA distribution pattern in *A. platensis* cultivated under both autotrophy and mixotrophy conditions with cheese whey supplementation. This study also found no significant changes in the levels of saturated (SFAs), monounsaturated (MUFAs), and polyunsaturated fatty acids (PUFAs) when *A. platensis* was grown under photoautotrophy (CTRL) and mixotrophy with salt stress (JBS). However, mixotrophy without salt stress (JB) resulted in increased levels of SFAs and MUFAs and a decrease in the total unsaturated fatty acids (UFAs) and PUFAs.

While C16:0 is an important energy source in infant nutrition, given its presence in breast milk at levels between 20 and 30%, heightened levels of free SFAs, including C16:0 and, to a lesser extent, C18:0, are correlated with an elevated risk of cardiovascular disease in adults [57]. This association is primarily due to vascular endothelial dysfunctions linked to oxidative stress, which arises from increased mitochondrial uncoupling [58]. Medical and pharmaceutical research highlights the importance of maintaining normal or non-elevated levels of C16:0 and C18:0 within the physiological range to prevent oxidative stress in the vascular endothelium. The FAME profile observed in this study is like that of olive oil, containing beneficial MUFAs known for their positive effects in preventing cardiovascular disease and addressing various risk factors. Therefore, it can be hypothesized that regular consumption of *A. platensis* biomass, where the levels of these two FAs do not exceed physiological limits, as seen in CTRL, JB, and JBS, could play a significant role in protecting cells against endothelial dysfunctions [58].

The α-linolenic acid (ALA, C18:3 ω-3) and γ-linolenic acid (GLA, C18:3 ω-6) are two PUFAs commonly found in microalgal and cyanobacterial oils [59]. Consistent with our results, various scientific reports have shown that GLA is the predominant isomer of this FA synthesized by *A. platensis* [60]. A recent review by [61] discussed several culture conditions (temperature, light intensity, nitrogen cell concentration, growth phase, and light/dark cycle) that enhance lipids and GLA production in *Spirulina*. It was noted that both lipid content and GLA levels are higher under mixotrophy compared to autotrophic conditions. However, our study did not align with this finding, as the addition of BWW to the culture medium reduced GLA content from 15.59% under photoautotrophy (CTRL) to 12.27% under mixotrophy (JB), with the addition of salt to JB only slightly decreasing the GLA content to 14.57%.

The content of ALA in *A. platensis* is generally negligible compared to GLA. In our study, ALA content increased significantly with the addition of BWW, rising from 1.19% in the CTRL to 1.43%, while the inclusion of salt led to a substantial decrease to 0.75%. ALA plays a critical biological role, particularly as a precursor for the synthesis of eicosapentaenoic acid C20:5 ω-3 (EPA) and docosahexaenoic acid C22:6 ω-3 (DHA). ALA is one of two essential FAs that humans must obtain through their diet for optimal health because it cannot be synthesized internally. The human body must convert ALA into EPA and DHA to harness its biological benefits, essential for the proper functioning of vital organs. This conversion process, although not highly efficient, allows the body to utilize 5% to 10% of the consumed ALA, with approximately 7% converting into EPA and from this, about 1% further converting into DHA [62].

The ratio of PUFA to SFA is a crucial parameter in assessing the nutritional quality of foods and their impact on cardiovascular health. PUFAs are known to lower low-density lipoprotein cholesterol (LDL-C) and overall serum cholesterol levels, whereas SFAs generally increase cholesterol levels. A higher PUFA:SFA ratio is thus considered beneficial. The British Department of Health recommends a PUFA:SFA ratio exceeding 0.45 in the human diet, and WHO/FAO guidelines suggest a balanced diet should have a PUFA:SFA ratio above 0.4, as the higher ratio is believed to mitigate the risk of cardiovascular and other chronic diseases [63]. In this study, the observed ratio exceeded 0.4 for both *A. platensis* cultivated under both autotrophy (0.85) and mixotrophy with (0.84) and without (0.74) the addition of salt.

The ratio of Σ hypocholesterolemic FAs/Σ hypercholesterolemic FAs (h/H) is another important factor in cholesterol metabolism. Higher h/H values are considered more favorable for human health, as they may accurately reflect the impact on cardiovascular disease. In our study, *A. platensis* showed an h/H ratio of 0.85 under photoautotrophic conditions and 0.74 under mixotrophic conditions with the addition of BWW to JM. When salt was added to JB, the value reverted to that of the control at 0.84. These h/H ratios under mixotrophy were higher than those (0.60–0.66) reported for *Spirulina* sp. by other authors [64,65]. However, various freshwater microalgae strains such as *Scenedesmus obliquus* (2.9), *Chlorella vulgaris* (2.8–2.04), *Porphyridium cruentum* (1.90), *Phaeodactylum tricornutum* (1.8–1.72), *Chlorococcum amblystomatis* (1.7), *Nannochloropsis oceanica* (1.7), *Nannochloropsis oculata* (1.44), and *Tetraselmis chui* (1.4) exhibited higher values [64,65]. It should be noted that variations in oil recovery methods (including the solvents used) from microalgae cells are not consistently specified in the literature.

3. Materials and Methods

3.1. Inoculums and Culture Media Preparation

The strain *Arthrospira platensis* SAG 21.99 used in this study was sourced not axenic from the culture collection of algae at the Gottingen University, Germany. The cell cultures were maintained and grown in a modified version of the Jourdan Medium (JM), composed as follows: in g L^{-1}: 5 $NaHCO_3$; 1.6 KOH, 5 $NaNO_3$; 0.027 $CaCl_2$ * $2H_2O$; 0.4 K_2SO_4, 2 K_2HPO_4; 1 NaCl; 0.4 $MgSO_4$ * $7H_2O$; 0.16 EDTA-Na_2; 0.01 $FeSO_4$ * $7H_2O$; and 1 mL of Trace elements. The Trace elements solution was prepared with the following composition (mg L^{-1}): 250 EDTA-Na_2; 57 H_3BO_3; 110 $ZnSO_4$ * $7H_2O$; 25.3 $MnCl_2$ * $4H_2O$; 8.05 $CoCl_2$ * $6H_2O$; 7.85 $CuSO_4$ * $5H_2O$; and 5.5 $Mo_7O_{24}(NH_4)_6$ * $4H_2O$. The original version of JM is detailed in Jourdan. Erlenmeyer flasks of 150 mL were filled with 50 mL of JM medium, inoculated with approximately 10 mL of microalgae, sealed with a cotton plug, and continuously illuminated at room temperature by white fluorescent lamps (Model T8 36 W IP20, CMI, Ense-Höingen, Germany) providing a light intensity of 50 μmol m^{-2} s^{-1}, measured with a luxmeter (Model HD 2302.0, Delta OHM, Padua, Italy). The inoculum was cultivated for about one week, until the end of exponential growth phase, before being used for experiments. BWW samples were collected from "Birrificio Emiliano", a brewery in Anzola, BO, Italy. The main chemical–physical parameters of these effluents are presented in Table 1.

Once collected, BWW was stored at 4 °C until use. It was then filtered using glass filter microfiber disks (GF/CTM 47 mm diameter, Whatman, Incofar Srl, Modena, MO, Italy), to remove solid materials, and sterilized at 121 °C and 0.1 MPa for 20 min before microalgae cultivation. Seawater, collected from the Adriatic Sea (N 42°67′07.58, E 14°02′43.14), was also sterilized at 121 °C and 0.1 MPa for 20 min.

3.2. Cultivation Conditions and Experimental Setup

A. platensis was cultivated in 20 L cylindrical PVC reactors (hereafter referred to as PBRs) with an outer diameter of 16 cm, inner diameter of 14 cm, and a height of 27 cm. Filtered compressed air, provided by an air pump (GIS Air Compressor, Carpi, MO, Italy), was supplied to the PBRs through a perforated rubber. PBRs were illuminated with a photoperiod of 12 h light/12 h dark at room temperature by four LEDs (model 2835, 30 W IP33, ImportLed, Desio, MB, Italy) providing a warm white light with an intensity of 150 μmol $m^{-2}s^{-1}$. A series of experiments were conducted with a working volume of 3 L to evaluate cell growth, biomass composition, FAME profile, and phycobiliproteins' (PBPs) production of *A. platensis*. In these experiments, JM was used as control (CTRL). In one setup, only BWW was added to JM (JB), and in another, both BWW and SW were added to JM (JBS). The PBRs operated in continuous mode with a daily removal and addition of specific percentages of JM, BWW, and SW according to the experiments setup detailed in Table 3. All tests were carried out at least in triplicate and lasted for 15 days. Microalgae growth was monitored by measuring optical density and biomass concentration. After cultivation, the final dry weight (g L^{-1}) and PBP content (mg g^{-1}_{DW}) were determined. In all experiments, the initial concentration of the inoculum was 0.3 g L^{-1}.

Table 3. Experimental setup showing the % (v v^{-1}) of BWW and SW added to the control.

	JM	BWW	SW
CTRL	2.5	-	-
JB	2.5	2	-
JBS	2.5	2	2

Note: JM = Jourdan medium, BWW = brewery wastewater, SW = seawater.

3.3. Cell Growth and Dry Weight Determination

A. platensis growth was evaluated by monitoring the absorbance (ABS) of the culture at 680 nm using a spectrophotometer (model ONDA V30 SCAN—UV VIS, ZetaLab, Padua, Italy). A regression equation was calculated to describe the relationship between dried biomass concentration and ABS. Dry biomass concentration was evaluated gravimetrically as follows: (a) a known volume (10 mL) of culture (V) was drawn from the PBRs, (b) the sample was filtered through a pre-weighted (W_1) glass microfiber filter (GF/C^{TM} 55 mm diameter, Whatman, Incofar Srl., Modena, Italy), the biomass retained on the filter was dried at 105 °C overnight to a constant weight (W_2), and (c) the filter paper had been previously dried in a forced-air oven (model 30, Memmert Gmbh, Scwabach, Germany) at 105 °C for 2 h, then cooled to room temperature in a desiccator, and weighed using an analytical scale (model M, Bel Engineering Srl, Monza, MI, Italy). The cell concentration (dry weight), X (g L^{-1}) was calculated using the following equation:

$$X = \frac{W_1 - W_2}{V} \tag{1}$$

where W = weight (g) of dried algal biomass, and V = volume (L) of the algae culture used for the test. The average biomass productivity (ΔX) was expressed as follows:

$$\Delta X = \frac{X_{max}}{t_{max}} \tag{2}$$

where X_{max} is the maximum biomass (g L^{-1}) obtained at (t_{max}). The specific growth rate (μ) was calculated according to the following equation:

$$\mu = \frac{ln(X_2 - X_1)}{(t_2 - t_1)} \tag{3}$$

where X_2 and X_1 correspond to dry biomass concentration (g L^{-1}) at time t_2 and t_1, respectively.

The pH of culture suspensions was measured by a pHmeter (model HI 2210, Hanna Instruments, Woonsocket, RI, USA).

3.4. Phycobilinproteins Extraction and Spectrophotometric Determination

The extraction of PBPs was conducted using an aqueous saline solution as described by [66]. Specifically, a known amount (10 g) of frozen *A. platensis* biomass was placed in 50 mL of a buffer solution consisting of water with 10 g L^{-1} of calcium dichloride (1%), then frozen and thawed repeatedly until the cells were completely broken up. This mixture was stirred for 30 to 45 min. This extraction process was performed twice, and the resulting phycobilins solution was separated by centrifugation at 8000 rpm for 10–15 min. The blue-colored supernatant was recovered and then used for optical readings with a spectrophotometer. The concentration of different PBPs, including C-phycocyanin (*PC*), allophycocyanin (*APC*), and phycoerythrin (*PE*), were determined by measuring the absorbance of each extract at three different wavelengths: 565, 620, and 650 nm.

The concentration of these *PBPs* was then calculated using the equations established by [67].

$$[PC] = \frac{A_{620} - 0.72\, A_{652}}{6.29} \tag{4}$$

$$[APC] = \frac{A_{652} - 0.191\,A_{620}}{5.79} \tag{5}$$

$$[PE] = \frac{A_{565} - 2.41 * [PC] - 1.40\,[APC]}{13.02} \tag{6}$$

The total *PBP* concentration was evaluated as the sum of *PE*, *PC*, and *APC* (mg mL^{-1}) of the collected supernatant as follows:

$$[PBPs] = [PC] + [APC] + [PE] \tag{7}$$

The extraction yield, estimated by relating the concentrations to the biomass of *A. platensis* used (expressed in mg of dry weight), was obtained according to Equation (8):

$$PBP\left(\frac{mg}{g}\right) = \frac{[PBPs]\,V_{ext}}{W_b.0.1} \tag{8}$$

In this equation, where V_{ext} (mL) represents the volume of the extract and W_b (g) denotes the weight of the wet biomass subjected to the extraction process, it is assumed based on experimental observations and literature data that the biomass pellet subjected to extraction contains 10% solid content.

The purity of PC was calculated according to the following equation:

$$PC\ purity\ (/) = \frac{A_{620}}{A_{280}} \tag{9}$$

3.5. FAMEs Determination

The fatty acids methyl ester (FAME) profile analysis in this study followed the method proposed by [68]. The 10 mg of biomass was weighed using a tube of glass after lyophilization. Subsequently, the biomass was suspended in a suitable volume of a methanol/chloroform (4:5 v v^{-1}) mixture along with 50 mg L^{-1} of the internal standard tritridecanoin (TAG 39:0, 13:0/13:0/13:0). The samples underwent eight rounds of vortexing for 60 s each, followed by sonication using an ultrasonic bath for 15 min at 5 °C. Subsequently, 2.5 mL of MilliQ water containing Amino-2-hydroxymethyl-propane-1,3-diol (Tris) (50 mg L^{-1}) and NaCl (1 M), with a pH adjusted to 7.0 using an HCl solution, were added. The solutions underwent sonication for 10 min at 5 °C. Subsequently, the samples were centrifuged for 10 min at 177 rcf at 5 °C and the chloroform phase was carefully transferred into a glass tube. The remaining samples were subjected to re-extraction with 1 mL of chloroform and were then further sonicated for 10 min at 5 °C. After an additional centrifugation step, the chloroform phase was mixed with the one formerly collected and then dried using a mild stream of nitrogen.

To obtain fatty acid methyl esters (FAMEs), the fatty acids (FAs) underwent transesterification. This process involved the addition of 3 mL of methanol, which contained 5% (v v^{-1}) sulfuric acid, to the tube containing the dried lipid samples. The mixture was then incubated at 70 °C for a duration of 3 h. After cooling, 3 mL of n-hexane and 3 mL of MilliQ water were added to the samples. The mixtures were vortexed three times for 1 min at 10 min intervals and then centrifuged at 177 RCF (Relative Centrifugal Force) at 5 °C for 10 min. For every sample, 2 mL of the hexane phase was recovered and washed with 2 mL of MilliQ water twice. The hexane phase with the FAMEs was then put into glass vials for subsequent GC-MS analysis. A gas chromatography trace 1300 equipped with a triple quadrupole mass spectrometry (TSQ 9000), a fused capillary column Agilent HP-5 (30 m × 0.32 i.d, 0.25 μm f.t.), and an automatic sampler (AI 1310) with a split-splitless injector were used (Waltham, MA, USA). The injector was set at 250 °C, and the carrier gas (helium) flow was 1.5 mL min^{-1}. The oven temperature was initially held at 50 °C for 1 min, followed by an increase from 50 to 175 °C at 10 °C min^{-1}. It was then kept at 175 °C for 10 min, augmented from 175 to 210 °C at 5 °C min^{-1}, kept at 210 °C for 10 min, and further increased at 5 °C min^{-1} from 210 °C to 230 °C. Afterwards, it was held for 9.5 min at 230 °C and then re-increased at 10 °C min^{-1} from 230 °C to 300 °C. The sample was injected in split mode (0.4 µL) with a split ratio set at 1:20. The mass spectrometry transfer line temperature and ion source were set at 250 °C and 300 °C, respectively. Ions were generated at 70 eV with electron ionization and recorded at 1.5 scans s^{-1} over the mass range m z^{-1} 50 to 550. Peak identification was conducted by comparing peak retention time with Supelco 37 component FAME Mix (Sigma Aldrich). Data are expressed as a mg g^{-1} of dry weight (mean ± standard deviation) and calculated using Equation (10) by [68]:

$$FA\left(\frac{mg}{g}\right) = IS \frac{\frac{A_{FAME_i}}{A_{C13:0} \; RRF_{FAME_i}}}{m_b} \tag{10}$$

where *IS* refers to the internal standard added; A_{FAME_i} represents the area of the *ith FAME*; *A*(C13:0) is the area of the *FAME C13:0*; RRF_{FAME_i} denotes the relative response factor of the *ith FAME*; and m_b is the mass of the processed biomass. The relative abundance of the generic *FA* was determined by dividing its concentration by the total *FA* content. The ratio between the sum of hypocholesterolemic *FAs* and the sum of hypercholesterolemic *FAs* (*h/H*) was calculated according to the equation suggested by [69]:

$$\frac{h}{H} = \frac{\sum_{j=1}^{n} C_j^{hypo}}{\sum_{k=1}^{m} C_k^{Hype}} \tag{11}$$

with *j* = 18:1 *n*-9, 18:1 *n*-7, 18:2 *n*-6, 18:3 *n*-6, 18:3 *n*-3, 20:3 *n*-6, 20:4 *n*-6, 20:5 *n*-3, 22:4 *n*-6, 22:5 *n*-3, and 22:6 *n*-3 while k = 14:0 and 16:0.

3.6. Statistical Analysis

Each experimental condition was investigated in triplicate. Statistical analysis was carried out using the software MetaboAnalysts 5.0 tuned by McGill University, Montreal, Canada. In particular, the difference among groups was evaluated by the one-way analysis of variance (ANOVA) followed by the honestly significant difference (HSD) test by Tukey. Variations were considered significant at 95% confidence.

4. Conclusions

This study highlighted the efficacy of using food industry wastes like BWW, which is rich in organic carbon, as a cost-effective and sustainable nutrient source for *A. platensis*. The application of BWW not only promoted higher biomass yields (3.70 g L^{-1}) but also addressed waste management issues by recycling industrial effluents. The mixotrophic conditions facilitated higher PC production, with BWW and SW supplements notably enhancing PC content (7.70 mg L^{-1} and 5.25 mg L^{-1}) compared to controls (3.28 mg L^{-1}). The observed increase in PC yield was attributed to the synergistic effects of mixotrophy and salt stress, which stimulated the metabolic pathways involved in PC biosynthesis. Additionally, the study explored the impact of salinity on the biochemical composition of *A. platensis* biomass. The addition of SW resulted in significant changes in protein, carbohydrate, and lipid fractions, suggesting that salinity stress can be leveraged to manipulate the biochemical profile of microalgal biomass. This adaptability of *A. platensis* to saline environments underscores its potential for large-scale cultivation in marine or brackish waters, thus expanding its applicability in diverse geographic regions. Mixotrophic cultivation with BWW and SW also influenced the FAME profile, enhancing the content of specific FAs such as C16:0 (43.77% and 42.48%) and C18:1 (7.85% and 6.49%). This modification in the lipid profile could have implications for the nutritional and commercial value of the biomass produced. The purity of the extracted PC under mixotrophic conditions was found to be suitable for various applications, including food and cosmetic industries, with BWW and SW supplementation yielding higher EP values (1.15 with JB and 0.92 with JBS, respectively). The economic analysis of PC production emphasized the potential market value of high-purity PC, which can reach significant commercial prices, highlighting the economic benefits of optimizing cultivation parameters for maximum PC yield. Overall, this study contributes to the understanding of the complex interactions between salinity, organic carbon sources, and *A. platensis* metabolism. It underscores the feasibility of using mixotrophic cultivation strategies to enhance biomass and metabolite production while promoting environmental sustainability through waste valorization. The findings pave the way for developing efficient and cost-effective bioprocesses for the profitable production of high-added-value compounds from microalgae, aligning with the principles of a circular bio-economy. Future research should focus on scaling up these processes and further exploring the biochemical mechanisms underlying the observed enhancements to optimize industrial applications.

Author Contributions: Conceptualization, S.T. and G.A.L.; Data curation, L.U., G.F., M.C., M.G., G.A.L., and A.C.; Formal analysis, M.G., G.F., M.G., M.C. and D.D.; Investigation, N.P.R. and M.B.; Methodology, L.U. and G.A.L.; Supervision, P.N., G.A.L., and A.C.; Writing—original draft, E.D., G.A.L., and A.C.; Writing—review and editing, G.F., G.A.L., and A.C. All authors have read and agreed to the published version of the manuscript.

Funding: The project MABIZO (COD. AGROIND_2022-7) funded by Sardegna Ricerca (Italy) in the framework of "Programma Regionale di Sviluppo 2020-2024 - Strategia 2 - Identità economica -Progetto 2.1 - Ricerca e innovazione tecnologica" is gratefully acknowledged for the economic support of some activities performed in this work.

Institutional Review Board Statement: Not applicable.

Data Availability Statement: Data will be available upon request.

Acknowledgments: One of the authors, M.C., performed his activity in the framework of the International PhD in Innovation Sciences and Technologies at the University of Cagliari, Italy.

Conflicts of Interest: Authors Luca Usai and Giovanni Antonio Lutzu are employed by the company Teregroup Srl, and Eya Damergi is employed by the company Algaltek SARL. The remaining authors declare that the research is conducted in the absence of any commercial or financial relationships that could be construed as a potential conflict of interest.

References

1. Barbosa, M.J.; Janssen, M.; Südfeld, C.; D'Adamo, S.; Wijffels, R.H. Hypes, Hopes, and the Way Forward for Microalgal Biotechnology. *Trends Biotechnol.* **2023**, *41*, 452–471. [CrossRef]
2. Usai, A.; Theodoropoulos, C.; Di Caprio, F.; Altimari, P.; Cao, G.; Concas, A. Structured Population Balances to Support Microalgae-Based Processes: Review of the State-of-Art and Perspectives Analysis. *Comput. Struct. Biotechnol. J.* **2023**, *21*, 1169–1188. [CrossRef] [PubMed]
3. Li, S.; Chang, H.; Zhang, S.; Ho, S.H. Production of Sustainable Biofuels from Microalgae with CO_2 Bio-Sequestration and Life Cycle Assessment. *Environ. Res.* **2023**, *227*, 115730. [CrossRef] [PubMed]
4. Lutzu, G.A.; Parsaeimehr, A.; Ozbay, G.; Ciurli, A.; Bacci, L.; Rao, A.R.; Ravishankar, G.A.; Concas, A. Microalgae and Cyanobacteria Role in Sustainable Agriculture: From Wastewater Treatment to Biofertilizer Production. *Algae Mediat. Bioremed. Ind. Prospect.* **2024**, *2*, 565–618.
5. Concas, A.; Lutzu, G.A.; Turgut Dunford, N. Experiments and Modeling of *Komvophoron* sp. Growth in Hydraulic Fracturing Wastewater. *Chem. Eng. J.* **2021**, *426*, 131299. [CrossRef]
6. Fernandes, T.; Cordeiro, N. Microalgae as Sustainable Biofactories to Produce High-Value Lipids: Biodiversity, Exploitation, and Biotechnological Applications. *Mar. Drugs* **2021**, *19*, 573. [CrossRef]
7. Lutzu, G.A.; Marin, M.A.; Concas, A.; Dunford, N.T. Growing *Picochlorum oklahomensis* in Hydraulic Fracturing Wastewater Supplemented with Animal Wastewater. *Water. Air. Soil Pollut.* **2020**, *231*, 457. [CrossRef]
8. Baunach, M.; Guljamow, A.; Miguel-Gordo, M.; Dittmann, E. Harnessing the Potential: Advances in Cyanobacterial Natural Product Research and Biotechnology. *Nat. Prod. Rep.* **2024**, *41*, 347–369. [CrossRef]
9. Citi, V.; Torre, S.; Flori, L.; Usai, L.; Aktay, N.; Dunford, N.T.; Lutzu, G.A.; Nieri, P. Nutraceutical Features of the Phycobiliprotein C-Phycocyanin: Evidence from *Arthrospira platensis* (Spirulina). *Nutrients* **2024**, *16*, 1752. [CrossRef]
10. Xiao, S.; Lu, Z.; Yang, J.; Shi, X.; Zheng, Y. Phycocyanobilin from *Arthrospira platensis*: A Potential Photodynamic Anticancer Agent. *Dye. Pigment.* **2023**, *219*, 111516. [CrossRef]
11. Fernandes, R.; Campos, J.; Serra, M.; Fidalgo, J.; Almeida, H.; Casas, A.; Toubarro, D.; Barros, A.I.R.N.A. Exploring the Benefits of Phycocyanin: From *Spirulina* Cultivation to Its Widespread Applications. *Pharmaceuticals* **2023**, *16*, 592. [CrossRef]

12. Miotti, T.; Pivetti, L.; Lolli, V.; Sansone, F.; Concas, A.; Lutzu, G.A. On the Use of Agro-Industrial Wastewaters to Promote Mixotrophic Metabolism in *Chlorella vulgaris*: Effect on FAME Profile and Biodiesel Properties. *Chem. Eng. Trans.* **2022**, *92*, 55–60. [CrossRef]
13. Proietti Tocca, G.; Agostino, V.; Menin, B.; Tommasi, T.; Fino, D.; Di Caprio, F. Mixotrophic and Heterotrophic Growth of Microalgae Using Acetate from Different Production Processes. *Rev. Environ. Sci. Biotechnol.* **2024**, *23*, 93–132. [CrossRef]
14. Shayesteh, H.; Laird, D.W.; Hughes, L.J.; Nematollahi, M.A.; Kakhki, A.M.; Moheimani, N.R. Co-Producing Phycocyanin and Bioplastic in *Arthrospira platensis* Using Carbon-Rich Wastewater. *BioTech* **2023**, *12*, 49. [CrossRef] [PubMed]
15. Pereira, I.; Rangel, A.; Chagas, B.; de Moura, B.; Urbano, S.; Sassi, R.; Camara, F.; Castro, C. Microalgae Growth under Mixotrophic Condition Using Agro-Industrial Waste: A Review. In *Biotechnological Applications of Biomass*; Intechopen: London, UK, 2021.
16. Cavallini, A.; Torre, S.; Usai, L.; Casula, M.; Fais, G.; Nieri, P.; Concas, A.; Lutzu, G.A. Effect of Cheese Whey on Phycobiliproteins Production and FAME Profile by *Arthrospira platensis* (*Spirulina*): Promoting the Concept of a Circular Bio-Economy. *Sustain. Chem. Pharm.* **2024**, *40*, 101625. [CrossRef]
17. Varandas, R.C.R.; Pereira, A.C.; da Silva Araújo, V.B.; de Moura Andrade, P.; da Silva Nonato, N.; da Costa, M.H.J.; de Pina, L.C.C.; Handam, V.P.T.; da Costa Sassi, C.F.; da Conceição, M.M. Utilização Do Resíduo de Malte Da Indústria Cervejeira Como Meio Alternativo Para o Cultivo Da *Spirulina platensis* e *Spirulina maxima*. *Res. Soc. Dev.* **2022**, *11*, e451111638249. [CrossRef]
18. Mirhosseini, N.; Davarnejad, R.; Hallajisani, A.; Cano-Europa, E.; Tavakoli, O. Sugarcane Molasses as a Cost-Effective Carbon Source on *Arthrospira maxima* Growth by Taguchi Technique. *Int. J. Eng. Trans. B Appl.* **2022**, *35*, 510–516. [CrossRef]
19. Al Mahrouqi, H.; Dobretsov, S.; Avilés, A.; Díaz, R.T.A. *Spirulina* Optimization Using Cane Molasses as the Cost-Effective Alternative of Sodium Bicarbonate. *Biol. Bull.* **2022**, *49*, S60–S68. [CrossRef]
20. Uzlasir, T.; Isik, O.; Uslu, L.H.; Selli, S.; Kelebek, H. Impact of Different Salt Concentrations on Growth, Biochemical Composition and Nutrition Quality of *Phaeodactylum tricornutum* and *Spirulina platensis*. *Food Chem.* **2023**, *429*, 136843. [CrossRef]
21. Markou, G.; Kougia, E.; Arapoglou, D.; Chentir, I.; Andreou, V.; Tzovenis, I. Production of *Arthrospira platensis*: Effects on Growth and Biochemical Composition of Long-Term Acclimatization at Different Salinities. *Bioengineering* **2023**, *10*, 233. [CrossRef]
22. Yu, C.; Hu, Y.; Zhang, Y.; Luo, W.; Zhang, J.; Xu, P.; Qian, J.; Li, J.; Yu, J.; Liu, J.; et al. Concurrent Enhancement of Biomass Production and Phycocyanin Content in Salt-Stressed *Arthrospira platensis*: A Glycine Betaine- Supplementation Approach. *Chemosphere* **2024**, *353*, 141387. [CrossRef]
23. Nosratimovafagh, A.; Fereidouni, A.E.; Krujatz, F. Modeling and Optimizing the Effect of Light Color, Sodium Chloride and Glucose Concentration on Biomass Production and the Quality of *Arthrospira platensis* Using Response Surface Methodology (RSM). *Life* **2022**, *12*, 371. [CrossRef]
24. Amenorfenyo, D.K.; Huang, X.; Zhang, Y.; Zeng, Q.; Zhang, N.; Ren, J.; Huang, Q. Microalgae Brewery Wastewater Treatment: Potentials, Benefits and the Challenges. *Int. J. Environ. Res. Public Health* **2019**, *16*, 1910. [CrossRef]
25. Budgen, J.; Le-Clech, P. Assessment of Brewery Wastewater Treatment by an Attached Growth Bioreactor. *H2Open J.* **2020**, *3*, 32–45. [CrossRef]
26. Jaiyeola, A.T.; Bwapwa, J.K. Treatment Technology for Brewery Wastewater in a Water-Scarce Country: A Review. *S. Afr. J. Sci.* **2016**, *112*, 8. [CrossRef] [PubMed]
27. Dias, C.; Santos, J.A.L.; Reis, A.; Lopes da Silva, T. The Use of Oleaginous Yeasts and Microalgae Grown in Brewery Wastewater for Lipid Production and Nutrient Removal: A Review. *Waste Biomass Valorization* **2023**, *14*, 1799–1822. [CrossRef]
28. Lutzu, G.A.; Zhang, W.; Liu, T. Feasibility of Using Brewery Wastewater for Biodiesel Production and Nutrient Removal by *Scenedesmus dimorphus*. *Environ. Technol.* **2016**, *37*, 1568–1581. [CrossRef] [PubMed]
29. Lois-Milevicich, J.; Casá, N.; Alvarez, P.; Mateucci, R.; Busto, V.; de Escalada Pla, M. *Chlorella vulgaris* Biomass Production Using Brewery Wastewater with High Chemical Oxygen Demand. *J. Appl. Phycol.* **2020**, *32*, 2773–2783. [CrossRef]
30. He, Y.; Lian, J.; Wang, L.; Su, H.; Tan, L.; Xu, Q.; Wang, H.; Li, Y.; Li, M.; Han, D.; et al. Enhanced Brewery Wastewater Purification and Microalgal Production through Algal-Bacterial Synergy. *J. Clean. Prod.* **2022**, *376*, 134361. [CrossRef]
31. Su, H.; Wang, K.; Lian, J.; Wang, L.; He, Y.; Li, M.; Han, D.; Hu, Q. Advanced Treatment and Resource Recovery of Brewery Wastewater by Co-Cultivation of Filamentous Microalga *Tribonema aequale* and Autochthonous Bacteria. *J. Environ. Manage.* **2023**, *348*, 119285. [CrossRef]
32. Akhere, M.A.; Ngbonyebi, E.C. Comparative study of the influence of brewery effluent on the growth of two marine microalgae. *Afr. J. Health Saf. Environ.* **2023**, *4*, 101–114. [CrossRef]
33. Miotti, T.; Sansone, F.; Lolli, V.; Concas, A.; Lutzu, G.A. Mixotrophic and Heterotrophic Metabolism in Brewery Wastewater by *Chlorella vulgaris*: Effect on Growth, FAME Profile and Biodiesel Properties. *Chem. Eng. Trans.* **2024**, *109*, 97–102.
34. Villaró, S.; García-Vaquero, M.; Morán, L.; Álvarez, C.; Cabral, E.M.; Lafarga, T. Effect of Seawater on the Biomass Composition of *Spirulina* Produced at a Pilot-Scale. *New Biotechnol.* **2023**, *4*, 101–114. [CrossRef]
35. Sassano, C.E.N.; Gioielli, L.A.; Almeida, K.A.; Sato, S.; Perego, P.; Converti, A.; Carvalho, J.C.M. Cultivation of *Spirulina platensis* by Continuous Process Using Ammonium Chloride as Nitrogen Source. *Biomass Bioenergy* **2007**, *31*, 593–598. [CrossRef]
36. Pavan, M.; Reinmets, K.; Garg, S.; Mueller, A.P.; Marcellin, E.; Köpke, M.; Valgepea, K. Advances in Systems Metabolic Engineering of Autotrophic Carbon Oxide-Fixing Biocatalysts towards a Circular Economy. *Metab. Eng.* **2022**, *71*, 117–141. [CrossRef]
37. Papadopoulos, K.P.; Economou, C.N.; Markou, G.; Nicodemou, A.; Koutinas, M.; Tekerlekopoulou, A.G.; Vayenas, D.V. Cultivation of *Arthrospira platensis* in Brewery Wastewater. *Water* **2022**, *14*, 1547. [CrossRef]

38. Bezerra, P.Q.M.; Moraes, L.; Cardoso, L.G.; Druzian, J.I.; Morais, M.G.; Nunes, I.L.; Costa, J.A.V. Spirulina Sp. LEB 18 Cultivation in Seawater and Reduced Nutrients: Bioprocess Strategy for Increasing Carbohydrates in Biomass. *Bioresour. Technol.* **2020**, *316*, 123883. [CrossRef]

39. Coelho, R.S.; Vidotti, A.D.S.; Reis, É.M.; Franco, T.T. High Cell Density Cultures of Microalgae under Fed-Batch and Continuous Growth. *Chem. Eng. Trans.* **2014**, *38*, 313–318. [CrossRef]

40. Sujatha, K.; Nagarajan, P. Effect of Salinity on Biomass and Biochemical Constituents of *Spirulina platensis* (Geitler). *Int. J. Plant Prot.* **2014**, *7*, 71–73.

41. Castillo, T.; Ramos, D.; García-Beltrán, T.; Brito-Bazan, M.; Galindo, E. Mixotrophic Cultivation of Microalgae: An Alternative to Produce High-Value Metabolites. *Biochem. Eng. J.* **2021**, *176*, 108183. [CrossRef]

42. Nur, M.M.A.; Rahmawati, S.D.; Sari, I.W.; Achmad, Z.; Setyoningrum, T.M.; Jaya, D.; Murni, S.W.; Djarot, I.N. Enhancement of Phycocyanin and Carbohydrate Production from *Spirulina platensis* Growing on Tofu Wastewater by Employing Mixotrophic Cultivation Condition. *Biocatal. Agric. Biotechnol.* **2023**, *47*, 102600. [CrossRef]

43. Occhipinti, P.S.; Del Signore, F.; Canziani, S.; Caggia, C.; Mezzanotte, V.; Ferrer-Ledo, N. Mixotrophic and Heterotrophic Growth of *Galdieria sulphuraria* Using Buttermilk as a Carbon Source. *J. Appl. Phycol.* **2023**, *35*, 2631–2643. [CrossRef]

44. Katari, J.K.; Uz Zama Khan, M.R.; Trivedi, V.; Das, D. Extraction, Purification, Characterization and Bioactivity Evaluation of High Purity C-Phycocyanin from *Spirulina* sp. NCIM 5143. *Process Biochem.* **2023**, *130*, 322–333. [CrossRef]

45. Khandual, S.; Sanchez, E.O.L.; Andrews, H.E.; de la Rosa, J.D.P. Phycocyanin Content and Nutritional Profile of *Arthrospira platensis* from Mexico: Efficient Extraction Process and Stability Evaluation of Phycocyanin. *BMC Chem.* **2021**, *15*, 24. [CrossRef]

46. Chini Zittelli, G.; Lauceri, R.; Faraloni, C.; Silva Benavides, A.M.; Torzillo, G. Valuable Pigments from Microalgae: Phycobiliproteins, Primary Carotenoids, and Fucoxanthin. *Photochem. Photobiol. Sci.* **2023**, *22*, 1733–1789. [CrossRef]

47. Fratelli, C.; Bürck, M.; Silva-Neto, A.F.; Oyama, L.M.; De Rosso, V.V.; Braga, A.R.C. Green Extraction Process of Food Grade C-Phycocyanin: Biological Effects and Metabolic Study in Mice. *Processes* **2022**, *10*, 1793. [CrossRef]

48. da Silva Figueira, F.; Moraes, C.C.; Kalil, S.J. C-Phycocyanin Purification: Multiple Processes for Different Applications. *Brazilian J. Chem. Eng.* **2018**, *35*, 1117–1128. [CrossRef]

49. Meticulous Research Phycocyanin Market to Be Worth $279.6 Million by 2030. Available online: https://www.meticulousresearch.com/pressrelease/30/phycocyanin-market-2030 (accessed on 26 June 2024).

50. Carvalho, F.; Prazeres, A.R.; Rivas, J. Cheese Whey Wastewater: Characterization and Treatment. *Sci. Total Environ.* **2013**, *445–446*, 385–396. [CrossRef] [PubMed]

51. Jiang, L.; Yu, S.; Chen, H.; Pei, H. Enhanced Phycocyanin Production from *Spirulina subsalsa* via Freshwater and Marine Cultivation with Optimized Light Source and Temperature. *Bioresour. Technol.* **2023**, *378*, 129009. [CrossRef]

52. Machalek, K.M.; Davison, I.R.; Falkowski, P.G. Thermal Acclimation and Photoacclimation of Photosynthesis in the Brown Alga *Laminaria saccharina*. *Plant, Cell Environ.* **1996**, *19*, 1005–1016. [CrossRef]

53. Nur, M.M.A.; Garcia, G.M.; Boelen, P.; Buma, A.G.J. Enhancement of C-Phycocyanin Productivity by *Arthrospira platensis* When Growing on Palm Oil Mill Effluent in a Two-Stage Semi-Continuous Cultivation Mode. *J. Appl. Phycol.* **2019**, *31*, 2855–2867. [CrossRef]

54. Schipper, K.; Fortunati, F.; Oostlander, P.C.; Al Muraikhi, M.; Al Jabri, H.M.S.J.; Wijffels, R.H.; Barbosa, M.J. Production of Phycocyanin by *Leptolyngbya* sp. in Desert Environments. *Algal Res.* **2020**, *47*, 101875. [CrossRef]

55. Liu, H.; Chen, H.; Wang, S.; Liu, Q.; Li, S.; Song, X.; Huang, J.; Wang, X.; Jia, L. Optimizing Light Distribution and Controlling Biomass Concentration by Continuously Pre-Harvesting *Spirulina platensis* for Improving the Microalgae Production. *Bioresour. Technol.* **2018**, *252*, 14–19. [CrossRef] [PubMed]

56. Pagels, F.; Guedes, A.C.; Amaro, H.M.; Kijjoa, A.; Vasconcelos, V. Phycobiliproteins from Cyanobacteria: Chemistry and Biotechnological Applications. *Biotechnol. Adv.* **2019**, *37*, 422–443. [CrossRef]

57. Maki, K.C.; Dicklin, M.R.; Kirkpatrick, C.F. Saturated Fats and Cardiovascular Health: Current Evidence and Controversies. *J. Clin. Lipidol.* **2021**, *15*, 765–772. [CrossRef]

58. Palomino, O.M.; Giordani, V.; Chowen, J.; Alfonso, S.F.; Goya, L. Physiological Doses of Oleic and Palmitic Acids Protect Human Endothelial Cells from Oxidative Stress. *Molecules* **2022**, *27*, 5217. [CrossRef]

59. Maltsev, Y.; Maltseva, K. Fatty Acids of Microalgae: Diversity and Applications. *Rev. Environ. Sci. Biotechnol.* **2021**, *20*, 515–547. [CrossRef]

60. Diraman, H.; Koru, E.; Dibeklioglu, H. Fatty Acid Profile of *Spirulina platensis* Used as a Food Supplement. *Isr. J. Aquac.-Bamidgeh* **2009**, *61*, 134–142. [CrossRef]

61. Choopani, A.; Poorsoltan, M.; Fazilati, M.; Latifi, A.M.; Salavati, H. *Spirulina*: A Source of Gamma-Linoleic Acid and Its Applications. *J. Appl. Biotechnol. Reports* **2016**, *3*, 483–488.

62. Barceló-Coblijn, G.; Murphy, E.J. Alpha-Linolenic Acid and Its Conversion to Longer Chain n-3 Fatty Acids: Benefits for Human Health and a Role in Maintaining Tissue n-3 Fatty Acid Levels. *Prog. Lipid Res.* **2009**, *48*, 355–374. [CrossRef]

63. Teixeira, A.; Fernandes, A.; Pereira, E. Fat Content Reduction and Lipid Profile Improvement in Portuguese Fermented Sausages Alheira. *Heliyon* **2020**, *6*, e05306. [CrossRef]

64. Conde, T.A.; Neves, B.F.; Couto, D.; Melo, T.; Neves, B.; Costa, M.; Silva, J.; Domingues, P.; Domingues, M.R. Microalgae as Sustainable Bio-Factories of Healthy Lipids: Evaluating Fatty Acid Content and Antioxidant Activity. *Mar. Drugs* **2021**, *19*, 357. [CrossRef] [PubMed]

65. Matos, Â.P.; Feller, R.; Moecke, E.H.S.; de Oliveira, J.V.; Junior, A.F.; Derner, R.B.; Sant'Anna, E.S. Chemical Characterization of Six Microalgae with Potential Utility for Food Application. *JAOCS J. Am. Oil Chem. Soc.* **2016**, *93*, 963–972. [CrossRef]
66. Herrera, A.; Boussiba, S.; Napoleone, V.; Hohlberg, A. Recovery of C-Phycocyanin from the Cyanobacterium *Spirulina maxima*. *J. Appl. Phycol.* **1989**, *1*, 325–331. [CrossRef]
67. Bryant, D.A.; Glazer, A.N.; Eiserling, F.A. Characterization and Structural Properties of the Major Biliproteins of *Anabaena* sp. *Arch. Microbiol.* **1976**, *110*, 61–75. [CrossRef] [PubMed]
68. Breuer, G.; Lamers, P.P.; Martens, D.E.; Draaisma, R.B.; Wijffels, R.H. Effect of Light Intensity, PH, and Temperature on Triacylglycerol ({TAG}) Accumulation Induced by Nitrogen Starvation in *Scenedesmus obliquus*. *Bioresour. Technol.* **2013**, *143*, 1–9. [CrossRef]
69. Fernández, M.; Ordóñez, J.A.; Cambero, I.; Santos, C.; Pin, C.; Hoz, L. de la Fatty Acid Compositions of Selected Varieties of Spanish Dry Ham Related to Their Nutritional Implications. *Food Chem.* **2007**, *101*, 107–112. [CrossRef]

marine drugs

Article

Enhanced Production of High-Value Porphyrin Compound Heme by Metabolic Engineering Modification and Mixotrophic Cultivation of *Synechocystis* sp. PCC6803

Kai Cao [1,2,†], Fengjie Sun [3,†], Zechen Xin [1], Yujiao Cao [2], Xiangyu Zhu [2], Huan Tian [1], Tong Cao [1], Jinju Ma [1], Weidong Mu [2], Jiankun Sun [1], Runlong Zhou [1], Zhengquan Gao [1,*] and Chunxiao Meng [1,*]

[1] School of Pharmacy, Binzhou Medical University, Yantai 264003, China; 17852032808@163.com (K.C.); 13287757622@163.com (Z.X.); tianhuan202310@163.com (H.T.); 19508010689@163.com (T.C.); 13862926695@163.com (J.M.); sjk203528@163.com (J.S.); xycrl21@163.com (R.Z.)

[2] School of Life Sciences and Medicine, Shandong University of Technology, Zibo 255049, China; c17122004049@163.com (Y.C.); zhuxyscut@163.com (X.Z.); muweidong2021@163.com (W.M.)

[3] Department of Biological Sciences, School of Science and Technology, Georgia Gwinnett College, Lawrenceville, GA 30043, USA; fsun@ggc.edu

[*] Correspondence: gaozhengquan@bzmc.edu.cn (Z.G.); mengchunxiao@bzmc.edu.cn (C.M.); Tel.: +86-187-6996-1859 (Z.G.); +86-135-8959-2609 (C.M.)

[†] These authors contributed equally to this work.

Abstract: Heme, as an essential cofactor and source of iron for cells, holds great promise in various areas, e.g., food and medicine. In this study, the model cyanobacteria *Synechocystis* sp. PCC6803 was used as a host for heme synthesis. The heme synthesis pathway and its competitive pathway were modified to obtain an engineered cyanobacteria with high heme production, and the total heme production of *Synechocystis* sp. PCC6803 was further enhanced by the optimization of the culture conditions and the enhancement of mixotrophic ability. The co-expression of *hemC*, *hemF*, *hemH*, and the knockout of *pcyA*, a key gene in the heme catabolic pathway, resulted in a 3.83-fold increase in the heme production of the wild type, while the knockout of *chlH*, a gene encoding a Mg-chelatase subunit and the key enzyme of the chlorophyll synthesis pathway, resulted in a 7.96-fold increase in the heme production of the wild type; further increased to 2.05 mg/L, its heme production was 10.25-fold that of the wild type under optimized mixotrophic culture conditions. *Synechocystis* sp. PCC6803 has shown great potential as a cell factory for photosynthetic carbon sequestration for heme production. This study provides novel engineering targets and research directions for constructing microbial cell factories for efficient heme production.

Keywords: heme; *Synechocystis* sp. PCC6803; cell factory; metabolic transformation

Citation: Cao, K.; Sun, F.; Xin, Z.; Cao, Y.; Zhu, X.; Tian, H.; Cao, T.; Ma, J.; Mu, W.; Sun, J.; et al. Enhanced Production of High-Value Porphyrin Compound Heme by Metabolic Engineering Modification and Mixotrophic Cultivation of *Synechocystis* sp. PCC6803. *Mar. Drugs* **2024**, *22*, 378. https://doi.org/10.3390/md22090378

Academic Editors: Cecilia Faraloni and Eleftherios Touloupakis

Received: 16 July 2024
Revised: 10 August 2024
Accepted: 21 August 2024
Published: 23 August 2024

1. Introduction

Heme ($C_{34}H_{33}FeN_4O_4$) is a type of iron porphyrin complex with a molecular weight of 616.49. As an important cofactor, heme is also involved in a variety of key life activities in the cell. For example, heme is responsible for oxygen transport/storage and its cytochromes facilitate electron transfer in the respiratory chain [1,2]. Heme also acts as a pro-oxidant and is involved in the production of reactive oxygen species (ROS), which is required for both growth and differentiation processes [3]. Heme is also a precursor of phycocyanin ($C_{33}H_{38}N_4O_6$), which is mainly present in cyanobacteria in the form of phycocyanin and is involved in efficient energy transfer [4].

In microorganisms, the heme biosynthesis pathway is divided into three modules, i.e., the 5-aminolevulinate (5-ALA) synthesis, the uroporphyrinogen (urogen) III syntheis, and the heme synthesis (Figure 1). The formation of 5-aminolevulinic acid is an important rate-limiting step in the heme biosynthesis pathway [5]. The C5 pathway for the formation of 5-ALA is predominantly found in algae, higher plants, and bacteria, while the C4 pathway is detected in animals, fungi, and non-sulfur photosynthesizing bacteria [6]. The C4 pathway primarily involves the conversion of CO_2 to organic acids (e.g., oxaloacetate, malate, and aspartate) through a series of enzymatic reactions in specific plants and some microorganisms that are subsequently involved in heme synthesis as intermediates. However, the C4 pathway is more associated with CO_2 fixation in plant photosynthesis than with direct involvement in heme biosynthesis. The C5 pathway is one of the direct pathways for heme synthesis and involves a series of enzymatic reactions, starting with precursors such as glycine and succinyl CoA, and culminating in the production of heme. The pathway from 5-ALA to uroporphyrinogen (urogen) III synthesis is highly conserved in microorganisms. The heme synthesis module includes the protoporphyrin-dependent (PPD) pathway, the cecal-porphyrin-dependent (CPD) pathway, and the siroheme-dependent (SHD) pathway. The PPD pathway is the most widely distributed and is present in Gram-negative bacteria and eukaryotes, the CPD pathway is mainly revealed in Gram-positive bacteria, and the SHD pathway is the most ancient but least common heme synthesis pathway [7]. The heme synthesis pathway in *Synechocystis* sp. PCC6803 consists of three modules: the C5 synthesis, the uroporphyrinogen (urogen) III synthesis, and the PPD synthesis modules [7]. Glu-tamyl-tRNA reductase, involved in the C5 synthesis module, is the key enzyme in the tetrapyrrole pathway, catalyzing the first reaction in the biosynthesis of tetrapyrrole, i.e., the conversion of tRNA^{Glu} to glutamate-1-semialdehyde. Glutamate-1-semialdehyde is relatively unstable and can be non-enzymatically converted to 5-ALA [8], which is sequentially converted to protoporphyrinogen IX by a group of enzymes, including cholecalciferol synthase (HemB), cholecalciferol deaminase (HemC), uroporphyrin III synthase (HemD), uroporphyrinogen decarboxylase (HemE), and fecal porphyrinogen oxidase (HemF and HemN), after which heme is finally synthesized via the PPD pathway. Our previous study found that the three enzymes in the heme synthesis pathway, i.e., HemC, HemF, and HemH, are the key regulators for heme synthesis in *Synechocystis* sp. PCC6803 [9]. However, whether or not the co-expression of these three key genes and the modification of key gene targets related to heme content outside the heme synthesis pathway can promote heme production in *Synechocystis* sp. PCC6803 has not yet been reported.

Currently, the heme on the market is mainly derived from plant materials (e.g., soybean roots) and chemical synthesis. However, there are many drawbacks in the actual production process, such as the collection, transportation, and storage of materials, the excessive cycle time, the cumbersome extraction processes, and environmental pollution [10–12]. The market demand for heme in many areas, such as food and pharmaceuticals, has been increasingly expanding, and the traditional heme production methods are no longer able to meet the increasing demand for heme applications. Recently, the development of microbe-based heme production methods has become a potential solution to this problem. In recent years, extensive research has been performed to construct heme-producing microbial cell factories using *Escherichia coli* [13], yeast [14], and *Corynebacterium glutamicum* [15] as chassis. All these studies have used heterotrophic microorganisms as chassis modified into cell factories. Although the heme production found in these studies is high, these methods lack the advantages of photosynthetic carbon fixation and autotrophic oxygenation, in addition to the fact that some of the bacteria themselves secrete endotoxins, severely limiting the application of the final heme product. Selecting cyanobacteria that can rapidly accumulate biomass through photosynthetic carbon sequestration and convert that biomass into high value-added products as the chassis to build a cell factory for high heme production is economically effective and environmentally friendly, meeting the demand of low carbon emission. *Synechocystis* sp. PCC6803 is a unicellular cyanobacteria capable of both photosynthetic autotrophic and heterotrophic growth. *Synechocystis* sp. PCC6803 co-cultured using both seawater and industrial wastewater has shown great potential in various areas, such as lipid [16] and amino acid [17] production. *Synechocystis* sp. PCC6803 is a model species of cyanobacterium, with the advantages of having a well-investigated genetic background, high resistance, and an anti-pollution ability, and which is also able to utilize mariculture and industrial flue gas cultivation to effectively solve the problems of occupying arable land and consuming freshwater resources [18]. In addition, *Synechocystis* sp. PCC6803 has shown a strong photosynthetic carbon sequestration efficiency which is more than 10-fold higher than that of terrestrial plants [19,20]. *Synechocystis* sp. PCC6803 is an excellent chassis organism that has been successfully transformed into cellular factories to produce a wide range of chemicals, such as acetone [21], ethylene [22], isoprene [23], β-hydroxybutyrate [24], and astaxanthin [25]. These studies indicate the significant potential of converting *Synechocystis* sp. PCC6803 into a light-driven, carbon-fixing heme cell factory to enhance the production of heme.

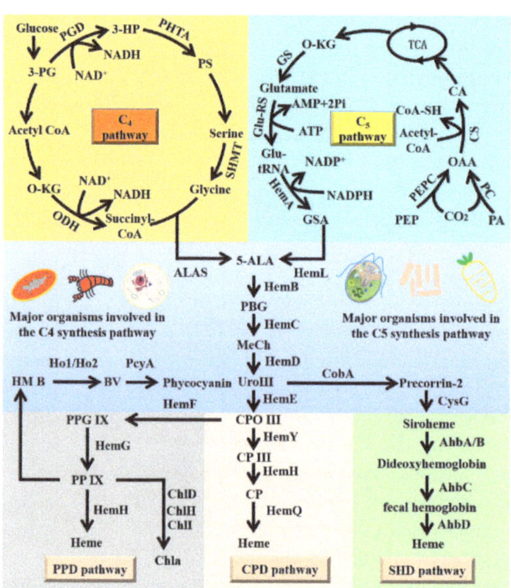

Figure 1. Heme synthesis pathway in microorganisms. 3-PG, 3-phosphoglycerate; 3-HP, 3-Hydroxypyruvate; PS, phosphatidyl serine; O-KG, α-ketoglutarate; PA, pyruvic acid; PEP, phosphoenolpyruvate; OAA, oxaloacetic acid; CA, citric acid; Glu-tRNA, glutamate-transfer RNA; GSA, glutamate-1-semialdehyde; 5-ALA, 5-aminolevulinic acid; PBG, porphobilinogen synthase; MeCh, methylocholine; UroIII, uroporphyrinogen III; CPO III, coproporphyrinogen III; CP III, coproporphyrin III; CP, coproporphyrin; PPG IX, protoporphyrinogen IX; PP IX, protoporphyrin IX; HM B, heme B; BV, biliverdin; HemB, bilirubinogen synthase; HemC, cholesterol deaminase; HemD, uroporphyrin III synthase; HemE, uroporphyrinogen decarboxylase; HemY, protoporphyrinogen oxidase; HemH, iron chelating enzyme; HemQ, fecal heme decarboxylase; HemF/HemN, fecal porphyrinogen oxidase; HemL, glutamate-1-semialdehyde 2,1-aminomutase; Ho1/Ho2, heme oxygenase; PcyA, phycocyanobilin:ferredoxin oxidoreductase; CysG, uroporphyrin-III C-methyltransferase; ChlI, magnesium chelatase subunit I; ChlH, magnesium chelatase subunit H; ChlD, magnesium chelatase subunit D. These are the siroheme-dependent (SHD) pathway, which is the most ancient but least common of the three; the coproporphyrin-dependent (CPD) pathway, which, with one known exception, is found only in Gram-positive bacteria; and the protoporphyrin-dependent (PPD) pathway, which is found in Gram-negative bacteria and all eukaryotes.

In this study, *Synechocystis* sp. PCC6803 was used as a microbial chassis for heme production. First, the competitive pathway for heme synthesis was knocked down, the heme repository was enhanced, and high-yielding strains were constructed both in combination with and modified in conjunction with the synthesis of heme in *Synechocystis* sp. PCC6803; then, the modules that could promote the heme synthesis of *Synechocystis* sp. PCC6803 were combined and modified to construct a high-yield cyanobacteria; finally, the mixotrophic fermentation of the recombinant cyanobacteria was optimized to further increase the heme production of *Synechocystis* sp. PCC6803. The objectives of this study were: (1) to construct an engineered cyanobacteria of *Synechocystis* sp. PCC6803 with high heme production, (2) to investigate effective methods to enhance the heterotrophic capacity of *Synechocystis* sp. PCC6803, and (3) to investigate the optimal glucose concentration for promoting the growth of *Synechocystis* sp. PCC6803 and optimize the mixotrophic fermentation of *Synechocystis* sp. PCC6803 to increase the heme production of *Synechocystis* sp. PCC6803. This study is highly expected to further broaden the application field of microalgae and provide a novel strategy to find new ways of increasing heme production.

2. Results

2.1. Modular Modification of the Heme-Producing Synthesis-Related Pathways of Synechocystis sp. PCC6803 and Its Enhanced Mixotrophic Capacity

The transformation of exogenous plasmid of *Synechocystis* sp. PCC6803 was carried out by the natural transformation method. The transformed *Synechocystis* sp. PCC6803 was incubated in an incubator, screened for resistance, and verified by colony PCR testing; singly picked single clones were streaked and passed four times and then transferred to a liquid medium for further culture.

Based on the plasmid transformation and transformant screening, the *Synechocystis* sp. PCC6803 mutant strain (CFH) with the co-expression of three *Synechocystis* sp. PCC6803 heme synthesis-related genes (*hemC*, *hemF*, and *hemH*) was obtained, and mutant strain ptsG was established with the overexpression of the *E. coli* glucose transporter protein gene *ptsG*. Another two *Synechocystis* sp. PCC6803 mutant strains, i.e., ΔChlH and ΔPcyA, were constructed based on homologous recombination. The knockout of the gene *ChlH*, which encodes a key enzyme for chlorophyll synthesis, and the gene *pcyA*, which encodes a key enzyme involved in phycocyanin synthesis, respectively, was achieved. The mutant strain Glbn-ptsG was established with the co-expression of both the *Synechocystis* sp. PCC6803 heme gene *glbn* and the *E. coli* glucose transporter protein gene *ptsG*.

2.1.1. Heme Content of Synechocystis sp. PCC6803

The results of the heme contents of *Synechocystis* sp. PCC6803 WT and the mutant strains showed that, compared to the WT, the heme contents of four mutant strains were increased, all except for the ptsG strain (Figure 2). Among these mutants, ΔPcyA showed the highest heme content (0.97 mg/L), reaching 4.41-fold higher content than the WT. Heme is a key precursor for phycocyanin synthesis in *Synechocystis* sp. PCC6803. Knocking out the key gene, *pcyA*, for phycocyanin synthesis can reduce the breakdown of heme into phycocyanin, thereby increasing the heme production in ΔPcyA. Zhao et al. effectively increased the accumulation of heme in *E. coli* by knocking down heme degrading enzymes [26]. This is consistent with our strategy of increasing the heme content in cells by reducing heme decomposition, and both have achieved positive results. The heme content of the mutant strain ΔChlH (0.29 mg/L) was 1.31-fold higher than that of the WT. ChlH drives the direct precursor of heme, protoporphyrin IX, towards the competitive pathway of heme synthesis, i.e., the chlorophyll synthesis pathway. Therefore, knocking out the gene encoding ChlH can promote an increase in protoporphyrin IX to synthesize more heme. The CFH strain (0.67 mg/mL) achieved a heme content that was 3.05-fold higher than that of the WT (0.22 mg/L). The co-expression of *hemC*, *hemF*, *hemH*, and other heme synthesis-related genes promoted heme synthesis in *E. coli* [27]. In addition, Lee et al. found that the expression of *hemC* enhanced the synthesis of the heme precursors ALA and uroporphyrin [28]. Both mutant strains, ptsG and Glbn-ptsG, showed no significant variations in heme content in comparison with the WT. This is due to the expression of *ptsG* encoding glucose transporter protein, the use of which aimed to enhance the biomass accumulation of *Synechocystis* sp. PCC6803 in cultures containing glucose, as previously reported [29]. In our study, we applied autotrophic culturing in glucose-free media. Therefore, the overexpression of *ptsG* showed no significant effect on the heme production of *Synechocystis* sp. PCC6803.

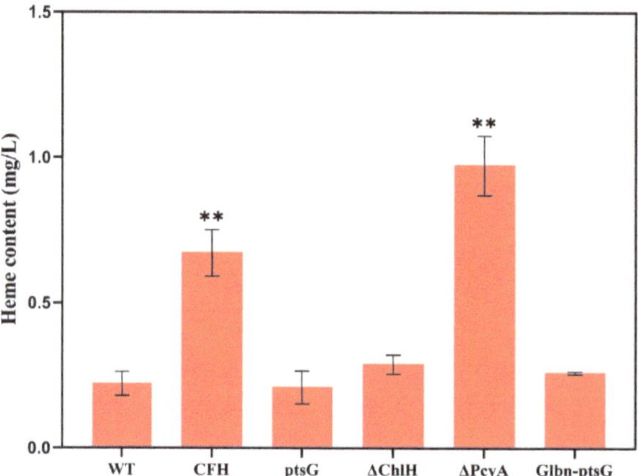

Figure 2. Heme contents of wild-type (WT) and five mutant strains of *Synechocystis* sp. PCC6803. Symbols "**" indicate significant difference compared with WT based on $p < 0.01$.

2.1.2. Growth Curves of *Synechocystis* sp. PCC6803

The WT and mutant strains of *Synechocystis* sp. PCC6803 were separately inoculated into BG11 liquid medium, with the absorbance detected at 730 nm by sampling at 48 h intervals after inoculation started to plot the growth curves (Figure 3A). The results showed that, up to 26 d after inoculation, the WT of *Synechocystis* sp. PCC6803 grew faster than all the mutant strains.

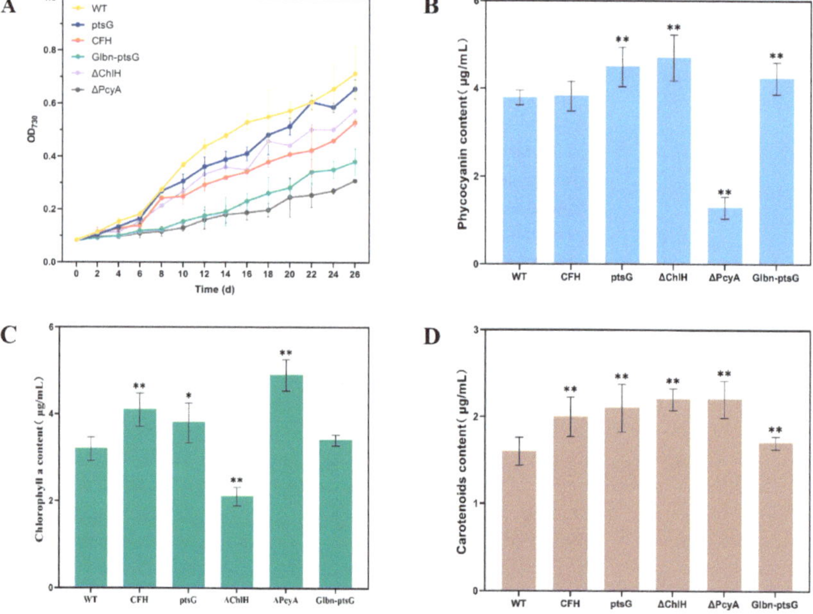

Figure 3. Growth and chemical characterization of wild-type (WT) and mutant strains of *Synechocystis* sp. PCC6803. (**A**) Growth curves. (**B**) Phycocyanin content. (**C**) Chlorophyll a content. (**D**) Carotenoid content. Symbols "*" and "**" indicate significant difference compared with WT based on $p < 0.05$ and $p < 0.01$, respectively.

2.1.3. Phycocyanin Content in *Synechocystis* sp. PCC6803

The phycocyanin contents of the WT and mutant strains of *Synechocystis* sp. PCC6803 revealed the highest phycocyanin content in the mutant strain ΔChlH (4.70 µg/mL), being 1.24-fold higher than that of the WT (3.79 µg/mL) (Figure 3B). Phycocyanin was still detected in the mutant strain ΔPcyA (1.28 µg/mL), with a 66.2% decrease in the phycocyanin content compared to the WT. The mutant strains ptsG and Glbn-ptsG showed a 1.18-fold (4.49 µg/mL) and 1.12-fold (4.23 µg/mL) increase, respectively, in the phycocyanin content compared to the WT, whereas no significant difference in the phycocyanin content was detected between the CFH strain and the WT.

2.1.4. Chlorophyll a Content in *Synechocystis* sp. PCC6803

Synechocystis sp. PCC6803 contains chlorophyll a but no chlorophyll b. Therefore, the chlorophyll a content of *Synechocystis* sp. PCC6803 was determined (Figure 3C). The results showed that the mutant strain ΔPcyA was revealed to have the highest chlorophyll a content (4.9 µg/mL), which was 1.53-fold higher than that of the WT (3.2 µg/mL), whereas the chlorophyll a content in the mutant strain ΔChlH (2.1 µg/mL) was reduced by 34.4% compared with that of the WT. The chlorophyll a content of the mutant strain CFH (4.1 µg/mL) was 1.28-fold higher than that of the WT, whereas the chlorophyll a content of both the ptsG and Glbn-ptsG strains showed no significant variation from that of the WT.

2.1.5. Carotenoid Content of *Synechocystis* sp. PCC6803

The results of the carotenoid content of *Synechocystis* sp. PCC6803 revealed a trivial increase in the carotenoid content of all mutant strains compared to the WT (1.6 µg/mL) (Figure 3D), with the Glbn-ptsG strain showing the lowest carotenoid content (1.7 µg/mL) among all transformants.

2.1.6. Mixotrophic Capacity of the *Synechocystis* sp. PCC6803 Mutant Strain ptsG

The growth curves of the WT and the mutant strain ptsG of *Synechocystis* sp. PCC6803 grown in mixotrophic medium containing different concentrations (0, 0.5, 1, 5, and 10 g/L) of glucose under low light culture conditions were established (Figure 4A). The results showed that both the WT and ptsG strain could grow in the mixotrophic medium with glucose concentrations of 0.5 and 1 g/L, and the growth rate of ptsG was higher than that of the WT; 1 g/L glucose was identified as the optimal concentration for the growth of *Synechocystis* sp. PCC6803 in the mixotrophic medium.

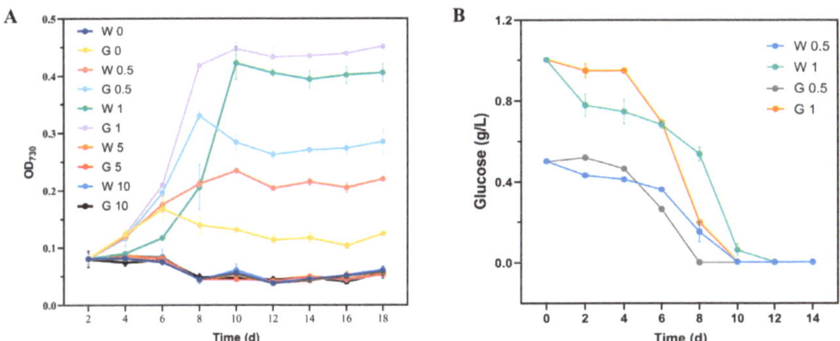

Figure 4. Growth of wild-type and mutant strain ptsG of *Synechocystis* sp. PCC6803, showing (**A**) growth curves and (**B**) residual amount of glucose in the culture medium containing glucose of different concentration. For example, "G 0.5" indicates transformant medium containing 0.5 g/L glucose and "W 0.5" stands for wild-type medium containing 0.5 g/L glucose.

Among the five glucose concentrations in the heterotrophic medium, only heterotrophic cultures with 0.5 and 1 g/L glucose caused increases in the concentrated biomass of *Syne-*

chocystis sp. PCC6803. Therefore, the changes in the glucose residual content in the heterotrophic culture medium of *Synechocystis* sp. PCC6803 with 0.5 and 1 g/L glucose were further investigated (Figure 4B). The results showed that the remaining glucose content in the culture medium was inversely proportional to the biomass of the *Synechocystis* sp. PCC6803. These results indicated that the addition of 1 g/L glucose to the culture medium promoted the growth of *Synechocystis* sp. PCC6803.

2.2. Multidimensional Engineering Modification of Heme-Producing Cell Factories Based on Synechocystis sp. PCC6803

The natural transformation method was performed based on multiple antibiotics corresponding to the plasmids that were added to the screening plate to obtain positive clones with successfully transformed multiple plasmids. After the second round of transformation and screening, two transformant strains of *Synechocystis* sp. PCC6803 were obtained that successfully transformed two plasmids, respectively, including (1) the *Synechocystis* sp. PCC6803 transformant with the co-expression of *hemC*, *hemF*, and *hemH*, as well as *pcyA* knockout (CFH-PcyA), and (2) the *Synechocystis* sp. PCC6803 transformant with the co-knockout of both the gene *chlH*, which encodes a key enzyme for chlorophyll synthesis, and the gene *pcyA*, which encodes a key enzyme involved in phycocyanin synthesis (PcyA-ChlH). After the third round of transformation and screening of the PcyA-ChlH strain, a transformant strain of *Synechocystis* sp. PCC6803 which successfully transformed two plasmids was obtained with the co-expression of *hemC*, *hemF*, *hemH*, *ptsG*, and *glbn* (glbn-ptsG-CFH). PCR testing of the insertion site and exogenous gene expression cassette was performed based on the above *Synechocystis* sp. PCC6803 transformants as well as the WT (Figure S2). The results of the electrophoresis and sequencing confirmed that these transformants contained the target exogenous gene expression cassette.

2.2.1. Heme Content of Combined Transformants of *Synechocystis* sp. PCC6803

The results of the heme contents of the combined transformants of *Synechocystis* sp. PCC6803 revealed the highest heme content of 1.83 mg/L in the mutant strain PcyA-ChlH, which was 7.96-fold higher than that of the WT (0.23 mg/L), and the heme content of the transformant CFH-PcyA (0.88 mg/L) was 3.83-fold higher than that of the WT (Figure 5). No significant variation was detected in the heme content between the transformant glbn-ptsG-CFH and the WT.

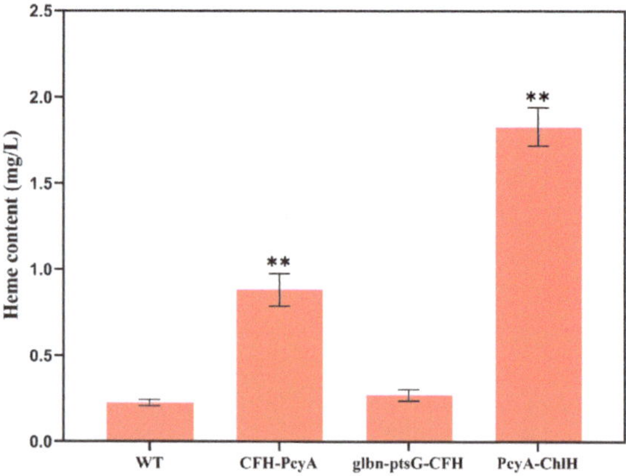

Figure 5. Heme content of wild-type (WT) and three transformant strains of *Synechocystis* sp. PCC6803. Symbols "**" indicate the significant difference compared with WT based on $p < 0.01$.

2.2.2. Growth of Combined Transformants of *Synechocystis* sp. PCC6803

Both the WT and mutant strains of *Synechocystis* sp. PCC6803 were inoculated into BG11 liquid medium and cultured under photoautotrophic conditions. Samples were taken at 48 h intervals starting from the inoculation to measure the absorbance at 730 nm and plot the growth curves (Figure 6A). The results showed that the growth rates of three combination-transformed *Synechocystis* sp. PCC6803 mutant strains were all lower than that of the WT, with the slowest growth rate being revealed in the CFH-PcyA strain.

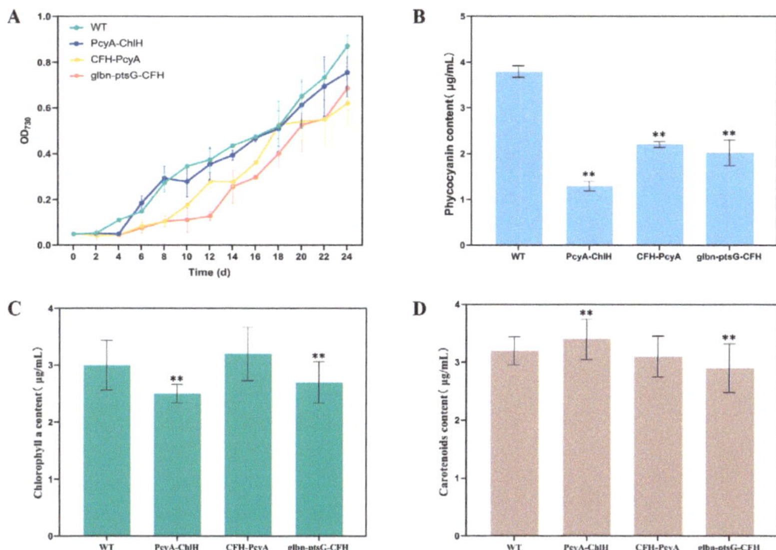

Figure 6. Growth and chemical characterization of wild-type (WT) and mutant strains of *Synechocystis* sp. PCC6803. (**A**) Growth curves. (**B**) Phycocyanin content. (**C**) Chlorophyll a content. (**D**) Carotenoid content. Symbols "******" indicate the significant difference compared with WT based on $p < 0.01$.

2.2.3. Phycocyanin Content in Combined Transformants of *Synechocystis* sp. PCC6803

The results of the phycocyanin contents revealed that, compared to the WT (3.79 µg/mL), all three transformants of *Synechocystis* sp. PCC6803, PcyA-ChlH (1.30 µg/mL), CFH-PcyA (2.02 µg/mL), and glbn-ptsG-CFH (2.20 µg/mL), showed significantly decreased phycocyanin content, by 65.7%, 46.7%, and 41.9%, respectively (Figure 6B).

2.2.4. Chlorophyll a Content in the Combined Transformants of *Synechocystis* sp. PCC6803

The results of the chlorophyll a content showed that, compared with the WT (3.00 µg/mL) of *Synechocystis* sp. PCC6803, the chlorophyll a content of the mutant strain CFH-PcyA was slightly increased to 3.20 µg/mL, whereas the chlorophyll a contents of both the mutants of PcyA-ChlH (2.50 µg/mL) and glbn-ptsG-CFH (2.70 µg/mL) were lower than that of the WT (Figure 6C).

2.2.5. Carotenoid Content of Combined Transformants of *Synechocystis* sp. PCC6803

The results of the carotenoid contents showed that, compared with the WT (3.20 µg/mL) of *Synechocystis* sp. PCC6803, the carotenoid content was slightly increased in the PcyA-ChlH strain (3.40 µg/mL) and slightly decreased in the Glbn-ptsG-CFH (2.90 µg/mL) strain, whereas no significant difference was revealed in the carotenoid contents between the CFH-PcyA strain and the WT (Figure 6D).

2.2.6. Expression of Heme Synthesis-Related Genes in *Synechocystis* sp. PCC6803 Combined Transformants

The expression patterns of heme synthesis-related genes, including *hemA*, *hemC*, *hemF*, and *hemH*, the hemoglobin-encoding gene *glbn*, the key gene for phycocyanin synthesis (*pcyA*), the key genes for chlorophyll synthesis (*chlH* and *chlM*), and a gene of glucose transporter protein (*ptsG*), were detected in the mutant strains of *Synechocystis* sp. PCC6803 using qRT-PCR testing (Figure 7). The results showed that, compared to the WT, most genes, except for *hemF* and *hemH*, were down-regulated in the glbn-ptsG-CFH strain. In the CFH-PcyA strain, *hemF*, *hemH*, *chlH*, and *chlM* were up-regulated, the expression of *ptsG* was not detected, and all other genes were down-regulated. In the PcyA-ChlH strain, *hemF*, *hemH*, *glbn*, and *chlM* were up-regulated, the expression of both *hemF* and *glbn* was substantially up-regulated by more than 5-fold, the expression of *ptsG* was not detected, and all other genes were down-regulated.

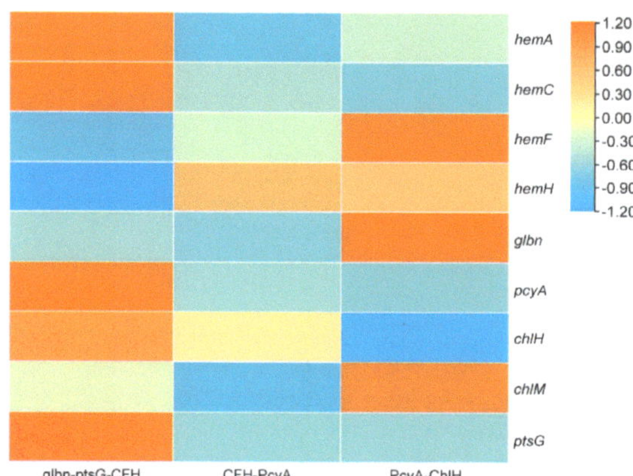

Figure 7. Expression profiles of nine heme synthesis-related genes in three mutant strains of *Synechocystis* sp. PCC6803, i.e., Glbn-ptsG-CFH, CFH-PcyA, and PcyA-ChlH.

2.2.7. Scanning Electron Microscopic Observation of the Morphological Characteristics of *Synechocystis* sp. PCC6803 Transformant Cells

Scanning electron microscopic observations of cells of both the WT and transformed strains of *Synechocystis* sp. PCC6803 revealed evident variations in extracellular adhesion, cell morphology, and cell size between the transformed strains and the WT (Figure 8). In both the PcyA-ChlH and CFH-PcyA strains, the algal cells were bound together by the extracellular secretion of large amounts of adhesive substances. Both the transformants CFH-PcyA and Glbn-ptsG-CFH showed smaller cell diameters and flatter cell shapes compared with the WT, and the Glbn-ptsG-CFH strain showed a decreased level of cell surface wrinkling.

Figure 8. *Cont.*

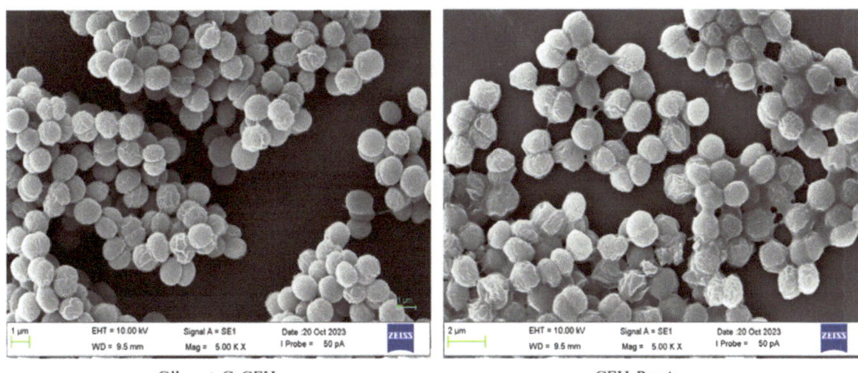

Glbn-ptsG-CFH   CFH-PcyA

Figure 8. Scanning electron micrographs of wild-type (WT) and three combined transformants of *Synechocystis* sp. PCC6803.

2.3. Fermentation Optimization of Synechocystis sp. PCC6803 Cell Factories

2.3.1. Heme Content of Combined Transformants of *Synechocystis* sp. PCC6803 Based on Optimized Fermentation Conditions

The *Synechocystis* sp. PCC6803 conditions with a glucose concentration of 1 g/L in the medium were analyzed. The results of the heme content of *Synechocystis* sp. PCC6803 revealed the highest heme content of 2.05 mg/L in the combined transformant PcyA-ChlH, which was 10.25-fold higher than that of the WT (0.20 mg/L) (Figure 9). The heme contents of both the combined transformants CFH-PcyA (0.14 mg/L) and Glbn-ptsG-CFH (0.23 mg/L) were slightly elevated, i.e., they were 0.70- and 1.15-fold higher than that of the WT, respectively.

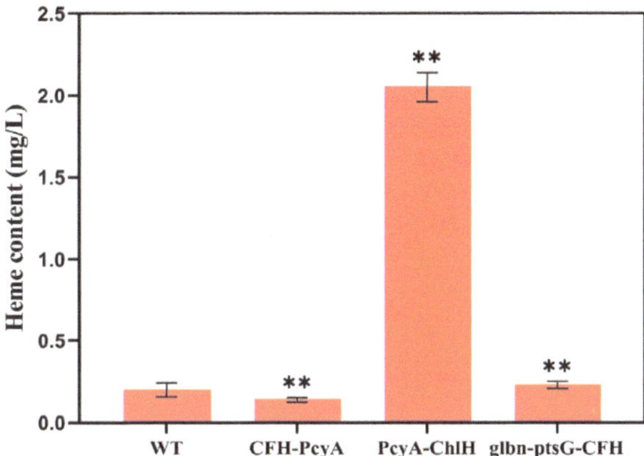

Figure 9. Heme content of wild-type (WT) and three combined transformants of *Synechocystis* sp. PCC6803 under optimized fermentation conditions. Symbols "*" indicate the significant difference compared with WT based on $p < 0.01$.

2.3.2. Growth of *Synechocystis* sp. PCC6803 Combined Transformants Based on Optimized Fermentation Conditions

Based on the results of Section 2.1.6, 1 g/L glucose was added to BG-11 culture medium, and the *Synechocystis* sp. PCC6803 combinatorial transformants with an OD_{730} value of 0.1 were initially inoculated and then cultivated in a thermostatic light-illuminated shaker at 100 rpm, with a light intensity of 2000 lux, a temperature of 30 °C, and a light/dark photoperiod of 12 h/12 h. The results of the growth curves showed that, unlike under photoautotrophic conditions, the engineered strains showed higher growth rates in the mixotrophic medium than the WT (Figure 10A). The highest heme production was achieved in the PcyA-ChlH strain, indicating that its growth disadvantage under autotrophic conditions was greatly improved. The maximum biomasses of the PcyA-ChlH and glbn-ptsG-CFH strains, as indicated by the OD_{730} values of 4.32 and 4.07, respectively, were obtained at day 15 of incubation.

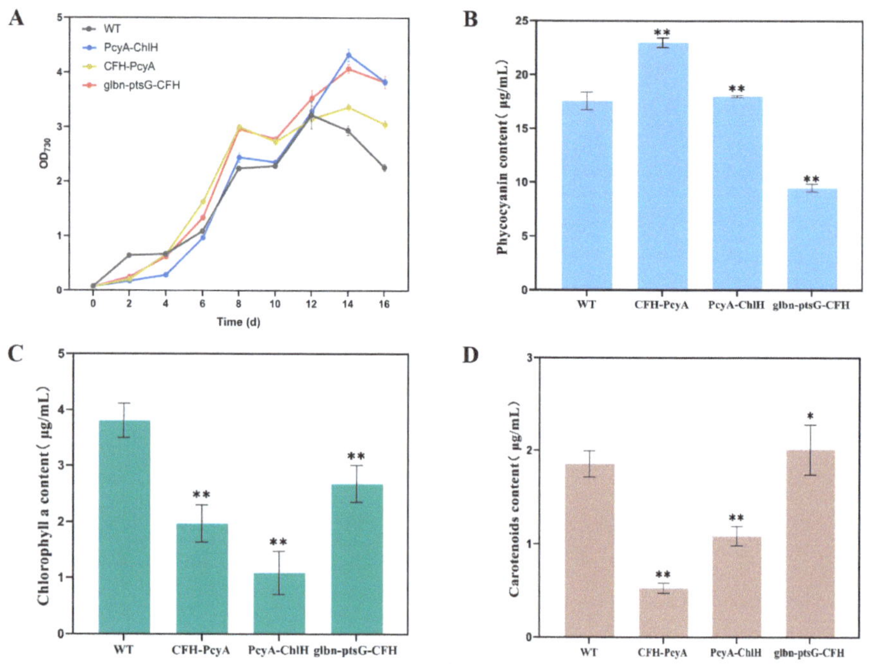

Figure 10. Growth and chemical characterization of wild-type (WT) and three mutant strains of *Synechocystis* sp. PCC6803. (**A**) Growth curves. (**B**) Phycocyanin content. (**C**) Chlorophyll a content. (**D**) Carotenoid content. Symbols "*" and "**" indicate the significant difference compared with WT based on $p < 0.05$ and $p < 0.01$, respectively.

2.3.3. Phycocyanin Content of *Synechocystis* sp. PCC6803 Combined Transformants Based on Optimized Fermentation Conditions

The results of the phycocyanin content revealed the highest content of 22.99 µg/mL in the combined transformant CFH-PcyA, which was 1.31-fold higher than that of the WT (17.57 µg/mL), followed by the combined transformant PcyA-ChlH, which showed slightly higher phycocyanin content (18.01 µg/mL) than that of the WT (Figure 10B). The phycocyanin content of the Glbn-ptsG-CFH strain (9.49 µg/mL) was decreased by 46.0% compared to the WT.

2.3.4. Chlorophyll a Content of *Synechocystis* sp. PCC6803 Combined Transformants Based on Optimized Fermentation Conditions

The chlorophyll a contents were decreased in all three combined transformants, i.e., glbn-ptsG-CFH (2.68 µg/mL), CFH-PcyA (1.97 µg/mL), and PcyA-ChlH (1.09 µg/mL), by 29.7%, 48.3%, and 71.4%, respectively, compared to the WT (3.81 µg/mL) (Figure 10C).

2.3.5. Carotenoid Content of *Synechocystis* sp. PCC6803 Combined Transformants after Fermentation Optimization

The results of the carotenoid content revealed a slightly higher carotenoid content in the Glbn-ptsG-CFH strain (2.01 µg/mL) than in the WT (1.85 µg/mL), whereas both the PcyA-ChlH (1.08 µg/mL) and CFH-PcyA (0.52 µg/mL) strains showed significantly decreased carotenoid contents, by 41.6% and 71.9%, respectively, compared to the WT (Figure 10D).

3. Discussion

The construction of microbial cell factories for heme production through metabolic engineering modifications has been widely successful in heterotrophic microorganisms, such as *E. coli* [13,26,28,30], *C. glutamicum* [15], and yeast [14]. We have also successfully constructed a heme-producing cell factory based on *B. subtilis* using metabolic engineering and synthetic biological techniques [11]. However, all the above chassis cells are heterotrophic microorganisms, which need to consume large amounts of organic cultures, lacking the advantages of photosynthetic carbon fixation and autotrophic oxygen release. In addition, *E. coli* and several other microbial taxa used as chassis secrete endotoxins [10], which severely limit the use of the high value-added products synthesized in these microbes. Microalgae hold strong promise as a source of bioavailable heme [31]. Although *Synechocystis* sp. PCC6803 contains a complete heme synthesis pathway, its heme yield is generally low. Previously, we found that the overexpression of biliverdinogen deaminase (HemC), fecal porphobilinogen oxidase (HemF), iron chelatase (HemH), and Heme GlbN outside of the heme synthesis pathway, respectively, promoted heme synthesis in *Synechocystis* sp. PCC6803 [32]. Based on this study, we utilized metabolic engineering and synthetic biology to engineer *Synechocystis* sp. PCC6803, targeting the heme synthesis pathway, competitive pathways, metabolic sinks, and biomass accumulation to obtain a high heme-producing *Synechocystis* sp. PCC6803 cell factory, i.e., PcyA-ChlH. In addition, we further explored methods to enhance the mixotrophic capacity of *Synechocystis* sp. PCC6803 by adding appropriate amounts of glucose (1 g/L) to the culture medium to promote the rapid accumulation of biomass of *Synechocystis* sp. PCC6803, ultimately increasing the heme production of *Synechocystis* sp. PCC6803. In our previous study, we screened for genes with positive effects on the accumulation of phycocyanin and heme in *Synechocystis* sp. PCC6803 by overexpressing 22 genes involved in the heme synthesis pathways of *Synechocystis* sp. PCC6803 and *Synechococcus elongatus* PCC7942, respectively, and individually in *Synechocystis* sp. PCC6803. In this study, we applied the combined overexpression of three key genes previously screened for heme synthesis to construct a heme-producing heme cell factory and investigated the effects of knocking out genes outside the heme synthesis pathway on the production of heme in *Synechocystis* sp. PCC6803. In addition, we further promoted heme production by introducing heterologous glucose transporter proteins and combining them with the optimization of mixotrophic culture conditions.

3.1. Improved Heme Production of Synechocystis sp. PCC6803 with Knockout of Genes pcyA and chlH and Co-Expression of Genes hemC, hemF, and hemH

In recent years, research on the use of microorganisms for heme production has mainly focused on heterotrophic microorganisms, such as *E. coli*, *B. subtilis*, and yeast. For example, Ge et al. constructed a high heme-producing *E. coli* strain, Ec-M13, which was engineered by the co-expression of the genes *hemE*, *hemF*, *hemG*, and *hemH* in *E. coli*, achieving a high heme production of 1.04 mg/L-OD$_{600}$ [33], and Zhao et al. reported that *E. coli* with the knockout of *ldhA*, *pta*, and *yfe* produced a total heme of 7.88 mg/L and 1.26 mg/L of extracellular heme. In addition, the engineered strains of *E. coli* with the overexpression of the heme exporter CcmABC for feed batch fermentation from glucose only and glucose supplemented with l-glutamic acid secreted extracellular heme amounts of 73.4 mg/L and 151.4 mg/L, respectively, which accounted for 63.5% and 63.3%, respectively, of the total heme production [26]. Yang et al. reported that, in *B. subtilis*, the knockout of *hemX*, *hemA*, and *rocG* resulted in an increase of 42.7% in heme production, the overexpression of *hemCDB* caused increased heme production by 39%, and the knockout of *nasF*, *hmoA*, and *hmoB* resulted in an increase of 52% in heme production [11]. The modified strain of *B. subtilis* produced 248.26 ± 6.97 mg/L of total heme and 221.83 ± 4.71 mg/L of extracellular heme during feed batch fermentation in a 10 L fermenter [11]. Kwon et al. reported the assembly of the entire heme biosynthesis pathway in a triple plasmid system and overexpression of the corresponding genes using *E. coli* as the host [27]; without further optimization, this method yielded significantly increased porphyrin production up to 90 μM, which was comparable to the industrial production levels of vitamin B12 [27]. Ishchuk et al. achieved a dramatic increase in the intracellular heme content in yeast cells by constructing a genome-scale metabolic model and combinatorically modifying the corresponding target genes based on computer simulations [18].

In this study, *Synechocystis* sp. PCC6803 was used as a chassis to overexpress a group of genes involved in heme synthesis, including the *hemC*, *hemF*, and *hemH* genes, the *Synechocystis* sp. PCC6803 hemoglobin gene, *glbn*, and the *E. coli* glucose transporter protein gene *ptsG*, all of which were co-expressed. The key enzyme ChlH of the chlorophyll synthesis gene (*chlH*) and the key enzyme PcyA of the phycocyanin synthesis gene (*pcyA*) were knocked out. A total of five mutant strains (ΔPcyA, ΔChlH, ptsG, CFH, and Glbn-ptsG) were successfully constructed. The heme contents of both the ΔPcyA and CFH strains were significantly increased by 4.41- and 3.05-fold, respectively, compared to the WT, with the highest heme content detected in the mutant strain ΔpcyA. This may be because, in *Synechocystis* sp. PCC6803, phycocyanin is synthesized using heme as a precursor, and the synthesis of phycocyanin causes a decrease in the heme content. Therefore, we further used homologous recombination knockout in *Synechocystis* sp. PCC6803 to knockout the key enzyme for phycocyanin synthesis to generate a mutant strain (PcyA-ChlH) with the heme content greatly increased. However, because of the important physiological functions that phycocyanin plays in the life activities of *Synechocystis* sp. PCC6803, it is still technically challenging to achieve a complete knockout of the phycocyanin synthesis pathway. A previous study found that overexpression of the genes *hemC*, *hemF*, and *hemH*, respectively, caused increased heme content in *Synechocystis* sp. PCC6803 [32]. In our study, the results indicated that co-expression of the genes *hemC*, *hemF*, and *hemH* was also effective in promoting heme content in mutant strains. Among the five *Synechocystis* sp. PCC6803 mutant strains constructed in this study (ΔPcyA, ΔChlH, CFH, ptsG, and Glbn-ptsG), a phycocyanin content of 4.70 μg/mL was obtained in strain ΔChlH, which was 1.24-fold higher than that of the WT. This may be because chlorophyll synthesis is a competing pathway of the phycocyanin synthesis pathway, and the knockout of *chlH* directs more metabolic fluxes to phycocyanin synthesis. In contrast, phycocyanin was still detected in the knockout mutant strain of the key enzyme for phycocyanin synthesis, PcyA, with a decreased phycocyanin content by 66.3% compared to the WT. This may be because phycocyanin is an important photosynthetic pigment of *Synechocystis* sp. PCC6803, playing an irreplaceable role in the life activities of *Synechocystis* sp. PCC6803. Therefore, it is not possible to obtain the mutant strain of *Synechocystis* sp. PCC6803 with complete knockout of phycocyanin.

Chlorophyll a contents of 4.90 μg/mL and 2.10 μg/mL were obtained in both the mutants ΔPCYA and ΔChlH of *Synechocystis* sp. PCC6803, respectively. The chlorophyll a content of the mutant strain ΔPcyA was 1.5-fold higher than that of the WT. This may be because phycocyanin synthesis is a competing pathway of the chlorophyll synthesis pathway, and the knockout of *pcyA* directs more metabolic fluxes to chlorophyll synthesis. However, the magnesium chelating enzyme ChlD/I, an isoform of ChlH, was present in *Synechocystis* sp. PCC6803. Therefore, although the knockout of *chlH* reduced the chlorophyll a content of *Synechocystis* sp. PCC6803 by 34.0% compared with that of the WT, chlorophyll a synthesis was not completely prevented.

Our results showed that the co-expression of three key enzymes (HemC, HemF, and HemH) of the heme synthesis pathway in *Synechocystis* sp. PCC6803 was also effective in promoting the content of heme in *Synechocystis* sp. PCC6803. However, the modification of a single engineered target played a limited role in enhancing the heme content in *Synechocystis* sp. PCC6803 and was prone to cause metabolic burden and cytotoxicity, ultimately adversely affecting the growth of *Synechocystis* sp. PCC6803. Further metabolic engineering modification and optimization are needed to solve these problems.

3.2. Highest Heme Yield Obtained by the Combined Transformant PcyA-ChlH of Synechocystis sp. PCC6803

The screened high heme-producing *Synechocystis* sp. PCC6803 mutant strain PcyA-ChlH was further combinatorically transformed with multiple heme synthesis-related plasmids to obtain three combined transformants (PcyA-ChlH, CFH-PcyA, and glbn-ptsG-CFH). Our findings revealed that the heme contents of all three combined transformants were higher than that of the WT, with the strain PcyA-ChlH showing a heme content that was 8.03-fold higher than that of the WT. The scanning electron microscopic observations revealed that the combined transformants underwent significant changes in extracellular adhesion as well as cell morphology and size. In particular, adhesion among algal cells was evident in the strain PcyA-ChlH. It is hypothesized that the increase in heme content caused significant changes in the oxidative stress and metabolic levels of the cyanobacterial cells, resulting in the secretion of large amounts of extracellular polysaccharides, which were accumulated to bind the cyanobacterial cells together.

Synechocystis sp. PCC6803 is rich in phycocyanin, with a well-established genetic transformation system [34]. Previous studies reported that the rate-limiting reaction in the synthesis of phycocyanin is catalyzed by Ho1 and PcyA [35–39]. In addition, PcyA plays a crucial role in the process of constructing heterologous hosts to synthesize phycocyanin [40]. The findings of our study revealed the highest heme content in the mutant strain ΔPcyA of *Synechocystis* sp. PCC6803 with the knockout of PcyA, a key enzyme for phycocyanin synthesis, and the lowest phycocyanin content in the combined transformants.

The results of biomass variations between the combined transformants and the WT revealed slower growth rates in all combined transformants than in the WT. Therefore, it is hypothesized that the increased contents of heme and its synthetic precursors caused metabolic burden and cytotoxicity to the cyanobacterial cells and, to a certain extent, adversely affected the growth of cyanobacterial cells. In addition, the contents of phycocyanin, chlorophyll a, and carotenoids of both the WT and combined transformants were investigated, and it was found that the increase in heme content showed the most pronounced effect on phycocyanin synthesis. In particular, the lowest content of phycocyanin was revealed in the strain PcyA-ChlH, showing a decrease of 65.7% compared to the WT. This could be due to the competition between the phycocyanin synthesis and heme synthesis for metabolic flux. It is noted that, although the heme content of *Synechocystis* sp. PCC6803 was improved in this study, it is still far from the industrial production level.

3.3. Addition of Glucose Enhanced Heme Content and Growth of Synechocystis sp. PCC6803 Combined Transformants under Mixotrophic Culture

Synechocystis sp. PCC6803, a mixotrophic cyanobacteria, can utilize glucose as the sole carbon source for biomass accumulation [41,42]. In this study, three metabolically modified heme-producing *Synechocystis* sp. PCC6803 combinatorial transformants were obtained. However, the growth rates of the three mutant strains under photoautotrophic conditions were all lower than that of the WT. This may be due to the increased intracellular free heme content causing an increase in the level of intracellular reactive oxygen species (ROS), which is detrimental to the metabolic activities and growth of *Synechocystis* sp. PCC6803 [43–45]. In order to promote the accumulation of biomass and heme synthesis in combinatorial transformants, we further explored the growth of *Synechocystis* sp. PCC6803 in a mixotrophic medium with additional glucose as a carbon source. The results revealed an enhanced growth rate and biomass accumulation of *Synechocystis* sp. PCC6803 in the mixotrophic medium containing 1 g/L glucose in comparison with glucose levels of 0, 0.5, 5, and 10 g/L, and, in particular, *Synechocystis* sp. PCC6803 could not grow in medium containing glucose levels of 5 and 10 g/L. In addition, compared to the WT, higher growth rates were revealed in the mutant strain ptsG in medium containing glucose levels of 0.5 g/L and 1 g/L.

Both the WT and combined transformants of *Synechocystis* sp. PCC6803 were cultured in the mixotrophic medium containing 1 g/L glucose. The biomass of two combined transformants, PcyA-ChlH and glbn-ptsG-CFH, reached the maximum level, as determined by the OD_{730} values of 4.32 and 4.07, respectively, at the 15th day of culture, which were higher than that of the WT (2.94). In addition, the contents of heme, phycocyanin, chlorophyll a, and carotenoid were assayed to investigate the variations in heme content and metabolic flow of *Synechocystis* sp. PCC6803. The highest intracellular and extracellular heme contents were revealed in the combined transformant PcyA-ChlH in mixotrophic culture (2.05 mg/L), which were both 10.25-fold higher than those of the WT. PcyA is a key enzyme involved in the synthesis of phycocyanin using heme as a precursor, while ChlH is a key enzyme in the chlorophyll synthesis pathway, which is a competing pathway of the heme synthesis pathway in *Synechocystis* sp. PCC6803. Thus, the knockout of these two enzymes facilitates the direction of more metabolic flow to heme synthesis. These results indicated that the mixotrophic conditions were effective in improving the poor growth of the combined transformants compared to the WT, and ultimately further increasing the heme content of the combined transformants of *Synechocystis* sp. PCC6803. This generates significant potential to help increase the heme content to the industrial production level.

4. Materials and Methods

4.1. Microbial Strains and Plasmids

All the microbial strains used in this study were listed in Table S1. *Escherichia coli* DH5α was used for DNA cloning as well as construction and preservation of plasmids. *Synechocystis* sp. PCC6803 was used as a template for amplification of the target genes.

4.2. Culture Condition

All strains of *Synechocystis* sp. PCC6803 were incubated at 30 °C in BG-11 medium (Table S2). BG-11 liquid culture medium was prepared as follows. The corresponding chemicals and reagents were accurately weighed and added to 1 L of dd H_2O for dissolution (Table S2). The $1000\times$ trace element solution was prepared (Table S3), with pH adjusted to 7.5, sterilized at 121 °C for 20 min, and stored at room temperature. BG-11 solid medium was prepared as follows. The liquid medium was sterilized by adding 1.5% agar powder at 121 °C for 20 min and stored at room temperature. *Escherichia coli* DH5α was cultured in LB medium at 37 °C and 200 rpm.

4.3. Construction of Recombinant Plasmids and Microbial Strains

DNA manipulation was performed using standard molecular genetic techniques [32]. The primers used for the construction of plasmids and cloning of target fragments were shown in Table S4. The 4UD homologous expression vector framework was obtained using primers 4UD-F and 4UD-R with the plasmid PBSK-4UD, stored in our laboratory, as a template. The primer pairs Prbcl-F and Prbcl-R, PpsbA2S-F and PpsbA2S-R, HemC-F and HemC-R, and HemF-F and HemH-R were used to amplify the promoters Prbcl and PpsbA2S, and the heme synthesis-associated genes (*hemC*, *hemF*, and *hemH*), respectively, using *Synechocystis* sp. PCC6803 genomic DNA as a template. The kanamycin resistance expression cassette was amplified using primers KAN-F and KAN-R with plasmid pBlunt-Kan, stored in our laboratory, as a template. Plasmid PBSK was used as a template to amplify the terminator T1T2. The promoter PPSBA2L was obtained using primers PPSBA2L-F and PPSBA2L-R with *Synechocystis* sp. PCC6803 genomic DNA as a template. The gene *ptsG*, encoding glucose transporter protein, was obtained using primers ptsG-F and ptsG-R with *E. coli* DH5α as a template. The terminator T1T2 was amplified using primers 6TT-F and 6TT-R with plasmid PBSK as a template. The gentamicin resistance expression cassette GM was amplified using primers GM-F with GM-R with the plasmid pBlunt-GM, stored in our laboratory, as a template. The 6UD homologous expression vector framework was obtained using primers 6UD-F and 6UD-R with *Synechocystis* sp. PCC6803 genomic DNA as a template. The *Synechocystis* sp. PCC6803 genome was amplified using primers PCYAU-F and PCYAU-R to obtain the PCYA upstream arm and using primers PCYAD-F and PCYAD-R to obtain the PCYA downstream arm. The pBlunt-Cm vector stored in our laboratory was amplified using primers Chl-F1 and Chl-R1 to obtain the chloramphenicol resistance gene (*Cm*). Primers PUC-F1 and PUC-R1 were used to amplify the PUC19 homologous vector expression frame stored in our laboratory.

Primer pairs ChlHU-F/ChlHU-R and ChlHD-F/ChlHD-R were used to amplify the *Synechocystis* sp. PCC6803 genome to obtain the ChlH upstream arm ChlHU and the ChlH downstream arm ChlHD, respectively. Primers CM-F1 and CM1-R1 were used to amplify the pBlunt-Gm vector stored in our laboratory to obtain the gentamicin resistance gene expression cassette Gm, and primers PUC-F2 and PUC-R2 were used to amplify the PUC19 homologous vector expression frame.

Linear plasmids and coding sequences were assembled using the ClonExpress II One Step Cloning Kit (Vazyme, Nanjing, China). Four recombinant plasmids, including PBSK-4UD-hem C/F/H (Figure S1A), PBSK-pstG-6UD (Figure S1B), pcyaUD-Chl (Figure S1C), and ChlH-GM (Figure S1D), were constructed for the transformation of *Synechocystis* sp. PCC6803, respectively, by using the methods described above.

4.4. Plasmid Transformation and Transformant Screening

The four recombinant plasmids, i.e., PBSK-4UD-hem C/F/H, PBSK-ptsG-6UD, pcyaUD-Chl, and ChlH-GM, were transformed into host cells of *Synechocystis* sp. PCC6803. A total of 8 mL of well-grown *Synechocystis* sp. PCC6803 cyanobacterial sap was collected and centrifuged at 3500 rpm for 10 min. The precipitate was collected and rinsed with 5 mL of sterile water, and then centrifuged at 3500 rpm for 10 min. The precipitate was added to 200 µL of BG-11 medium to resuspend, followed by adding a total of 20 µL of homologous recombinant vector, incubated at 30 °C under low light for 6 h. Then, the mixture was evenly spread in BG-11 plates with mixed cellulose ester membranes for 24 h, which were then transferred to the BG-11 plates containing antibiotics. Transformed *Synechocystis* sp. PCC6803 was incubated for two weeks at 30 °C in an incubator with a light intensity of 2000 lux and a light/dark photoperiod of 12 h/12 h to cultivate the transformants. After the unicellular algae grew on the mixed cellulose ester membranes, the single algal colonies were picked with a receiving ring and zoned on solid BG-11 medium containing the appropriate antibiotics, with the antibiotic concentration increased generation by generation. After four passages, the single algal colonies were transferred to 96-well plates for incubation, and when the OD_{730} value of the algal solution reached 0.2 or more, the algal solution in the 96-well plates was transferred to conical flasks containing culture medium for expanded culture. Algal colony PCR was performed to compare the size of the target bands, and then the target bands were recovered for sequencing to determine the acquisition of *Synechocystis* sp. PCC6803 transformants.

4.5. Determination of the Growth Curve of Synechocystis sp. PCC6803

Synechocystis sp. PCC6803 was inoculated into the BG-11 liquid medium, with three parallel control groups performed to minimize error, and samples were collected at 24 h intervals from the time of inoculation. A total of 200 µL *Synechocystis* sp. PCC6803 cyanobacterial solution was spiked into a 96-well plate, with the OD_{730} values determined using an enzyme labeler. Growth curves were plotted after the sampling cycle of a total of 26 d was completed.

4.6. Measurement of Heme Content

Synechocystis sp. PCC6803 cells were harvested by centrifugation of cyanobacterial fluid at 12,000× *g*/min for 5 min with the supernatant removed. The precipitate was collected and washed thrice with sterile water and each wash was followed by centrifugation at 12,000× *g*/min for 5 min to collect the precipitate. Each experiment was performed with three biological replicates. The collected *Synechocystis* sp. PCC6803 cells were lyophilized in a freeze dryer. The lyophilized powder of *Synechocystis* sp. PCC6803 was accurately weighed (0.01 g) and added with neutral acetone (80%) to extract chlorophyll a [15]. The above mixture was centrifuged to remove the supernatant and the extraction was repeated using neutral acetone (80%) until the supernatant was no longer green. The heme content in the cyanobacterial cells was later detected using a fluorescence method [46]. The precipitate was resuspended using 500 μL of 20 mM oxalic acid solution and then placed in a refrigerator at 4 °C for 16 h. After sonication and fragmentation, 500 μL of 2 M oxalic acid solution was added to the mixture, which was then divided into two equal portions. One portion was heated at 98 °C and the other was kept at room temperature. In 30 min, the above solution was centrifuged, and the supernatant was collected and added to a black 96-well plate to measure the fluorescence values under excitation light at 400 nm and emission light at 620 nm (Microplate reader, SpectraMax M2, Molecular Devices, Shanghai, China). The fluorescence value of heme was obtained by subtracting the fluorescence value measured at room temperature from the fluorescence value measured at 98 °C. The heme standard was added to a solution containing 1% (*w/v*) bovine serum albumin and 0.01 M KOH, dissolved, added to oxalic acid solution, and heated to prepare the standard, then the standard curve was plotted.

The content of phycocyanin was determined using the method previously reported [47]. *Synechocystis* sp. PCC6803 cyanobacterial solution was centrifuged at 6000 rpm for 5 min, the supernatant was removed, and the precipitate was collected and dried for 12 h in a freeze dryer. Then, a total of 10 mg of *Synechocystis* sp. PCC6803 dry powder was accurately weighed, added to 1 mL of PBS solution, and placed at −70 °C for 1 d for freezing. The frozen solution was thawed in a refrigerator at 4 °C and sonicated, followed by centrifugation at 10,000 rpm and 4 °C for 10 min. Then, a total of 200 μL of the supernatant was collected and placed in a 96-well plate, with the absorbance measured at 615 nm and 652 nm, respectively, using an enzyme labeler (Molecular Devices, Shanghai, China). The phycocyanin content was calculated according to the following equation: phycocyanin content (mg/mL) = $(OD_{615} - 0.474 \times OD_{652})/5.34$.

4.7. Determination of Chlorophyll a Content

A total of 1 mL of *Synechocystis* sp. PCC6803 cyanobacterial solution was collected and centrifuged at 10,000 rpm for 6 min, the supernatant was removed, and the precipitate was resuspended using 1 mL of neutral acetone (80%). The sample was heated at 55 °C for 40 min and centrifuged at 10,000 rpm for 6 min. Then, a total of 200 μL of supernatant was collected and placed in a 96-well plate, with the absorbance measured at 665 nm and 720 nm, respectively, using an enzyme labeler. The content of chlorophyll a was calculated according to the following equation: chlorophyll a content (μg/mL) = $12.9447 \times (OD_{665} - OD_{720})$.

4.8. Determination of Carotenoid Content

For the determination of carotenoid content of *Synechocystis* sp. PCC6803, the sample treatment was the same as that for the determination of chlorophyll a content (above), and the absorbance of the supernatant was measured at 470 nm and 720 nm, respectively, using an enzyme meter. The carotenoid content was calculated according to the following equation: carotenoid content (μg/mL) = $[1000 \times (OD_{470} - OD_{720}) - 2.86 (Chla [\mu g/mL])]/221$.

4.9. qRT-PCR Assay

Quantitative reverse transcription PCR (qRT-PCR) analysis of heme synthesis gene expression was performed using the CFX Connect real-time PCR system (Bio-Rad, Shanxi, China). The internal reference gene *rnpB* was amplified using primers rnpB-F and rnpB-R. Primers were designed to amplify the heme synthesis genes (Table S5), and qPCR was performed according to the manufacturer's instructions (Ruiboxingke, Beijing, China). The total volume of each qPCR was 20 μL, containing 10 μL of qPCR Master Mix, 200 nM of each primer, and 1 μL of 10-fold diluted cDNA template. The reaction was performed in an 8-well optical grade PCR plate with the amplification program provided in Table S6. The melting curve was generated using a Bio-Rad CFX Maestro 2.2 version 5.2.008.0222 (Bio-Rad, Hercules, CA, USA). Each experiment was repeated with three biological replicates. Negative controls were performed using sterile water instead of cDNA template. Relative expression levels of heme synthesis-related genes were determined using the relative $2^{-\Delta\Delta Ct}$ method [48].

4.10. Scanning Electron Microscopic Observation of Cell Morphology of of Synechocystis sp. PCC6803

The sample (1 mL) of *Synechocystis* sp. PCC6803 cyanobacterial solution was collected and centrifuged at 10,000 rpm for 6 min, with the supernatant removed. Then, 1 mL PBS buffer was used to wash the sample surface impurities thrice. Next, 0.5 mL of 3% glutaraldehyde was added to the sample for 6 h of fixation. The sample was treated with gradient dehydration in ethanol solutions of 30%, 40%, 50%, 60%, 70%, 80%, and 90% for 10 min each. Dehydration in 100% anhydrous ethanol for 10 min was repeated thrice. Displacement with tert-butanol was performed three times, each for 30 min. The sample was mounted to the sample stage with double-sided tape. Scanning electron microscopy (Zeiss, Oberkochen, German) analysis was performed after coating to observe the morphological features of the cell surface at a magnification of 5000×.

4.11. Statistical Analysis

Data were expressed as mean ± standard deviation (SD) based on three biological replicates. Data were statistically analyzed using SPSS 16.0 (IBM SPSS, Chicago, IL, USA). A one-way analysis of variance (ANOVA) was performed, with significant differences determined at $p < 0.05$ (*) and highly significant differences determined at $p < 0.01$ (**), respectively.

5. Conclusions

In this study, we successfully modularized *Synechocystis* sp. PCC6803 to screen for functional elements and modules of genes capable of increasing the heme content of *Synechocystis* sp. PCC6803. The co-expression of the key heme synthesis genes *hemC*, *hemF*, and *hemH* in *Synechocystis* sp. PCC6803, and the knockout of *pcyA*, a key gene in the phycocyanin synthesis pathway, and *chlH*, a key gene in the chlorophyll synthesis pathway, were all able to effectively increase the heme content in *Synechocystis* sp. PCC6803. The highest heme content (0.97 mg/L) was obtained in the mutant strain of *Synechocystis* sp. PCC6803 with the knockout of *pcyA* (ΔPcyA), which was 4.41-fold higher than that of the WT. A total of three combined transformants of *Synechocystis* sp. PCC6803 were obtained by natural transformation and had improved heme production compared with the WT; the highest heme content (1.83 mg/L) was revealed in the combined transformant PcyA-ChlH with the knockout of both *pcyA* and *chlH*, which was 8.03-fold higher than that of the WT. In the combined transformant PcyA-ChlH, *hemF*, *hemH*, *glbn*, and *chlM* were up-regulated, with *hemF* and *glbn* substantially up-regulated, showing an increased expression by more than 5-fold. Finally, we investigated the effects of different concentrations (0, 0.5, 1, 5, and 10 g/L) of glucose on the growth of *Synechocystis* sp. PCC6803 to optimize its culture conditions. The optimal growth of *Synechocystis* sp. PCC6803 was obtained in medium containing 1 g/L glucose. The combined transformant PcyA-ChlH in mixotrophic medium produced a heme content of 2.05 mg/L, which was 10.25-fold higher than that of the WT. This study provides novel engineering strategies, key target genes, and chassis selection for the construction of microbial cell factories for high heme production, as well as strong experimental evidence to support the metabolic modification of microbes for the production of porphyrin-like high-value compounds.

Supplementary Materials: The following supporting information can be downloaded at: https://www.mdpi.com/article/10.3390/md22090378/s1. Figure S1. plasmid mapping. Figure S2. PCR detection of *Synechocystis* sp. PCC6803 wild-type and combined transformants. Table S1. Strains used in this study. Table S2. BG-11 Media Formulation. Table S3. 1000× trace elements. Table S4. Primers used for construction of plasmids and cloning of target fragments. Table S5. qPCR primers and their sequences. Table S6. PCR amplification procedure.

Author Contributions: Conceptualization, Z.G., C.M., K.C. and F.S.; methodology, K.C., F.S., Z.X., Y.C., X.Z. and T.C.; validation, Z.G., C.M., K.C. and F.S.; investigation, K.C., F.S., X.Z., Y.C. and J.M.; resources, Z.G. and C.M.; data curation, H.T., K.C. and Z.X.; writing—original draft preparation, K.C., Z.X. and Y.C.; writing—review and editing, F.S., K.C., X.Z., J.M., W.M., J.S., R.Z. and Y.C.; funding acquisition, Z.G. and C.M. All authors have read and agreed to the published version of the manuscript.

Funding: This work was supported by the National Natural Science Foundation of China (31972815 and 42176124), the Natural Science Foundation of Shandong Province (ZR2019ZD17, ZR2020ZD23, ZR2021QD137, ZR2021MC051, and ZR2023ZD30), the Scientific Research Fund of Binzhou Medical University (BY2021KYQD18, BY2021KYQD25, and BY2021KYQD28), and the Yantai Economic-Technological Development Area Innovation Pilot Project (2021RC007).

Institutional Review Board Statement: Not applicable.

Data Availability Statement: All raw data are readily available upon request.

Acknowledgments: We thank Zilong Wang and Yu Wang for their great technical support.

Conflicts of Interest: The authors declare no conflicts of interest.

References

1. Chiabrando, D.; Vinchi, F.; Fiorito, V.; Mercurio, S.; Tolosano, E. Heme in pathophysiology: A matter of scavenging, metabolism and trafficking across cell membranes. *Front. Pharmacol.* **2014**, *5*, 61. [CrossRef]
2. Beas, J.Z.; Videira, M.A.; Saraiva, L.M. Regulation of bacterial haem biosynthesis. *Coord. Chem. Rev.* **2022**, *452*, 214286. [CrossRef]
3. Sassa, S.; Nagai, T. The role of heme in gene expression. *Int. J. Hematol.* **1996**, *63*, 167–178. [CrossRef] [PubMed]
4. Emerson, R.; Lewis, C.M. The photosynthetic efficiency of phycocyanin in chroococcus, and the problem of carotenoid participation in photosynthesis. *J. Gen. Physiol.* **1942**, *25*, 579–595. [CrossRef]
5. Fan, T.; Roling, L.; Hedtke, B.; Grimm, B. FC2 stabilizes POR and suppresses ALA formation in the tetrapyrrole biosynthesis pathway. *New Phytol.* **2023**, *239*, 624–638. [CrossRef] [PubMed]
6. Chung, J.; Chen, C.; Paw, B.H. Heme metabolism and erythropoiesis. *Curr. Opin. Hematol.* **2012**, *19*, 156–162. [CrossRef]
7. Dailey, H.A.; Medlock, A.E. A primer on heme biosynthesis. *Biol. Chem.* **2022**, *403*, 985–1003. [CrossRef] [PubMed]
8. Koreny, L.; Oborník, M.; Horáková, E.; Waller, R.F.; Lukes, J. The convoluted history of haem biosynthesis. *Biol. Rev.* **2022**, *97*, 141–162. [CrossRef]
9. Rivero-Müller, A.; Lajić, S.; Huhtaniemi, I. Assisted large fragment insertion by Red/ET-recombination (ALFIRE)—An alternative and enhanced method for large fragment recombineering. *Nucleic Acids Res.* **2007**, *35*, e78. [CrossRef]
10. Zhao, X.; Zhou, J.; Du, G.; Chen, J. Recent Advances in the Microbial Synthesis of Hemoglobin. *Trends Biotechnol.* **2021**, *39*, 286–297. [CrossRef]
11. Yang, S.; Wang, A.; Li, J.; Shao, Y.; Sun, F.; Li, S.; Cao, K.; Liu, H.; Xiong, P.; Gao, Z. Improved biosynthesis of heme in *Bacillus subtilis* through metabolic engineering assisted fed-batch fermentation. *Microb. Cell Factories* **2023**, *22*, 1–12. [CrossRef]
12. Pizarro, F.; Olivares, M.; Valenzuela, C.; Brito, A.; Weinborn, V.; Flores, S.; Arredondo, M. The effect of proteins from animal source foods on heme iron bioavailability in humans. *Food Chem.* **2016**, *196*, 733–738. [CrossRef] [PubMed]
13. Geng, Z.X.; Ge, J.X.; Cui, W.; Zhou, H.; Deng, J.Y.; Xu, B.C. Efficient De Novo Biosynthesis of Heme by Membrane Engineering in *Synechocystis* sp. PCC 6803. *Int. J. Mol. Sci.* **2022**, *23*, 15524. [CrossRef]
14. Ishchuka, O.P.; Domenzain, I.; Sánchez, B.J.; Muñiz-Paredes, F.; Martínez, J.L.; Nielsen, J.; Petranovic, D. Genome-scale modeling drives 70-fold improvement of intracellular heme production in *Synechocystis* sp. PCC 6803. *Proc. Natl. Acad. Sci. USA* **2022**, *119*, e2108245119.
15. Espinas, N.A.; Kobayashi, K.; Takahashi, S.; Mochizuki, N.; Masuda, T. Evaluation of Unbound Free Heme in Plant Cells by Differential Acetone Extraction. *Plant Cell Physiol.* **2012**, *53*, 1344–1354. [CrossRef] [PubMed]
16. Cai, T.; Ge, X.M.; Park, S.Y.; Li, Y.B. Comparison of *Synechocystis* sp. PCC6803 and for lipid production using artificial seawater and nutrients from anaerobic digestion effluent. *Bioresour. Technol.* **2013**, *144*, 255–260. [CrossRef] [PubMed]
17. Liu, Y.; Cui, Y.; Chen, J.; Qin, S.; Chen, G. Metabolic engineering of *Synechocystis* sp. PCC6803 to produce astaxanthin. *Algal Res.* **2019**, *44*, 101679. [CrossRef]
18. Iijima, H.; Nakaya, Y.; Kuwahara, A.; Hirai, M.Y.; Osanai, T. Seawater cultivation of freshwater cyanobacterium *Synechocystis* sp. PCC 6803 drastically alters amino acid composition and glycogen metabolism. *Front. Microbiol.* **2015**, *6*, 326. [CrossRef]
19. Chen, C.-Y.; Yeh, K.-L.; Aisyah, R.; Lee, D.-J.; Chang, J.-S. Cultivation, photobioreactor design and harvesting of microalgae for biodiesel production: A critical review. *Bioresour. Technol.* **2011**, *102*, 71–81. [CrossRef]

20. Santos-Merino, M.; Torrado, A.; Davis, G.A.; Röttig, A.; Bibby, T.S.; Kramer, D.M.; Ducat, D.C. Improved photosynthetic capacity and photosystem I oxidation via heterologous metabolism engineering in cyanobacteria. *Proc. Natl. Acad. Sci. USA* **2021**, *118*, e2021523118. [CrossRef]

21. Zhou, J.; Zhang, H.; Zhang, Y.; Li, Y.; Ma, Y. Designing and creating a modularized synthetic pathway in cyanobacterium Synechocystis enables production of acetone from carbon dioxide. *Metab. Eng.* **2012**, *14*, 394–400. [CrossRef] [PubMed]

22. Mo, H.; Xie, X.; Zhu, T.; Lu, X. Effects of global transcription factor NtcA on photosynthetic production of ethylene in recombinant *Synechocystis* sp. PCC 6803. *Biotechnol. Biofuels* **2017**, *10*, 145. [CrossRef] [PubMed]

23. Lindberg, P.; Park, S.; Melis, A. Engineering a platform for photosynthetic isoprene production in cyanobacteria, using Synechocystis as the model organism. *Metab. Eng.* **2010**, *12*, 70–79. [CrossRef]

24. Carpine, R.; Du, W.; Olivieri, G.; Pollio, A.; Hellingwerf, K.J.; Marzocchella, A.; Branco dos Santos, F. Genetic engineering of *Synechocystis* sp. PCC6803 for poly-β-hydroxybutyrate overproduction. *Algal Res.* **2017**, *25*, 117–127. [CrossRef]

25. Diao, J.; Song, X.; Zhang, L.; Cui, J.; Chen, L.; Zhang, W. Tailoring cyanobacteria as a new platform for highly efficient synthesis of astaxanthin. *Metab. Eng.* **2020**, *61*, 275–287. [CrossRef]

26. Zhao, X.R.; Choi, K.R.; Lee, S.Y. Metabolic engineering of for secretory production of free haem. *Nat. Catal.* **2018**, *1*, 720–728. [CrossRef]

27. Kwon, S.J.; de Boer, A.L.; Petri, R.; Schmidt-Dannert, C.; Sj, K. High-Level Production of Porphyrins in Metabolically Engineered *Escherichia coli*: Systematic Extension of a Pathway Assembled from Overexpressed Genes Involved in Heme Biosynthesis. *Appl. Environ. Microbiol.* **2003**, *69*, 4875–4883. [CrossRef] [PubMed]

28. Lee, M.J.; Kim, H.J.; Lee, J.Y.; Kwon, A.S.; Jun, S.Y.; Kang, S.H.; Kim, P. Effect of gene amplifications in porphyrin pathway on heme biosynthesis in a recombinant *Escherichia coli*. *J. Microbiol. Biotechnol.* **2013**, *23*, 668–673. [CrossRef]

29. Krämer, L.C.; Wasser, D.; Haitz, F.; Sabel, B.; Büchel, C. Heterologous expression of HUP1 glucose transporter enables low-light mediated growth on glucose in *Phaeodactylum tricornutum*. *Algal Res.* **2022**, *64*, 102719. [CrossRef]

30. Zhang, J.L.; Kang, Z.; Chen, J.; Du, G.C. Optimization of the heme biosynthesis pathway for the production of 5-aminolevulinic acid in *Synechocystis* sp. PCC6803. *Sci. Rep.* **2015**, *5*, 8584.

31. Lithi, U.J.; Laird, D.W.; Ghassemifar, R.; Wilton, S.D.; Moheimani, N.R. Microalgae as a source of bioavailable heme. *Algal Res.* **2024**, *77*, 103363. [CrossRef]

32. Cao, K.; Wang, X.; Sun, F.; Zhang, H.; Cui, Y.; Cao, Y.; Yao, Q.; Zhu, X.; Yao, T.; Wang, M.; et al. Promoting Heme and Phycocyanin Biosynthesis in *Synechocystis* sp. PCC 6803 by Overexpression of Porphyrin Pathway Genes with Genetic Engineering. *Mar. Drugs* **2023**, *21*, 403. [CrossRef] [PubMed]

33. Ge, J.Z.; Wang, X.L.; Bai, Y.G.; Wang, Y.; Wang, Y.; Tu, T.; Qin, X.; Su, X.; Luo, H.; Yao, B.; et al. Engineering for efficient assembly of heme proteins. *Microb. Cell Fact.* **2023**, *22*, 59. [CrossRef] [PubMed]

34. Veetil, V.P.; Angermayr, S.A.; Hellingwerf, K.J. Ethylene production with engineered *Synechocystis* sp PCC 6803 strains. *Microb Cell Fact.* **2017**, *16*, 34. [CrossRef] [PubMed]

35. Brown, S.B.; Houghton, J.D.; Vernon, D.I. New trends in photobiology biosynthesis of phycobilins. Formation of the chromophore of phytochrome, phycocyanin and phycoerythrin. *J. Photochem. Photobiol. B Biol.* **1990**, *5*, 3–23. [CrossRef]

36. Plank, T.; Anderson, L.K. Heterologous assembly and rescue of stranded phycocyanin subunits by expression of a foreign cpcBA operon in *Synechocystis* sp. strain 6803. *J. Bacteriol.* **1995**, *177*, 6804–6809. [CrossRef]

37. Shang, M.-H.; Sun, J.-F.; Bi, Y.; Xu, X.-T.; Zang, X.-N. Fluorescence and antioxidant activity of heterologous expression of phycocyanin and allophycocyanin from Arthrospira platensis. *Front. Nutr.* **2023**, *10*, 1127422. [CrossRef]

38. Yilmaz, M.; Kang, I.; Beale, S.I. Heme oxygenase 2 of the cyanobacterium *Synechocystis* sp. PCC 6803 is induced under a microaerobic atmosphere and is required for microaerobic growth at high light intensity. *Photosynth. Res.* **2010**, *103*, 47–59. [CrossRef] [PubMed]

39. Heck, S.; Sommer, F.; Zehner, S.; Schroda, M.; Gehringer, M.M.; Frankenberg-Dinkel, N. Expanding the toolbox for phycobiliprotein assembly: Phycoerythrobilin biosynthesis in *Synechocystis* sp. PCC6803. *Physiol. Plantarum.* **2024**, *176*, e14137. [CrossRef]

40. Sun, D.G.; Xiao, D.F.; Zang, X.N.; Wu, F.; Jin, Y.; Cao, X.; Guo, Y.; Liu, Z.; Wang, H.; Zhang, X. Cloning of the pcyA gene from Arthrospira platensis FACHB314 and its function on expression of a fluorescent phycocyanin in heterologous hosts. *J. Appl. Phycol.* **2019**, *31*, 3665–3675. [CrossRef]

41. Schmetterer, G.R. Sequence conservation among the glucose transporter from the cyanobacterium *Synechocystis* sp. PCC 6803 and mammalian glucose transporters. *Plant Mol. Biol.* **1990**, *14*, 697–706. [CrossRef] [PubMed]

42. Hu, J.; Nagarajan, D.; Zhang, Q.; Chang, J.-S.; Lee, D.-J. Heterotrophic cultivation of microalgae for pigment production: A review. *Biotechnol. Adv.* **2018**, *36*, 54–67. [CrossRef]

43. Roumenina, L.T.; Rayes, J.; Lacroix-Desmazes, S.; Dimitrov, J.D. Heme: Modulator of Plasma Systems in Hemolytic Diseases. *Trends Mol. Med.* **2016**, *22*, 200–213. [CrossRef] [PubMed]

44. Wang, L.; Yang, T.; Pan, Y.; Shi, L.; Jin, Y.; Huang, X. The Metabolism of Reactive Oxygen Species and Their Effects on Lipid Biosynthesis of Microalgae. *Int. J. Mol. Sci.* **2023**, *24*, 11041. [CrossRef] [PubMed]

45. Ding, X.T.; Fan, Y.; Jiang, E.Y.; Shi, X.-Y.; Krautter, E.; Hu, G.-R.; Li, F.-L. Expression of the hemoglobin gene in regulates intracellular oxygen balance under high-light. *J. Photochem. Photobiol. B Biol.* **2021**, *221*, 112237. [CrossRef]

46. Sinclair, P.R.; Gorman, N.; Jacobs, J.M. Measurement of Heme Concentration. *Curr. Protoc. Toxicol.* **1999**, 8.3.1–8.3.7. [CrossRef]

47. Bennett, A.; Bogorad, L. Complementary chromatic adaptation in a filamentous blue-green alga. *J. Cell Biol.* **1973**, *58*, 419–435. [CrossRef]
48. Livak, K.J.; Schmittgen, T.D. Analysis of relative gene expression data using real-time quantitative PCR and the 2(-Delta Delta C(T)) Method. *Methods* **2001**, *25*, 402–408. [CrossRef]

marine drugs

Article

Comparative Transcriptomic Analysis on the Effect of Sesamol on the Two-Stages Fermentation of *Aurantiochytrium* sp. for Enhancing DHA Accumulation

Xuewei Yang [1,*], Liyang Wei [1], Shitong Liang [1], Zongkang Wang [2] and Shuangfei Li [1,*]

[1] Guangdong Technology Research Center for Marine Algal Bioengineering, Guangdong Key Laboratory of Plant Epigenetics, College of Life Sciences and Oceanography, Shenzhen University, Shenzhen 518060, China; 2200251002@szu.edu.cn (L.W.); 2060251019@szu.edu.cn (S.L.)

[2] Ecological Fertilizer Research Institute, Shenzhen Batian Ecological Engineering Co., Ltd., Shenzhen 518057, China; wangzongkang712@163.com

* Correspondence: yangxw@szu.edu.cn (X.Y.); sfli@szu.edu.cn (S.L.); Tel.: +86-18565693989 (X.Y.)

Abstract: *Aurantiochytrium* is a well-known long-chain polyunsaturated fatty acids (PUFAs) producer, especially docosahexaenoic acid (DHA). In order to reduce the cost or improve the productivity of DHA, many researchers are focusing on exploring the high-yield strain, reducing production costs, changing culture conditions, and other measures. In this study, DHA production was improved by a two-stage fermentation. In the first stage, efficient and cheap soybean powder was used instead of conventional peptone, and the optimization of fermentation conditions (optimal fermentation conditions: temperature 28.7 °C, salinity 10.7‰, nitrogen source concentration 1.01 g/L, and two-nitrogen ratio of yeast extract to soybean powder 2:1) based on response surface methodology resulted in a 1.68-fold increase in biomass concentration. In the second stage, the addition of 2.5 mM sesamol increased the production of fatty acid and DHA by 93.49% and 98.22%, respectively, as compared to the optimal culture condition with unadded sesamol. Transcriptome analyses revealed that the addition of sesamol resulted in the upregulation of some genes related to fatty acid synthesis and antioxidant enzymes in *Aurantiochytrium*. This research provides a low-cost and effective culture method for the commercial production of DHA by *Aurantiochytrium* sp.

Keywords: *Aurantiochytrium*; DHA; thraustochytrid; two-stage fermentation; transcriptomics

Citation: Yang, X.; Wei, L.; Liang, S.; Wang, Z.; Li, S. Comparative Transcriptomic Analysis on the Effect of Sesamol on the Two-Stages Fermentation of *Aurantiochytrium* sp. for Enhancing DHA Accumulation. *Mar. Drugs* **2024**, *22*, 371. https://doi.org/10.3390/md22080371

Academic Editors: Cecilia Faraloni and Eleftherios Touloupakis

Received: 8 July 2024
Revised: 6 August 2024
Accepted: 12 August 2024
Published: 16 August 2024

1. Introduction

Docosahexaenoic acid (DHA, C22:6) belongs to omega-3 long-chain polyunsaturated fatty acids (PUFAs), and it is an essential structural element of the brain, retina, and neuron cell membrane [1]. DHA has demonstrably beneficial effects on depression [2], lowering cardiovascular risk [3], suppressing inflammation [4] and atherosclerosis [5], improving nervous system and retina development [6], and decreasing cancer risk [7]. The human body is unable to produce DHA; hence, it must come from food. Since marine fish and fish oil is cheap and contains a lot of DHA, it is generally thought to be the greatest source of DHA in the human diet [8]. However, marine pollution and overfishing limit the increasing demand for DHA [9]. Creating substitute sources to fulfill the demand for DHA is a significant task. Marine oleaginous microorganisms, as the original source of DHA production in marine fish, have garnered significant attention globally in the search for the sustainable production of DHA because of their potential applications in the biopharmaceutical and nutraceutical sectors [10,11]. The single-celled eukaryotic thraustochytrid, which includes the *Thraustochytrium*, *Aurantiochytrium*, and *Schizochytrium*, is currently the main focus of research. These organisms are typically found in marine environments and are capable of synthesizing large levels of lipids and carotenoids. Among them, it has been previously documented that *Aurantiochytrium* sp. can accumulate significant lipid content, particularly DHA [12,13].

Thraustochytrids are a prospective new market for DHA synthesis, and increasing DHA production is now a focus of intense study. The optimization of fermentation conditions (such as varying the source of nitrogen, C/N ratio, temperature, salinity etc.) as a crucial tactic to boost biomass and fatty acids production has been an important focus for thraustochytrid biotechnological research [14]. Fermentation costs are largely influenced by the carbon and nitrogen substrates used, and since nitrogen sources are often more expensive than carbon sources, less expensive options are being considered. Both organic (such as soybean meal hydrolysate, yeast extract, peptone, and monosodium glutamate) and inorganic (such as nitrate and ammonium) nitrogen can be used by thraustochytrids [15]. To reduce production costs, traditional nitrogen sources may be replaced with agricultural products and the food industry by-products [16]. The two-stage strategy, which divides the biomass growth phase from the lipid accumulation phase, is an efficient cultivation system [17]. The purpose of the first stage is to produce as much biomass as possible when nutrition is adequate, while the second stage—which is typically nitrogen-starved but has an excess of the carbon source—is meant to accumulate lipids [18].

An increasing number of research studies have shown that incorporating antioxidants such as vitamin C, plant hormone, and melatonin into the culture medium could enhance the ability of oleaginous microorganisms to produce lipids [19]. In a prior investigation, the utilization of *Dioscorea zingiberensis*'s phenolic-rich starch saccharification liquid markedly enhanced *Schizochytrium* sp.'s DHA yield and antioxidant capability [20]. One of the primary naturally occurring phenols in sesame, sesamol, has a potent antioxidant capacity and enhancement of radical scavenging [21], and it is frequently employed as an inexpensive, non-toxic antioxidant to stop lipid peroxidation in food and medicine. Sesamol supplementation has shown increased DHA production in *Schizochytrium*. The addition of 1 mM sesamol exogenously to the fermentation medium increased *Schizochytrium* sp. H016's yield of DHA and lipids by 53.52% and 78.30%, respectively [22].

This study optimizes the type of nitrogen source in the culture medium in terms of cost and utilization efficiency. According to the four factors of temperature, salinity, nitrogen source concentration and the ratio of two nitrogen sources (yeast extract and soybean powder), the best level was selected by a one-factor test. And then the culture conditions of four factors and three levels were optimized to enhance the biomass yield of *Aurantiochytrium* sp. DECR-KO (2,4-dienyl-CoA reductase-knockout) [23] on response surface methodology (RSM). The effects of various sesamol concentrations on the biomass, lipid accumulation, and the synthesis of fatty acids of *Aurantiochytrium* sp. DECR-KO were examined in this study. This study elucidated the possible mechanisms of lipid metabolism regulation through antioxidant supplementation through transcriptome analysis.

2. Results

2.1. Screening of Different Nitrogen Source Components

The N content of the six nitrogen source components was determined by an elemental analyzer, and the results are shown in Table S1. All six nitrogen sources had more than 10% of N content, among which the peptone contained the highest N, up to 13.35%. In order to reflect the types of nitrogen sources that can be efficiently utilized by *Aurantiochytrium* sp. DECR-KO, we added the six nitrogen sources at an addition rate of 2.5 g/L into the M4 medium without yeast extract and peptone to examine the effects of various nitrogen source components on the biomass concentration of *Aurantiochytrium* sp. DECR-KO.

The results of the initial screening of nitrogen source types are shown in Figure 1. Among the biomass concentration results from the 65 h incubation under the same nitrogen source addition concentration, the best biomass concentration was obtained from the culture by yeast extract, which amounted to 5.28 ± 0.07 g/L, and the second one was obtained from the soybean powder, with the obtained biomass concentration of 4.13 ± 0.12 g/L. The lowest biomass concentration obtained in culture was peptones, only 2.69 ± 0.06 g/L, which was 0.51 times that of yeast extract and 0.65 times that of soybean powder. Combining the effects of the six nitrogen sources on the biomass concentration of *Aurantiochytrium* sp. DECR-KO and its market price, the low-cost and high-efficiency soybean powder was selected to replace the original high-cost and low-utility peptone. The two nitrogen sources, yeast extract and soybean powder, will be used in subsequent studies for the compounding and optimization of culture conditions.

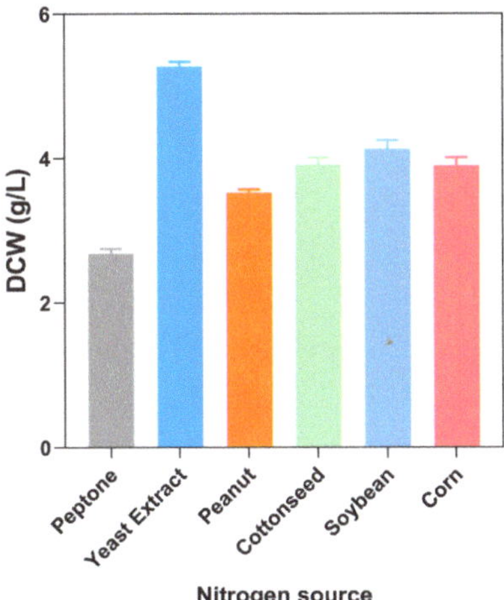

Figure 1. Effects of various nitrogen sources on the biomass concentration of *Aurantiochytrium* sp. DECR-KO. Cell dry weight (DCW) was obtained by culture for 65 h with 2.5 g/L different nitrogen source.

2.2. Response Surface Methodology (RSM) to Optimize the Biomass Yield

Using a single-factor experiment and an RSM central composite design based on the fermentation factors of *Aurantiochytrium* sp. DECR-KO, the fermentation factors were optimized in this study to increase the biomass concentration on the first stage. The results show that the temperature of 26 °C, salinity 10‰, nitrogen source concentration 0.90 g/L and N ratio 2:1 (yeast extract:soybean powder) were the optimal single factor levels for biomass concentration (Figure 2A–D). The Box–Behnken experiment designed a total of 29 sites, including 5 central sites and 24 factorial points (Table 1). The central experimental point is the central point of the partition, which can be used as the calibration of the points. With cell dry weight (DCW) as the response value (Y), and temperature (A), salinity (B), the ratio of two nitrogen sources (C), and the nitrogen source concentration (D) as independent variables, the response surface data as follows:

$$\text{DCW (g/L)} = 5.61313 - 0.12102 \times A + 0.134542 \times B + 0.0265687 \times C + 0.134915 \times D + 0.121345 \times AB + 0.21728 \times AC - 0.0554881 \times AD - 0.309067 \times A^2 - 0.604891 \times B^2 - 0.14348 \times C^2 - 0.128604 \times D^2$$

The model was built using the software Design Expert 12 to obtain a multiple quadratic regression response surface model for biomass yield, and the model obtained from fitting the experimental results was subjected to ANOVA (Table 2) to verify the usability of the regression model. The ANOVA yielded a model with a *p*-value of <0.0001, which is extremely significant; where the lack of fit had a *p*-value of 0.1640, which is greater than 0.05, and the lack of fit was not significant. These two items indicate that the model obtained by fitting this response surface analysis is accurate in its predictions. The contour plots and response surface curves of the interaction terms were obtained by simulation with Design Expert 12 software (Figure 2E–J). According to the *F*-value of each factor, the order of influence on biomass concentration of *Aurantiochytrium* sp. DECR-KO was D (nitrogen concentration) > B (salinity) > A (temperature) > C (ratio of two nitrogen sources). In this experiment, the differences in the effects on biomass concentration were extremely significant for the secondary term B^2, highly significant for the secondary term A^2, and reached significance for the primary terms A, B, D, and the interaction term AC.

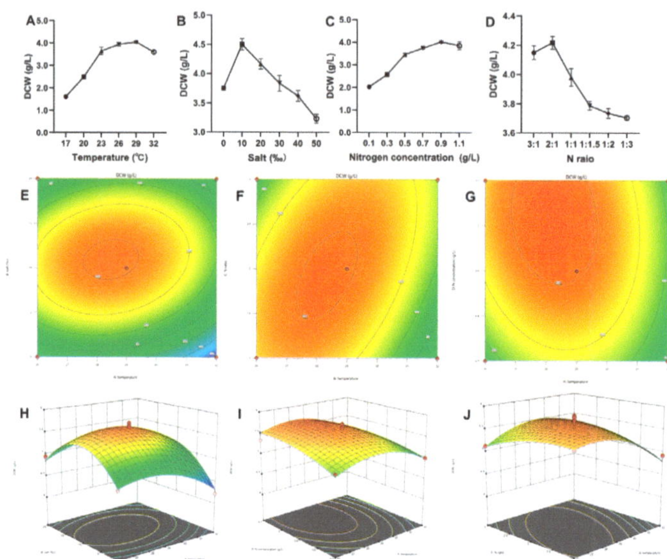

Figure 2. (**A**) The effects of six different temperature levels; (**B**) the effects of six different salinity levels; (**C**) the effects of six different nitrogen sources; (**D**) the effect of the ratio of two nitrogen sources (yeast extract: soybean powder). Contour plots showing the effect of (**E**) temperature and salt (**F**), temperature and nitrogen ratio (**G**), temperature and nitrogen source concentration to dry cell weight (DCW). Response surface plots show the effect of (**H**) temperature and salt, (**I**) temperature and N ratio, (**J**) temperature and nitrogen source concentration to DCW.

Table 1. The Box–Behnken design matrix for real values and coded values (in parentheses).

Run	A: Temperature (°C)	B: Salinity (%)	C: N Ratio	D: N Concentration (g/L)	DCW (g/L)
1	26 (−1)	10 (0)	3 (+1)	0.9 (0)	5.12
2	29 (0)	20 (+1)	1 (−1)	0.9 (0)	5.20
3	32 (+1)	10 (0)	1 (−1)	0.9 (0)	4.92
4	29 (0)	0 (−1)	2 (0)	0.7 (−1)	4.73
5	29 (0)	10 (0)	2 (0)	0.9 (0)	5.75
6	29 (0)	20 (+1)	3 (+1)	0.9 (0)	4.79
7	29 (0)	10 (0)	2 (0)	0.9 (0)	5.49
8	29 (0)	0 (−1)	2 (0)	1.1 (+1)	5.17
9	29 (0)	0 (−1)	1 (−1)	0.9 (0)	4.58
10	29 (0)	20 (+1)	2 (0)	0.7 (−1)	4.90
11	26 (−1)	10 (0)	2 (0)	1.1 (+1)	5.33
12	29 (0)	10 (0)	2 (0)	0.9 (0)	5.69
13	32 (+1)	0 (−1)	2 (0)	0.9 (0)	4.22
14	29 (0)	10 (0)	2 (0)	0.9 (0)	5.49
15	29 (0)	10 (0)	3 (+1)	1.1 (+1)	5.79
16	32 (+1)	10 (0)	2 (0)	1.1 (+1)	4.92
17	26 (−1)	0 (−1)	2 (0)	0.9 (0)	4.78
18	29 (0)	0 (−1)	3 (+1)	0.9 (0)	4.59
19	29 (0)	10 (0)	1 (−1)	0.7 (−1)	5.03
20	29 (0)	20 (+1)	2 (0)	1.1 (+1)	5.01
21	32 (+1)	10 (0)	3 (+1)	0.9 (0)	5.44
22	29 (0)	10 (0)	2 (0)	0.9 (0)	5.65
23	26 (−1)	20 (+1)	2 (0)	0.9 (0)	4.94

Table 1. *Cont.*

Run	A: Temperature (°C)	B: Salinity (%)	C: N Ratio	D: N Concentration (g/L)	DCW (g/L)
24	29 (0)	10 (0)	3 (+1)	0.7 (−1)	5.17
25	32 (+1)	20 (+1)	2 (0)	0.9 (0)	4.86
26	26 (−1)	10 (0)	1 (−1)	0.9 (0)	5.47
27	26 (−1)	10 (0)	2 (0)	0.7 (−1)	5.17
28	32 (+1)	10 (0)	2 (0)	0.7 (−1)	4.99
29	29 (0)	10 (0)	1 (−1)	1.1 (+1)	5.38

Table 2. ANOVA analysis of response surface experiment.

Source	Sum of Squares	df	Mean Square	F-Value	p-Value	
Model	3.5100	11	0.3191	9.45	<0.0001	significant
A—temperature	0.1757	1	0.1757	5.20	0.0357	
B—salinity	0.2172	1	0.2172	6.43	0.0213	
C—the ratio of two nitrogen sources	0.0085	1	0.0085	0.2508	0.6229	
D—nitrogen source concentration	0.2184	1	0.2184	6.47	0.0210	
AB	0.0589	1	0.0589	1.74	0.2042	
AC	0.1888	1	0.1888	5.59	0.0302	
AD	0.0123	1	0.0123	0.3646	0.5539	
A^2	0.6196	1	0.6196	18.34	0.0005	
B^2	2.3700	1	2.37	70.27	<0.0001	
C^2	0.1335	1	0.1335	3.95	0.0631	
D^2	0.1073	1	0.1073	3.18	0.0926	
Residual	0.5742	17	0.0338			
Lack of Fit	0.5176	13	0.0398	2.82	0.1640	not significant
Pure Error	0.0566	4	0.0141			
Cor Total	4.0800	28				

Meanwhile, the reliability analysis of the BBD model was also obtained through the fitting and analysis of the software. As shown in Table 2, the coefficient of variation in the model fitted in this experiment is 3.59, which is within the normal range. The regression coefficient R^2 is 85.94% > 85%, which indicates that the equation created by the fitted model fits well. Adjusted R^2 is the correction coefficient, and its value is 0.7684. Predicted R^2 is the prediction coefficient, and its value is 0.5766. The difference between the correction coefficient and prediction coefficient is less than 0.2, which is within a reasonable range. The signal-to-noise ratio of this model is 10.919 according to "Adeq Precision", which is greater than 4, indicating that the signal is sufficient and the model is reliable.

The culture conditions for the highest biomass yield were fitted by Design-Expert 12 software as follows: temperature 28.7 °C, salinity 10.7‰, nitrogen source concentration 1.01 g/L, and two-nitrogen ratio of yeast extract to soybean powder 2:1 (i.e., 11.62 g/L artificial sea salt, 3.16 g/L yeast extract, and 1.58 g/L water-soluble soybean powder). The predicted biomass yield that could be obtained under this optimal condition was 5.66 g/L of DCW. Three parallel experiments were conducted to validate under the predicted optimal culture conditions. The DCW results of the experiments were 5.56 g/L, 5.34 g/L, and 5.05 g/L, with an average value of 5.32 g/L, which was approximately 6% different from the predicted value.

The experimental group which was cultured using response surface methodology was named YS (yeast extract–soybean powder, Figure 3A). As shown in Figure 3B, the biomass yields of the different culture methods of YS and M4 were investigated at three time points including the exponential growth period (42 h), stationary period (63 h), and decline period (90 h) of the culture [23]. It can be observed that the difference in biomass concentration of YS and M4 cultures at the mid-late stage of the culture decreased gradually with the gradually decreasing with increasing incubation time: at 42 h, the biomass concentration of the YS culture was 1.68 times more than that of the M4 culture; at 63 h, the biomass concentration of the YS culture was 1.51 times more than that of the M4 culture; and at 90 h, the biomass concentration of the YS culture was 1.23 times higher than that of the M4 culture. At the decline period of the *Aurantiochytrium* sp. DECR-KO, the biomass concentration of the M4 cultures was still rising, while the YS cultures declined, but the biomass yields obtained from the YS cultures were greater than those obtained from the M4 cultures throughout the mid-late phase of the culture.

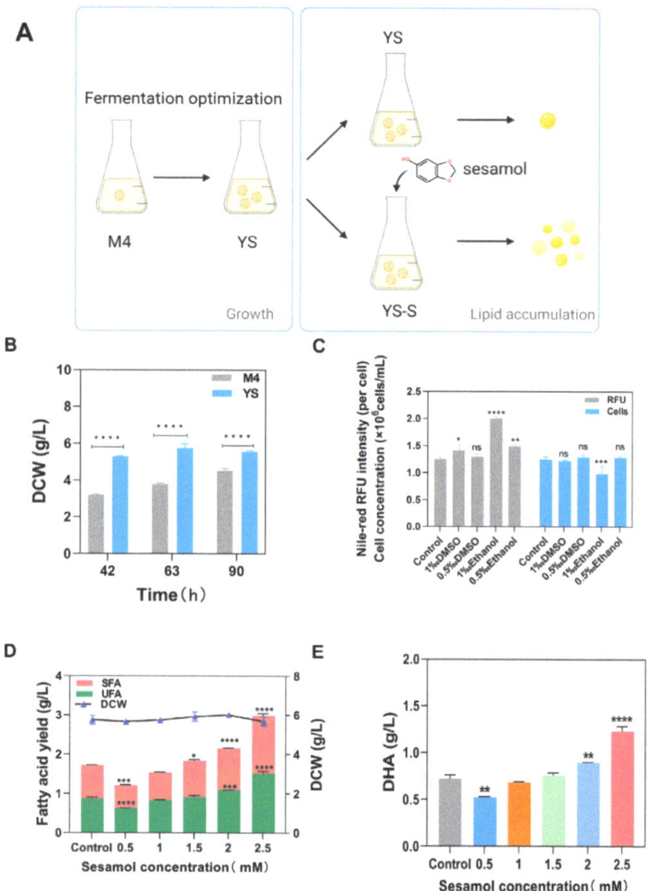

Figure 3. (**A**) Schematic diagram of enhancing lipid production in *Aurantiochytrium* sp. DECR-KO. (**B**) The difference of biomass concentration in YS and M4 culture after exponential phase. (**C**) Effects of DMSO and ethanol on cell concentration and neutral lipid content of *Aurantiochytrium* sp. DECR-KO at different volume concentrations. Solvent addition amount is expressed as volume concentration (*v/v*), and YS culture without adding any solvent was used as the control group. Cell concentration was shown as the number of cells per milliliter of culture medium, and neutral lipid content was represented by the relative fluorescence intensity of Nile red in each cell. (**D,E**): Effects of varying sesamol concentrations on DECR-KO strains' fatty acid yield and DHA synthesis. M4: experimental group before fermentation optimization; YS: experimental group cultured with yeast extract and soybean powder after fermentation optimization; YS-S: experimental group YS treated with 2.5 mM sesamol. DHA: docosahexaenoic acid; UFAs: unsaturated fatty acids; SFAs: saturated fatty acids. ns: not significant, * $p < 0.05$, ** $p < 0.01$, *** $p < 0.001$, and **** $p < 0.0001$.

2.3. Effect of Sesamol on the Fatty Acid Production Capacity of Aurantiochytrium sp. DECR-KO

As sesamol is insoluble in water, pre-experiments were carried out on both DMSO and ethanol in order to rule out the impact of the solvents on the biomass concentration and neutral lipid content of *Aurantiochytrium* sp. DECR-KO. There was no significant difference between the 0.5‰ (*v*/*v*) addition of DMSO and the control (Figure 3C), indicating that 0.5‰ of DMSO had no effect on the cell concentration of *Aurantiochytrium* sp. DECR-KO as well as the neutral lipid content. On the other hand, the neutral lipid content or cell concentration was significantly impacted by the volume ratio of 1‰ of DMSO and the ethanol of 0.5‰ and 1‰. Therefore, the subsequent sesamol will be solubilized using DMSO, and the volume ratio of the added solution will be controlled to be less than or equal to 0.5‰.

The addition of 0.5 mM to 2.5 mM sesamol had no visible effect on DCW compared to the control group (Figure 3D). The fatty acid yield of *Aurantiochytrium* sp. DECR-KO increased as the concentration of sesamol increased under 0.5–2.5 mM sesamol treatment (Figure 3D). However, in contrast to the control group, the fatty acid production decreased slightly when treated with low concentrations (0.5 to 1 mM) of sesamol, indicating that low concentrations of sesamol inhibited the production of fatty acids. When the concentration of sesamol treatment was greater than 1 mM, the fatty acid production was more than the control group. With the increase in the concentration of sesamol, the total fatty acid production (including saturated fatty acids and unsaturated fatty acids) also increased gradually. The DHA yield showed the same trend as the total fatty acid production (Figure 3E). Based on the results of the 0.5–2.5 mM sesamol treatments explored in this experiment, 2.5 mM was the optimal sesamol treatment concentration that most improved fatty acid yield. Compared to the control group, the 2.5 mM sesamol-treated group showed a 78.79% increase in total fatty acid yield and 69.83% increase in DHA yield.

2.4. Biomass Concentration and Fatty Production Analysis of Fermentation-Optimized

In order to better compare the optimization effect produced by fermentation optimization, the experimental group before fermentation optimization was named as M4, the experimental group using response surface methodology to optimize the biomass culture was named as YS (yeast extract and soybean powder), and the group treated with 2.5 mM sesamol was named as YSS (YS with sesamol treatment) (Figure 3A). Figure 4 displays the results of a comparison of the biomass concentration, fatty acid yield, and DHA yield of the three culture conditions. Compared with the M4 group, the biomass concentrations of the YS and YSS groups were increased by 50.06% and 49.85%, respectively. The addition of 2.5 mM sesamol treatment did not significantly affect the biomass concentration in the YS group during the stationary period (Figure 4A). Fatty acid production is shown in Figure 4B, and it can be seen that although the YS group increased the biomass concentration, its total fatty acid production was only increased by 8.23% compared to that of the M4 group. The addition of 2.5 mM sesamol increased the total fatty acid yield by 93.49% and 78.79% compared to the M4 and YS groups, respectively. A comparison of DHA yields among the three groups is shown in Figure 4C, which showed an increase of 16.71% and 98.22% in the YS and YSS groups, respectively, compared to the M4 group. The addition of 2.5 mM sesamol treatment to the experimental group increased DHA production by 69.83% compared to no sesamol addition.

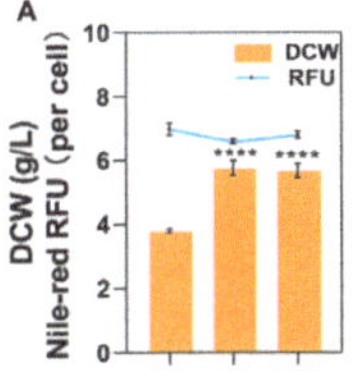
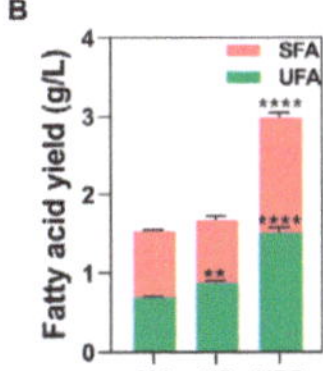
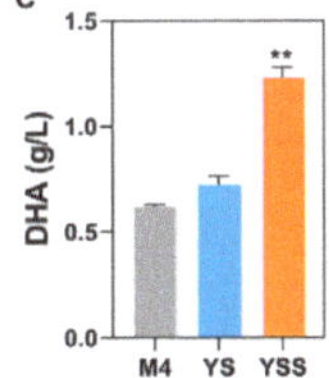

Figure 4. Difference of (**A**) biomass concentration, (**B**) fatty acids and (**C**) DHA yield in fermentation optimization experiments. ** $p < 0.01$ and **** $p < 0.0001$.

2.5. Transcriptome Profiling of Aurantiochytrium sp. DECR-KO with Sesamol Treatment

The mechanism of *Aurantiochytrium* sp. DECR-KO's response to sesamol treatment was ascertained by utilizing the Illumina RNA-seq method. There were 2677 differentially expressed genes (DEGs) in total identified (p-adjust < 0.05 and $|\log2FC| \geq 1$). A total of 1411 genes were upregulated and 1266 genes were downregulated in the YSS compared to the YS group. KEGG was used to examine the biological roles and interactions of the discovered DEGs. According to KEGG, notable enrichment pathways involved in lipid metabolism include fatty acid degradation and elongation, fatty acid metabolism, peroxisome and the biosynthesis of unsaturated fatty acids (Figure S1).

In oleaginous microorganisms, acetyl-CoA is an important central metabolite of carbon metabolism and one of the key precursors for fatty acid synthesis [24]. The process of glycolysis is one of the sources of acetyl coenzyme A. In glycolysis, 6-phosphofructokinase (PFK) is the rate-limiting enzyme, and PFK was upregulated by 1.07-fold in the YSS group (Table 3). Additionally, other key enzyme genes, including triosephosphate isomerase (TPI) and glyceraldehyde 3-phosphate dehydrogenase (GAPDH), were upregulated by 1.00-fold and 1.21-fold, respectively. In pyruvate metabolism, malate dehydrogenase was upregulated by 1.00-fold, which catalyzes the conversion of malate to pyruvate. Increasing the expression of the above genes results in more pyruvate production, and pyruvate produces acetyl-CoA through pyruvate dehydrogenase.

Table 3. Transcriptomics of the expression of genes under sesamol treatment.

Gene ID	Name	Description	YSS vs. YS ($\log_2$ Fold Change)
		Fatty Acid Synthesis	
TRINITY_DN338_c0_g1_i2-SM4	ACC	acetyl-CoA carboxylase	0.81
TRINITY_DN7076_c0_g1_i1-SM4	MCAT	malonyl-CoA:ACP transacylase	1.07
TRINITY_DN14219_c0_g1_i1-SM4	ME	malate dehydrogenase (oxaloacetate-decarboxylating)	1.00
TRINITY_DN11008_c0_g1_i1-YS	FAS	fatty acid synthase	0.92
TRINITY_DN15725_c0_g1_i1-YSS	KS	3-ketoacyl-synthase	0.68
TRINITY_DN4969_c0_g2_i1-YS	KR	ketoreductase	1.33
TRINITY_DN2556_c0_g1_i1-YSS	PFK	6-phosphofructokinase	1.07
TRINITY_DN897_c2_g1_i1-YSS	TPI	triosephosphate isomerase	1.00
TRINITY_DN10506_c0_g1_i1-YSS	GAPDH	glyceraldehyde 3-phosphate dehydrogenase	1.21
		Fatty Acid Degradation	
TRINITY_DN2005_c6_g1_i1-YSS	HADH	3-hydroxyacyl-CoA dehydrogenase	1.45
TRINITY_DN13554_c0_g1_i1-AM4	ECH	enoyl-CoA hydratase	1.38
TRINITY_DN11028_c0_g1_i1-YS	ACD	acyl-CoA dehydrogenase	1.17
TRINITY_DN2439_c0_g1_i1-YSS	KAT	3-ketoacyl-CoA thiolase	0.90
		antioxidant system	
TRINITY_DN12527_c0_g1_i1-YSS	GST	glutathione S-transferase	1.18
TRINITY_DN11074_c0_g1_i1-AM4	SOD	superoxide dismutase	1.27

Acetyl-CoA carboxylase (ACC) catalyzes the conversion of acetyl-CoA to malonyl-CoA, a direct substrate for the synthesis of fatty acids, limiting the rate of fatty acid synthesis [25]. Acetyl-CoA carboxylase was upregulated by 0.81-fold in the YSS group. Malonyl-CoA:ACP transacylase (MCAT) was upregulated by 1.07-fold in the YSS group, converting malonyl-CoA to malonyl-ACP to initiate the elongation cycle. Concentrations of both SFA and UFA increased under sesamol treatment. In the fatty acid synthase (FAS) pathway, FAS was upregulated by 0.92-fold in the YSS group, which led to the accumulation of the SFA. In the polyketide synthase (PKS) pathway, 3-ketoacyl-CoA synthase (KS) and ketoreductase (KR) were upregulated by 0.68-fold and 1.33-fold. However, other co-catalyzing enzymes associated with the PKS pathway, dehydrase/isomerase (DH/I) and enoyl reductase (ER), were not identified in this study, which were similarly not identified in previous studies [26]. The above results indicated that fatty acid synthesis was enhanced under 2.5 mM sesamol treatment, which was in agreement with the experimental results. Except for fatty acid synthesis, fatty acid degradation is an important factor affecting lipid content. The fatty acid β-oxidation pathway is an important pathway for fatty acid degradation [24]. In the fatty acid β-oxidation pathway, acyl-CoA dehydrogenase (ACD), enoyl-CoA hydratase (ECH), 3-hydroxyacyl-CoA dehydrogenase (HADH), and 3-ketoacyl-CoA thiolase (KAT) were significantly more expressed in the YSS group than in the YS group (Table 3).

Superoxide dismutase (SOD) is one of the major intracellular enzymes that protects cells from oxidative damage [27]. SOD was upregulated 1.27-fold in the YSS group. Glutathione S-transferase (GST) further enhances the antioxidant capacity by catalyzing the binding of glutathione to electrophilic substrates [28], which was upregulated 1.18-fold in YSS. Therefore, sesamol treatment improved the total antioxidant capacity of *Aurantiochytrium* sp. DECR-KO, which was beneficial for accumulating more PUFAs.

2.6. Detection of the Gene Expression through qRT-PCR

The expression profiles of eight genes related to fatty acid synthesis were analyzed to validate the transcriptome analysis data (Figure 5). The results of reverse transcriptase quantitative PCR (RT-qPCR) were consistent with the transcriptome sequencing (RNA-Seq) results.

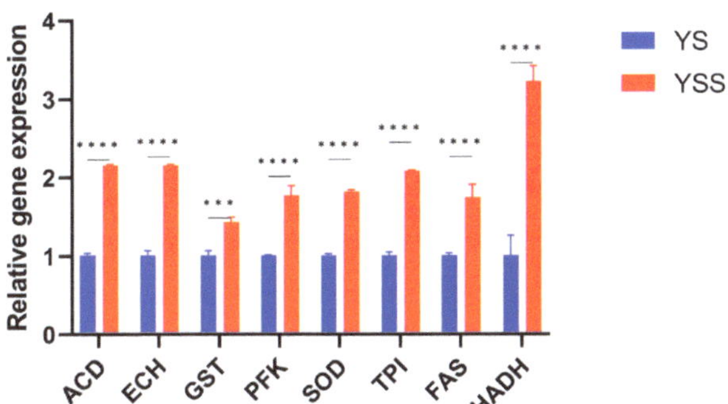

Figure 5. The relative expression levels of related enzyme genes by qRT-PCR. *** $p < 0.001$, and **** $p < 0.0001$.

3. Discussion

3.1. Effect of Fermentation Optimization for Growth

The results of elemental N content showed a different trend from that of biomass concentration obtained from culture, which may be related to the solubility rate of each organic nitrogen source, and the efficiency of its availability to *Aurantiochytrium* sp. DECR-KO. The phenomenon that peptone had the highest N content but the least biomass concentration obtained from culture may be due to the fact that peptone was least consistent with the amino acid composition of *Aurantiochytrium* sp. DECR-KO, and it is difficult to be utilized by *Aurantiochytrium* sp. DECR-KO in the pre- and mid-fermentation stages, which led to its accumulation of less biomass concentration [29].

Medium composition and fermentation conditions significantly affect fatty acid accumulation in *Aurantiochytrium*. Influential factors include the selection of carbon and nitrogen sources, culture strategy, dissolved oxygen concentration, salinity, pH and temperature [30]. To improve the lipid production capacity of *Aurantiochytrium*, a thorough evaluation of the effects of different fermentation conditions on it is required. The *Auranti-*

ochytrium's development and metabolism are significantly impacted by temperature. The temperature between 20 and 30 °C seems to be the optimum incubation temperature [31]. Low temperature had been shown to stimulate DHA production to maintain membrane fluidity and permeability, but at the expense of biomass, resulting in lower overall DHA production [32]. *Aurantiochytrium* is found from mangroves and other sea areas, while the average salinity of natural seawater is 35‰, and the optimal salinity for growth and tolerance level varies according to the strains [33]. High salinity stress can stimulate lipid accumulation in thraustochytrids [32], while high salinity can corrode equipment and increase costs. In *Schizochytrium limacinum* OUC88, the lipid content and biomass was significantly reduced when the salinity was less than 18 g/L (51% of seawater). Both inorganic (such as nitrate and ammonium) and organic nitrogen (such as yeast extract, peptone and corn steep liquor) can be utilized by *Aurantiochytrium* [34–36]. The combination of organic nitrogen proved to be more supportive of production because of the non-specific growth factors (vitamins, trace elements) it provided [15]. For thraustochytrids, balancing the relationship between biomass concentration and lipid content per cell is important to increase the total lipid yield. A single increase in lipid content may lead to a reduction in biomass concentration [17]. Therefore, a staged culture strategy was used to separate the biomass increase stage from the lipid accumulation stage, increasing the final lipid yield. Response surface methodology allows for the identification of optimal culture conditions that will improve the growth of *Aurantiochytrium* [37]. In this study, the optimal conditions for maximum growth under the first phase were determined by response surface methodology optimization.

3.2. The Effect of Sesamol Additionon Lipid Accumulation

Sesamol is a naturally occurring phenolic molecule that is added to foods and medicines as a cheap and safe antioxidant [38]. Despite the antioxidant activity of sesamol, the presence of 0.5 mM sesamol decreased the fatty acids content of *Crypthecodinium cohnii* by 25.24% [39]. In *Schizochytrium* sp., the addition of 1 mM sesamol caused a 59.06% increase in lipid yield [22]. The fatty acid content of *Aurantiochytrium* was also significantly reduced in this study by 0.5 mM sesamol. When sesamol was added at a concentration greater than 1 mM, the yield of total fatty acids increased with increasing concentration. Compared to other oil-producing microorganisms, sesamol induces lipid synthesis in *Aurantiochytrium* sp. possibly due to the presence of a specific fatty acid synthesis system in *Aurantiochytrium* sp.

Two independent pathways for polyunsaturated fatty acid (PUFA) synthesis pathways were reported in *Aurantiochytrium*. In the fatty acid synthase (FAS) pathway, firstly in the action of fatty acid synthase, acetyl-CoA and malonyl-CoA are used to produce palmitic acid (C16:0), and then PUFAs are produced from C16:0 through a sequence of desaturases and elongases [40]. In the polyketide synthase (PKS) pathway, PUFAs can be generated more efficiently starting from acetyl-ACP without oxygen dependence [41]. In fatty acid production, NADPH is the crucial precursor in the FAS and PKS pathway [12]. ME (malic enzyme) and G6PD(glucose-6-phosphate 1-dehydrogenase) are crucial enzymes in the production of NADPH. The overexpression of G6PD and ME increased NADPH supply, resulting in a 10.6% and >105% increase in PUFA and SFA, respectively [42,43]. Sesamol can reduce NADPH supply by inhibiting ME, leading to a reduction in lipid accumulation in oleaginous microorganisms [44]. Previous studies have shown that the addition of sesamol leads to a decrease in ME activity and an increase in G6PD activity in *Schizochytrium* sp.H016 [22]. However, there was no significant difference in the expression of G6PD by the addition of sesamol in the present study, while ME was upregulated 1.00-fold in the YSS group. The increase in ME expression may have compensated for the decrease in its activity to ensure the supply of NADPH. For the production of another precursor, acetyl-CoA, there was no ATP citrate lyase (ACL) identified in this study. In the FAS pathway, fatty acid synthase was upregulated, leading to the accumulation of SFA (Figure 6). In the PKS pathway, the synergistic activity of β-ketoacyl synthase (KS), β-ketoreductase (KR), dehydration, and enoyl-reductase (ER) lead to the synthesis of PUFA [45]. The overexpression of PKS pathway genes increased the accumulation of DHA in the YSS group. Interestingly, genes associated with fatty acid degradation were significantly upregulated. The β-oxidation of fatty acids is generally considered to be detrimental to fatty acid accumulation. However, β-oxidation provides acetyl coenzyme A and ATP, which are also necessary for the synthesis of fatty acid. The upregulation of the fatty acid degradation pathway may result from the consumption of large amounts of short-chain fatty acids for cell division and other life activities [46]. Acetyl coenzyme A generated from the breakdown of short-chain fatty acids can enter the TCA cycle or serve as a precursor substance for unsaturated fatty acids. PUFAs have a high degree of unsaturation, which makes them easily oxidized. Due to the increased accumulation of polyunsaturated fatty acids, there is an increased risk of lipid peroxidation, which is accompanied by increased levels of ROS [47]. In this study, the antioxidant system mitigates oxidative stress damage through enzymatic (superoxide dismutase) and non-enzymatic mechanisms (glutathione S-transferase) [48]. Superoxide dismutase (SOD) can catalyze superoxide anions, the precursors of most ROS, to oxygen and hydrogen peroxide. Glutathione S-transferase (GST) quenches reactive molecules by adding glutathione, assisting in the elimination of hydrogen peroxide and other oxidative stress metabolites [49].

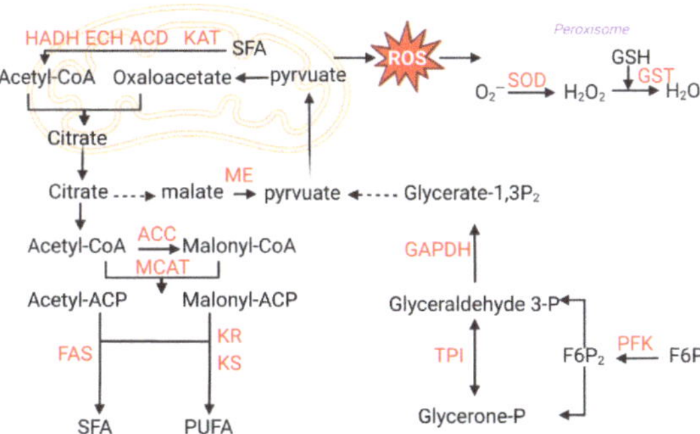

Figure 6. Schematic map of transcriptional analysis of the pathways associated with lipid and carbon metabolism in *Aurantiochytrium* sp. DECR-KO. HADH: 3-hydroxyacyl-CoA dehydrogenase; ECH: enoyl-CoA hydratase; ACD: acyl-CoA dehydrogenase; KAT: 3-ketoacyl-CoA thiolase; SOD: superoxide dismutase; GST: glutathione S-transferase; ACC: acetyl-CoA carboxylase; MCAT: malonyl-CoA:ACP transacylase; ME: malic enzyme; FAS: fatty acid synthase; KS: 3-ketoacyl-synthase; KR: ketoreductase; PFK: 6-phosphofructokinase; TPI: triosephosphate isomerase; GAPDH: glyceraldehyde 3-phosphate dehydrogenase.

4. Materials and Methods

4.1. Nitrogen Content Determination

The nitrogen elemental composition of six nitrogen sources (yeast extract, bacteriological peptone, cottonseed powder, peanut powder, corn powder, and soybean powder) was determined using an elemental analyzer and sulfanilamide as a standard.

4.2. Strain and Cultural Methods

The *Aurantiochytrium* sp. DECR knockout engineered strain (*Aurantiochytrium* sp. DECR-KO) is stored in the China Center for Type Culture Collection (CCTCC M 2022545) and obtained from prior study [23,50]. Routine culture conditions are as follows: M4 culture medium (1 g/L yeast extract, 20 g/L glucose, 0.025 g/L potassium dihydrogen phosphate and 1.5 g/L peptone dissolved in artificial seawater with a salinity of 30‰), 23 °C and 200 rpm. The strain was cultured in a 250 mL shake flask with 100 mL M4 culture medium.

4.3. Determination of Cell Dry Weight and Neutral Lipids

Cultured cells were collected by centrifugation at 8000 rpm for 10 min and then freeze-drying using a freeze-dryer for 48 h. The cell dry weight (DCW) was used as the biomass.

Neutral lipids were stained with Nile red fluorescent dye (Rhawn, Shanghai, China). First, 0.1 mg/mL Nile red (in acetone) is used for staining 200 mL of cells. After 20 min of a dark incubation at 200 rpm, Nile red fluorescence was measured using a fluorescence microplate reader (Synergy H1, Bio-Tek, Winooski, VT, USA) with an excitation wavelength of 488 nm and emission wavelength of 592 nm [51].

4.4. Experimental Design

To maximize the biomass production of *Aurantiochytrium* sp. DECR-KO by optimizing culture conditions, this study selected four factors (temperature, salinity, nitrogen concentration and nitrogen ratio) to carry out a one-way experimental design to observe their effects on the biomass production. *Aurantiochytrium* sp. DECR-KO was inoculated into 250 mL shake flasks with 100 mL of M4 medium and incubated at 23 °C and 200 rpm for 42 h as a seed culture. For temperature, the seed culture was inoculated into M4 medium at six temperatures: 17 °C, 20 °C, 23 °C, 26 °C, 29 °C, and 32 °C. For salinity, seed culture was inoculated into M4 medium with different salinities (0‰, 10‰, 20‰, 30‰, 40‰, 50‰) at 23 °C. The nitrogen source concentration of the M4 medium was calculated to be about 0.31 g/L by the percentage of N content. For nitrogen concentration, the medium with different nitrogen source concentrations (0.1, 0.3, 0.5, 0.7, 0.9, and 1.1 g/L) was prepared at a ratio of 1:1.5 between yeast extract and soya bean powder, respectively. The seed culture was inoculated into M4 medium with different concentrations of nitrogen sources at 23 °C. The medium with different ratios of yeast extract and soybean flour (3:1, 2:1, 1:1, 1:1.5, 1:2, 1:3) was prepared separately. Seed culture was inoculated into M4 medium with different nitrogen source ratios at 23 °C. All experiments were inoculated into 250 mL culture flasks containing 100 mL of medium and incubated at 200 rpm with shaking for 42 h. Cells were collected by centrifugation and weighed after freeze-drying. Then, we carried out a Box–Behnken design (BBD) of experiments based on the results and finally obtained the optimized culture conditions with the highest biomass production.

A Box–Behnken experimental design with four factors, namely, temperature (A), salinity (B), nitrogen concentration (C) and nitrogen ratio (D), as independent variables, with −1, 0, and +1 levels for each factor, and cell dry weight (DCW) as the response value was carried out as shown in Table 4. The experimental design was assisted by Design-Expert 12 software. The design of the central experimental site was 5, which required a total of 29 experiments, and each experiment was incubated for 42 h. The experimental results were entered into Design-Expert software for response surface analysis.

Table 4. Factors and levels of Box–Behnken for the optimization of the culture conditions of the *Aurantiochytrium* sp. DECR-KO.

Level	A Temperature (°C)	B Salinity (‰)	C nitrogen concentration (g/L)	D nitrogen ratio *
−1	26	0	0.7	1:1
0	29	10	0.9	2:1
1	32	20	1.1	3:1

* The nitrogen ratio is the weight ratio of yeast extract to water-soluble soybean powder.

DMSO and ethanol were used as solvents, and 1‰ and 0.5‰ (*v/v*) of different solvents were added to the strain culture medium that had been cultured for 42 h. The dry cell weight and Nile red relative fluorescence intensity were determined after 24 h of incubation. A gradient concentration of 0, 0.5, 1, 1.5, 2, and 2.5 mM sesamol was supplemented into the *Aurantiochytrium* sp. DECR-KO that had been cultured for 42 h in optimized culture conditions (temperature 28.7 °C, 11.62 g/L artificial sea salt (salinity of 10.7‰), 3.16 g/L yeast extract, 1.58 g/L water-soluble soybean powder, 20 g/L glucose, 0.025 g/L potassium dihydrogen phosphate). The culture was continued under optimized culture conditions for 21 h to stationary phase (63 h) [23].

4.5. Lipid Extraction and Fatty Acid Analysis

Lipids were extracted by a chloroform–methanol (2:1, v/v) method as previously described [52,53]. First, 500 mg of the freeze-dried cells was mixed with chloroform–methanol and extracted for 72 h at 65 °C in a Soxhlet extractor (AG-SXT-06, OUGE, Shanghai, China). The crude total lipids were obtained by evaporating the solvent at 65 °C. To the crude total lipids, 4 mL of 4% sulfuric acid in methanol was added to obtain fatty acid methyl esters (FAMEs) at 65 °C for 1 h. The FAMEs were treated with hexane and deionized water, which was followed by volatilization of the hexane off in a stream of nitrogen to gain the methyl esterified fatty acids (MEFs). Then, 1 mL of dichloromethane was used for the dissolution of the MEFs. Compositional and content analyses of MEFs were performed by gas chromatography–mass spectrometry (GC-MS, 7890-5975 Agilent, Santa Clara, CA, USA). Chromatographic conditions were set as claimed in previous studies [53]. The mass spectrometry library of the National Institute of Standards and Technology (NIST) was used to identify the fatty acids. Methyl nonadecylate (Solarbio, Beijing, China) was used as an internal standard, and the content was determined by comparing the internal standard peak areas.

4.6. RNA Extraction, Transcriptomic Analysis, and Real-Time Quantitative PCR (RT-qPCR) Analysis

For transcriptome analysis, samples were collected at 63 h for RNA extraction. Then, samples were sent to the BioTechnology Genomics Institute, Shenzhen, China for transcriptome sequencing. Under the condition of fold change ≥ 2 and adjusted p-value ≤ 0.001, DEseq2 was used to conduct differential gene analysis between groups [54]. According to the gene ontology (GO) and Kyoto Encyclopedia of Genes and Genomes (KEGG) annotation results and official classification, the differentially expressed genes were functionally classified, and the phyper in R software was used for KEGG enrichment analysis. Genes satisfying Q-value ≤ 0.05 were defined as significantly enriched.

The total RNA was extracted and collected from both the experimental and control sample using a Trizol reagent. RNA concentration and purity measurements were performed by NanoDrop2000 (Thermo Scientific, Waltham, MA, USA). The PrimeScript™ RT reagent Kit was used for cDNA synthesis. According to the manufacturer's protocol, TB Green® Premix Ex Taq™ II and ABI QuantStudio 6 Flex (Applied Biosystems, Foster, CA, USA) were used for RT-qPCR. Primer sequences used for RT-qPCR (Table S2) were designed by Primer Premier 5.0. The relative gene expression was calculated as $2^{-\Delta\Delta ct}$ using 18S rDNA as the internal standard [26].

4.7. Statistical Analysis

All the experimental data were expressed as the mean $\pm$ standard deviation (S.D.) of at least three independent experiments. Design Expert 12 software was used to perform response surface experiments. GraphPad Prism (version 8.0.2) was used to analyze data. Two-way ANOVA and t-tests were used to determine differences between groups at a confidence level of $p < 0.05$. A different number of asterisks (*) on each column indicates the significance of the difference, * $p < 0.05$, ** $p < 0.01$, *** $p < 0.001$, and **** $p < 0.0001$.

5. Conclusions

In the present study, the lipid production of *Aurantiochytrium* sp. DECR-KO was enhanced by a two-phase strategy. In the first stage, the biomass concentration of *Aurantiochytrium* sp. DECR-KO was significantly increased by response surface methodology optimization. In the second stage, the fatty acid yield was increased by adding the antioxidant sesamol. Compared to the M4 culture condition, the biomass concentration, total fatty acid yield and DHA yield of 2.5 mM sesamol treatment were increased by 49.85%,

93.49% and 98.22%, respectively. The treatment of sesamol induced the gene expression related to fatty acid synthesis (FAS, KS, KR) and the antioxidant system (SOD, GST). This research provides a methodological basis for the use of *Aurantiochytrium* sp. DECR-KO as a feedstock for the industrial production of DHA.

Supplementary Materials: The following supporting information can be downloaded at https://www.mdpi.com/article/10.3390/md22080371/s1, Figure S1: Differential gene KEGG pathway enrichment bubble map; Table S1: Nitrogen content of different nitrogen sources; Table S2: The primer pairs used in the cloning experiment.

Author Contributions: Conceptualization, X.Y. and L.W.; methodology, X.Y.; validation, L.W. and S.L. (Shitong Liang); formal analysis, X.Y., L.W. and S.L. (Shitong Liang); resources, X.Y., Z.W. and S.L. (Shuangfei Li); data curation, X.Y. and L.W.; writing—original draft preparation, L.W.; writing—review and editing, X.Y. and S.L. (Shuangfei Li); visualization, X.Y.; supervision, X.Y. and S.L. (Shuangfei Li); project administration, X.Y., S.L. (Shuangfei Li) and Z.W.; funding acquisition, X.Y., S.L. (Shuangfei Li) and Z.W. All authors have read and agreed to the published version of the manuscript.

Funding: This study was supported by the following projects: the National Key Research and Development Programme (2018YFA0902500), the National Key Research and Development Programme of China (2020YFD0901002), the Natural Science Foundation of Guangdong Province (2021A1515012557), the Shenzhen Science and Technology Programme (KCXST20221021111206015 and KCXFZ20201221173404012), and the Shenzhen Agricultural Development Special Funds (Fishery) Agricultural High-Tech Project (2021-928).

Institutional Review Board Statement: Not applicable.

Data Availability Statement: The data presented in this study are available on request from the corresponding author.

Acknowledgments: We would like to thank the Instrument Analysis Center of Shenzhen University and the Public Service Platform for Large Instruments and Equipment of the College of Life and Marine Science of Shenzhen University for equipment sharing.

Conflicts of Interest: Zongkang Wang is employed by Shenzhen Batian Ecological Engineering Co., Ltd., and the other authors declare that there are no potential conflicts of interest. Shenzhen Batian Ecological Engineering Co., Ltd. has no role in the study design, collection, analysis, interpretation of data, the writing of this article or the decision to submit it for publication. The authors declare no conflicts of interest.

References

1. Wang, Y.; Liu, W.J.; Chen, X.D.; Selomulya, C. Micro-encapsulation and stabilization of DHA containing fish oil in protein-based emulsion through mono-disperse droplet spray dryer. *J. Food Eng.* **2016**, *175*, 74–84. [CrossRef]
2. Wang, C.C.; Wang, J.Y.; Shi, H.H.; Zhao, Y.C.; Yang, J.Y.; Wang, Y.M.; Yanagita, T.; Xue, C.H.; Zhang, T.T. DHA-Enriched Phospholipids Exhibit Anti-Depressant Effects by Immune and Neuroendocrine Regulation in Mice: A Study on Dose- and Structure-Activity Relationship. *Mol. Nutr. Food Res.* **2023**, *67*, e2200089. [CrossRef] [PubMed]
3. Lee, J.H.; O'Keefe, J.H.; Lavie, C.J.; Harris, W.S. Omega-3 fatty acids: Cardiovascular benefits, sources and sustainability. *Nat. Rev. Cardiol.* **2009**, *6*, 753–758. [CrossRef]
4. Che, H.; Li, H.; Song, L.; Dong, X.; Yang, X.; Zhang, T.; Wang, Y.; Xie, W. Orally Administered DHA-Enriched Phospholipids and DHA-Enriched Triglyceride Relieve Oxidative Stress, Improve Intestinal Barrier, Modulate Inflammatory Cytokine and Gut Microbiota, and Meliorate Inflammatory Responses in the Brain in Dextran Sodium Sulfate Induced Colitis in Mice. *Mol. Nutr. Food Res.* **2021**, *65*, e2000986. [CrossRef]
5. Liu, L.; Hu, Q.; Wu, H.; Xue, Y.; Cai, L.; Fang, M.; Liu, Z.; Yao, P.; Wu, Y.; Gong, Z. Protective role of n6/n3 PUFA supplementation with varying DHA/EPA ratios against atherosclerosis in mice. *J. Nutr. Biochem.* **2016**, *32*, 171–180. [CrossRef]
6. Anderson, R.E. Lipids of ocular tissues. IV. A comparison of the phospholipids from the retina of six mammalian species. *Exp. Eye Res.* **1970**, *10*, 339–344. [CrossRef]
7. Williams, C.D.; Whitley, B.M.; Hoyo, C.; Grant, D.J.; Iraggi, J.D.; Newman, K.A.; Gerber, L.; Taylor, L.A.; McKeever, M.G.; Freedland, S.J. A high ratio of dietary n-6/n-3 polyunsaturated fatty acids is associated with increased risk of prostate cancer. *Nutr. Res.* **2011**, *31*, 1–8. [CrossRef]
8. Zhou, Q.; Wei, Z. Food-grade systems for delivery of DHA and EPA: Opportunities, fabrication, characterization and future perspectives. *Crit. Rev. Food Sci. Nutr.* **2023**, *63*, 2348–2365. [CrossRef] [PubMed]

9. Patel, A.; Liefeldt, S.; Rova, U.; Christakopoulos, P.; Matsakas, L. Co-production of DHA and squalene by thraustochytrid from forest biomass. *Sci. Rep.* **2020**, *10*, 1992. [CrossRef]
10. Prabhakaran, P.; Nazir, M.Y.M.; Thananusak, R.; Hamid, A.A.; Vongsangnak, W.; Song, Y. Uncovering global lipid accumulation routes towards docosahexaenoic acid (DHA) production in *Aurantiochytrium* sp. SW1 using integrative proteomic analysis. *Biochim. Biophys. Acta (BBA) Mol. Cell Biol. Lipids* **2023**, *1868*, 159381. [CrossRef]
11. Xu, X.; Huang, C.; Xu, Z.; Xu, H.; Wang, Z.; Yu, X. The strategies to reduce cost and improve productivity in DHA production by *Aurantiochytrium* sp.: From biochemical to genetic respects. *Appl. Microbiol. Biotechnol.* **2020**, *104*, 9433–9447. [CrossRef]
12. Aasen, I.M.; Ertesvåg, H.; Heggeset, T.M.; Liu, B.; Brautaset, T.; Vadstein, O.; Ellingsen, T.E. Thraustochytrids as production organisms for docosahexaenoic acid (DHA), squalene, and carotenoids. *Appl. Microbiol. Biotechnol.* **2016**, *100*, 4309–4321. [CrossRef]
13. Kadalag, N.L.; Pawar, P.R.; Prakash, G. Co-cultivation of *Phaeodactylum tricornutum* and *Aurantiochytrium limacinum* for polyunsaturated omega-3 fatty acids production. *Bioresour. Technol.* **2022**, *346*, 126544. [CrossRef] [PubMed]
14. Fossier Marchan, L.; Lee Chang, K.J.; Nichols, P.D.; Mitchell, W.J.; Polglase, J.L.; Gutierrez, T. Taxonomy, ecology and biotechnological applications of thraustochytrids: A review. *Biotechnol. Adv.* **2018**, *36*, 26–46. [CrossRef] [PubMed]
15. Chi, G.; Xu, Y.; Cao, X.; Li, Z.; Cao, M.; Chisti, Y.; He, N. Production of polyunsaturated fatty acids by Schizochytrium (*Aurantiochytrium*) spp. *Biotechnol. Adv.* **2022**, *55*, 107897. [CrossRef]
16. Humhal, T.; Kastanek, P.; Jezkova, Z.; Cadkova, A.; Kohoutkova, J.; Branyik, T. Use of saline waste water from demineralization of cheese whey for cultivation of *Schizochytrium limacinum* PA-968 and *Japonochytrium marinum* AN-4. *Bioprocess Biosyst. Eng.* **2017**, *40*, 395–402. [CrossRef]
17. Du, F.; Wang, Y.-Z.; Xu, Y.-S.; Shi, T.-Q.; Liu, W.-Z.; Sun, X.-M.; Huang, H. Biotechnological production of lipid and terpenoid from thraustochytrids. *Biotechnol. Adv.* **2021**, *48*, 107725. [CrossRef]
18. Jakobsen, A.N.; Aasen, I.M.; Josefsen, K.D.; Strøm, A.R. Accumulation of docosahexaenoic acid-rich lipid in thraustochytrid *Aurantiochytrium* sp. strain T66: Effects of N and P starvation and O2 limitation. *Appl. Microbiol. Biotechnol.* **2008**, *80*, 297–306. [CrossRef]
19. Zhao, Y.; Wang, H.P.; Yu, C.; Ding, W.; Han, B.; Geng, S.; Ning, D.; Ma, T.; Yu, X. Integration of physiological and metabolomic profiles to elucidate the regulatory mechanisms underlying the stimulatory effect of melatonin on astaxanthin and lipids coproduction in *Haematococcus pluvialis* under inductive stress conditions. *Bioresour. Technol.* **2021**, *319*, 124150. [CrossRef] [PubMed]
20. Bao, Z.; Zhu, Y.; Zhang, K.; Feng, Y.; Chen, X.; Lei, M.; Yu, L. High-value utilization of the waste hydrolysate of Dioscorea zingiberensis for docosahexaenoic acid production in *Schizochytrium* sp. *Bioresour. Technol.* **2021**, *336*, 125305. [CrossRef]
21. Ramachandran, S.; Rajendra Prasad, N.; Karthikeyan, S. Sesamol inhibits UVB-induced ROS generation and subsequent oxidative damage in cultured human skin dermal fibroblasts. *Arch. Dermatol. Res.* **2010**, *302*, 733–744. [CrossRef]
22. Bao, Z.; Zhu, Y.; Feng, Y.; Zhang, K.; Zhang, M.; Wang, Z.; Yu, L. Enhancement of lipid accumulation and docosahexaenoic acid synthesis in *Schizochytrium* sp. H016 by exogenous supplementation of sesamol. *Bioresour. Technol.* **2022**, *345*, 126527. [CrossRef]
23. Liang, S.; Yang, X.; Zhu, X.; Ibrar, M.; Liu, L.; Li, S.; Li, X.; Tian, T.; Li, S. Metabolic Engineering to Improve Docosahexaenoic Acid Production in Marine Protist *Aurantiochytrium* sp. by Disrupting 2,4-Dienoyl-CoA Reductase. *Front. Mar. Sci.* **2022**, *9*, 939716. [CrossRef]
24. Ou, Y.; Li, Y.; Feng, S.; Wang, Q.; Yang, H. Transcriptome Analysis Reveals an Eicosapentaenoic Acid Accumulation Mechanism in a *Schizochytrium* sp. Mutant. *Microbiol. Spectr.* **2023**, *11*, e00130-23. [CrossRef] [PubMed]
25. Krivoruchko, A.; Zhang, Y.; Siewers, V.; Chen, Y.; Nielsen, J. Microbial acetyl-CoA metabolism and metabolic engineering. *Metab. Eng.* **2015**, *28*, 28–42. [CrossRef]
26. Song, Y.; Hu, Z.; Xiong, Z.; Li, S.; Liu, W.; Tian, T.; Yang, X. Comparative transcriptomic and lipidomic analyses indicate that cold stress enhanced the production of the long C18–C22 polyunsaturated fatty acids in *Aurantiochytrium* sp. *Front. Microbiol.* **2022**, *13*, 915773. [CrossRef]
27. Du, H.; Liao, X.; Gao, Z.; Li, Y.; Lei, Y.; Chen, W.; Chen, L.; Fan, X.; Zhang, K.; Chen, S.; et al. Effects of Methanol on Carotenoids as Well as Biomass and Fatty Acid Biosynthesis in *Schizochytrium limacinum* B4D1. *Appl. Environ. Microbiol.* **2019**, *85*, e01243-19. [CrossRef]
28. Chatterjee, A.; Gupta, S. The multifaceted role of glutathione S-transferases in cancer. *Cancer Lett.* **2018**, *433*, 33–42. [CrossRef]
29. Song, X.; Zang, X.; Zhang, X. Production of High Docosahexaenoic Acid by *Schizochytrium* sp. Using Low-cost Raw Materials from Food Industry. *J. Oleo Sci.* **2015**, *64*, 197–204. [CrossRef] [PubMed]
30. Morabito, C.; Bournaud, C.; Maës, C.; Schuler, M.; Aiese Cigliano, R.; Dellero, Y.; Maréchal, E.; Amato, A.; Rébeillé, F. The lipid metabolism in thraustochytrids. *Prog. Lipid Res.* **2019**, *76*, 101007. [CrossRef]
31. Zeng, Y.; Ji, X.-J.; Lian, M.; Ren, L.-J.; Jin, L.-J.; Ouyang, P.-K.; Huang, H. Development of a Temperature Shift Strategy for Efficient Docosahexaenoic Acid Production by a Marine Fungoid Protist, *Schizochytrium* sp. HX-308. *Appl. Biochem. Biotechnol.* **2011**, *164*, 249–255. [CrossRef]
32. Sun, X.-M.; Ren, L.-J.; Bi, Z.-Q.; Ji, X.-J.; Zhao, Q.-Y.; Jiang, L.; Huang, H. Development of a cooperative two-factor adaptive-evolution method to enhance lipid production and prevent lipid peroxidation in *Schizochytrium* sp. *Biotechnol. Biofuels* **2018**, *11*, 65. [CrossRef]

33. Yokochi, T.; Honda, D.; Higashihara, T.; Nakahara, T. Optimization of docosahexaenoic acid production by *Schizochytrium limacinum* SR21. *Appl. Microbiol. Biotechnol.* **1998**, *49*, 72–76. [CrossRef]
34. Wang, L.R.; Zhang, Z.X.; Wang, Y.Z.; Li, Z.J.; Huang, P.W.; Sun, X.M. Assessing the potential of *Schizochytrium* sp. HX-308 for microbial lipids production from corn stover hydrolysate. *Biotechnol. J.* **2022**, *17*, e2100470. [CrossRef]
35. Jiang, X.; Zhang, J.; Zhao, J.; Gao, Z.; Zhang, C.; Chen, M. Regulation of lipid accumulation in *Schizochytrium* sp. ATCC 20888 in response to different nitrogen sources. *Eur. J. Lipid Sci. Technol.* **2017**, *119*, 1700025. [CrossRef]
36. Sun, L.; Ren, L.; Zhuang, X.; Ji, X.; Yan, J.; Huang, H. Differential effects of nutrient limitations on biochemical constituents and docosahexaenoic acid production of *Schizochytrium* sp. *Bioresour. Technol.* **2014**, *159*, 199–206. [CrossRef] [PubMed]
37. Nazir, Y.; Shuib, S.; Kalil, M.S.; Song, Y.; Hamid, A.A. Optimization of Culture Conditions for Enhanced Growth, Lipid and Docosahexaenoic Acid (DHA) Production of *Aurantiochytrium* SW1 by Response Surface Methodology. *Sci. Rep.* **2018**, *8*, 8909. [CrossRef]
38. Zhou, S.; Zou, H.; Huang, G.; Chen, G. Preparations and antioxidant activities of sesamol and it's derivatives. *Bioorganic Med. Chem. Lett.* **2021**, *31*, 127716. [CrossRef] [PubMed]
39. Liu, B.; Liu, J.; Sun, P.; Ma, X.; Jiang, Y.; Chen, F. Sesamol Enhances Cell Growth and the Biosynthesis and Accumulation of Docosahexaenoic Acid in the Microalga *Crypthecodinium cohnii*. *J. Agric. Food Chem.* **2015**, *63*, 5640–5645. [CrossRef] [PubMed]
40. Ratledge, C. Fatty acid biosynthesis in microorganisms being used for Single Cell Oil production. *Biochimie* **2004**, *86*, 807–815. [CrossRef]
41. Song, Y.; Yang, X.; Li, S.; Luo, Y.; Chang, J.S.; Hu, Z. Thraustochytrids as a promising source of fatty acids, carotenoids, and sterols: Bioactive compound biosynthesis, and modern biotechnology. *Crit. Rev. Biotechnol.* **2023**, *44*, 618–640. [CrossRef]
42. Cui, G.-Z.; Ma, Z.; Liu, Y.-J.; Feng, Y.; Sun, Z.; Cheng, Y.; Song, X.; Cui, Q. Overexpression of glucose-6-phosphate dehydrogenase enhanced the polyunsaturated fatty acid composition of *Aurantiochytrium* sp. SD116. *Algal Res.* **2016**, *19*, 138–145. [CrossRef]
43. Cui, G.; Wang, Z.; Hong, W.; Liu, Y.-J.; Chen, Z.; Cui, Q.; Song, X. Enhancing tricarboxylate transportation-related NADPH generation to improve biodiesel production by *Aurantiochytrium*. *Algal Res.* **2019**, *40*, 101505. [CrossRef]
44. Wynn, J.P.; Kendrick, A.; Ratledge, C. Sesamol as an inhibitor of growth and lipid metabolism in Mucor circinelloides via its action on malic enzyme. *Lipids* **1997**, *32*, 605–610. [CrossRef]
45. Meesapyodsuk, D.; Qiu, X. Biosynthetic mechanism of very long chain polyunsaturated fatty acids in *Thraustochytrium* sp. 26185. *J. Lipid Res.* **2016**, *57*, 1854–1864. [CrossRef]
46. Xie, Y.; Wang, G. Mechanisms of fatty acid synthesis in marine fungus-like protists. *Appl. Microbiol. Biotechnol.* **2015**, *99*, 8363–8375. [CrossRef] [PubMed]
47. Johansson, M.; Chen, X.; Milanova, S.; Santos, C.; Petranovic, D. PUFA-induced cell death is mediated by Yca1p-dependent and -independent pathways, and is reduced by vitamin C in yeast. *FEMS Yeast Res.* **2016**, *16*, fow007. [CrossRef]
48. Bi, Z.-Q.; Ren, L.-J.; Hu, X.-C.; Sun, X.-M.; Zhu, S.-Y.; Ji, X.-J.; Huang, H. Transcriptome and gene expression analysis of docosahexaenoic acid producer *Schizochytrium* sp. under different oxygen supply conditions. *Biotechnol. Biofuels* **2018**, *11*, 249. [CrossRef]
49. Zou, Y.; Cao, S.; Zhao, B.; Sun, Z.; Liu, L.; Ji, M. Increase in glutathione S-transferase activity and antioxidant damage ability drive resistance to bensulfuron-methyl in *Sagittaria trifolia*. *Plant Physiol. Biochem.* **2022**, *190*, 240–247. [CrossRef]
50. Liu, L.; Hu, Z.; Li, S.; Yang, H.; Li, S.; Lv, C.; Zaynab, M.; Cheng, C.H.K.; Chen, H.; Yang, X. Comparative Transcriptomic Analysis Uncovers Genes Responsible for the DHA Enhancement in the Mutant *Aurantiochytrium* sp. *Microorganisms* **2020**, *8*, 529. [CrossRef]
51. Wang, X.; Liu, S.F.; Qin, Z.H.; Balamurugan, S.; Li, H.Y.; Lin, C.S.K. Sustainable and stepwise waste-based utilisation strategy for the production of biomass and biofuels by engineered microalgae. *Environ. Pollut.* **2020**, *265*, 114854. [CrossRef] [PubMed]
52. Folch, J.; Lees, M.; Sloane Stanley, G.H. A simple method for the isolation and purification of total lipides from animal tissues. *J. Biol. Chem.* **1957**, *226*, 497–509. [CrossRef] [PubMed]
53. Li, S.; Hu, Z.; Yang, X.; Li, Y. Effect of Nitrogen Sources on Omega-3 Polyunsaturated Fatty Acid Biosynthesis and Gene Expression in *Thraustochytriidae* sp. *Mar. Drugs* **2020**, *18*, 612. [CrossRef] [PubMed]
54. Anders, S.; Huber, W. Differential expression analysis for sequence count data. *Genome Biol.* **2010**, *11*, R106. [CrossRef]

marine drugs

MDPI

Article

Production of Fucoxanthin from Microalgae *Isochrysis galbana* of Djibouti: Optimization, Correlation with Antioxidant Potential, and Bioinformatics Approaches

Fatouma Mohamed Abdoul-Latif [1,*], Ayoub Ainane [2], Laila Achenani [2], Ali Merito Ali [1], Houda Mohamed [1,3], Ahmad Ali [4], Pannaga Pavan Jutur [5] and Tarik Ainane [2,*]

[1] Medicinal Research Institute, Center for Research and Study of Djibouti, Djibouti City P.O. Box 486, Djibouti
[2] Superior School of Technology, University of Sultan Moulay Slimane, P.O. Box 170, Khenifra 54000, Morocco; laila.achenani@usms.ma (L.A.)
[3] Peltier Hospital of Djibouti, Djibouti City P.O. Box 2123, Djibouti
[4] University Department of Life Sciences, University of Mumbai, Vidyanagari, Santacruz (East), Mumbai 400098, India
[5] Omics of Algae Group, Industrial Biotechnology, International Centre for Genetic Engineering and Biotechnology, Aruna Asaf Ali Marg, New Delhi 110067, India; pavan.jutur@icgeb.org
* Correspondence: fatoumaabdoulatif@gmail.com (F.M.A.-L.); t.ainane@usms.ma (T.A.)

Citation: Mohamed Abdoul-Latif, F.; Ainane, A.; Achenani, L.; Merito Ali, A.; Mohamed, H.; Ali, A.; Jutur, P.P.; Ainane, T. Production of Fucoxanthin from Microalgae *Isochrysis galbana* of Djibouti: Optimization, Correlation with Antioxidant Potential, and Bioinformatics Approaches. *Mar. Drugs* **2024**, *22*, 358. https://doi.org/10.3390/md22080358

Academic Editors: Cecilia Faraloni and Eleftherios Touloupakis

Received: 9 July 2024
Revised: 30 July 2024
Accepted: 3 August 2024
Published: 6 August 2024

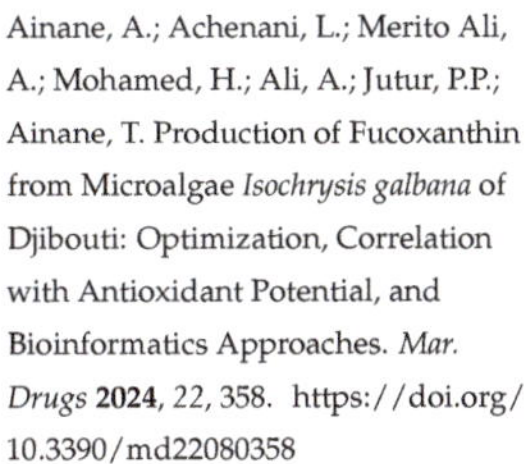

Abstract: Fucoxanthin, a carotenoid with remarkable antioxidant properties, has considerable potential for high-value biotechnological applications in the pharmaceutical, nutraceutical, and cosmeceutical fields. However, conventional extraction methods of this molecule from microalgae are limited in terms of cost-effectiveness. This study focused on optimizing biomass and fucoxanthin production from *Isochrysis galbana*, isolated from the coast of Tadjoura (Djibouti), by testing various culture media. The antioxidant potential of the cultures was evaluated based on the concentrations of fucoxanthin, carotenoids, and total phenols. Different nutrient formulations were tested to determine the optimal combination for a maximum biomass yield. Using the statistical methodology of principal component analysis, Walne and Guillard F/2 media were identified as the most promising, reaching a maximum fucoxanthin yield of 7.8 mg/g. Multiple regression models showed a strong correlation between antioxidant activity and the concentration of fucoxanthin produced. A thorough study of the optimization of *I. galbana* growth conditions, using a design of experiments, revealed that air flow rate and CO_2 flow rate were the most influential factors on fucoxanthin production, reaching a value of 13.4 mg/g. Finally, to validate the antioxidant potential of fucoxanthin, an in silico analysis based on molecular docking was performed, showing that fucoxanthin interacts with antioxidant proteins (3FS1, 3L2C, and 8BBK). This research not only confirmed the positive results of *I. galbana* cultivation in terms of antioxidant activity, but also provided essential information for the optimization of fucoxanthin production, opening up promising prospects for industrial applications and future research.

Keywords: antioxidant; carotenoids; fucoxanthin; in silico studies; microalgae; production; statistical approaches

1. Introduction

Microalgae are emerging as a game-changer in the realm of natural ingredients [1]. These microscopic organisms hold immense potential for various industrial applications, particularly in the food and pharmaceutical sectors [2–4]. Their remarkable biodiversity translates into producing a diverse range of valuable intracellular metabolites. These include proteins, carbohydrates, lipids, and most importantly, carotenoids—a class of compounds renowned for their potent antioxidant properties that significantly contribute to human health [5,6].

Carotenoids are a diverse group of pigments found in all photosynthetic organisms and some non-photosynthetic ones, including microalgae [7]. These vibrant molecules, responsible for the yellow, orange, and red hues in fruits, flowers, and even our skin, are built on a C_{40} isoprenoid backbone. Depending on the specific structure, this backbone can be acyclic or cyclic, with various modifications at its ends [8–10]. Common and well-studied carotenoids include lycopene, β-carotene, α-carotene, lutein, zeaxanthin, and β-cryptoxanthin [11,12]. These compounds are not just visually striking; they play an essential role in human health by acting as precursors to vitamin A and providing essential antioxidant protection against cellular damage [13,14]. In recent years, fucoxanthin has garnered significant interest due to its therapeutic potential and ability to address nutritional deficiencies. However, traditional methods for extracting fucoxanthin from brown algae (macroalgae) face limitations. These methods often involve complex and lengthy growth cycles, making large-scale production impractical and driving up costs [15,16].

Microalgae, on the other hand, offer a promising solution for the commercial production of fucoxanthin. These diverse photoautotrophs have the advantage of efficiently accumulating fucoxanthin and achieving high biomass productivity [17]. Unlike macroalgae, their cultivation cycles are significantly shorter and more manageable, paving the way for sustainable and cost-effective production [18]. Among the various fucoxanthin-producing microalgae, *I. galbana*, a member of the Prymnesiophyceae class, takes center stage. Its small size, high digestibility, and rich content of essential nutrients make it a valuable food source for the larvae of shellfish like mussels and clams [19–21]. Furthermore, *I. galbana* demonstrates significant potential for successful industrialization due to its ease of cultivation and rapid growth. These advantages have propelled this microalga to the forefront of research and development efforts focused on commercial applications [22–24].

Several previous works have explored the optimization of fucoxanthin production from various microalgae, including *Isochrysis galbana*, *Phaeodactylum tricornutum*, and *Tisochrysis lutea* [25,26]. These studies analyzed the impact of different culture media, abiotic parameters (such as temperature, irradiance, and pH), and nutrient sources on fucoxanthin productivity. Other studies have also used sophisticated approaches [27,28], such as response surface methodology and machine learning algorithms, to optimize fucoxanthin production. However, there are still gaps in understanding the complex interactions between the various factors influencing fucoxanthin production and the underlying molecular mechanisms. This research study aims to lay the groundwork for enhancing the cultivation efficiency of *I. galbana* and explore its potential in the pharmaceutical and nutraceutical industries. The primary objective is to optimize fucoxanthin production and the biomass's overall antioxidant activity. This will be achieved through controlled cultivation experiments examining the influence of different enrichment solutions in the culture media. We will explore the relationships between these factors and their impact on the microalgae.

The first step of our investigation focused on analyzing the correlations between antioxidant activity, fucoxanthin production, carotenoid content, and total phenol levels. Biostatistical tools, such as principal component analysis (PCA) and multiple regression, were used to identify these connections within different culture media formulations designed explicitly for *I. galbana*. Building on these results, the second step consisted of implementing experimental models to optimize fucoxanthin production. The influence of modifiable growth conditions, including choice of growing medium, temperature, pH, lighting intensity, air flow rate, and CO_2 flow rate, was explored. These parameters were meticulously adjusted to maximize fucoxanthin production while simultaneously assessing their impact on the antioxidant properties of the microalgae. Finally, the last step consisted of an in silico study based on molecular docking to study the molecular interaction between fucoxanthin and proteins with renowned antioxidant potentials in cellular processes.

2. Results

2.1. Optimization of Growth Parameters

The optimal enriched culture medium for maximizing the biomass production of *I. galbana*, as well as the content of fucoxanthin, carotenoids, total phenols, and antioxidant activity, was selected from several media: 3N-BBM+V medium, ASP-M medium, CHU-10 medium, Conway medium, Erdschreiber medium, Guillard F/2 medium, PM medium, Walne medium, and WC aquatic culture medium. Natural seawater served as a negative control due to its lack of nutrient supplements compared with enriched solutions.

Under standard operating conditions (temperature T = 25 °C, pH = 6.5, light intensity LI = 100 μmol/m^2/s, air flow AFR = 0.5 L/min, and CO_2 flow CO_2FR from 0.1 to 0.2 L/min) over a culture period of 15 days, the results are summarized in Table 1. This table presents data on biomass, antioxidant activity of each extract, and biomolecule contents of each culture medium.

Table 1. Dry biomass, antioxidant activity, carotenoid content, total phenolic content, and fucoxanthin content of cultures of the microalga *I. galbana*.

Culture Medium	Dry Weight Biomass (g/L)	DPPH IC_{50} (μg/mL)	Carotenoids Content (mg/g)	Total Phenolic Content (mg/100 g)	Fucoxanthin (mg/g)
Natural Seawater (Djibouti)	0.65 ± 0.12 [a]	481 ± 25 [a]	9.9 ± 0.2 [a]	12.0 ± 1.1 [a]	2.8 ± 0.5 [a]
3N-BBM+V Medium	0.88 ± 0.17 [a,b,c]	438 ± 20 [a,b]	11.5 ± 1.3 [b]	20.0 ± 1.8 [b]	4.6 ± 0.8 [b]
ASP-M Medium	0.82 ± 0.16 [a,b,c]	435 ± 25 [a,b]	11.7 ± 1.2 [b]	19.2 ± 1.7 [b]	4.8 ± 0.7 [b]
CHU-10 Medium	0.91 ± 0.18 [b,c]	440 ± 25 [a,b]	11.5 ± 1.1 [b]	20.1 ± 1.8 [b]	5.0 ± 0.8 [b]
Conway Medium	0.92 ± 0.18 [b,c]	415 ± 25 [b,c]	13.1 ± 1.5 [b,c]	20.3 ± 2.0 [b]	5.0 ± 0.6 [b]
Erdschreiber Medium	0.95 ± 0.19 [b,c]	405 ± 20 [c]	12.8 ± 1.8 [b,c]	20.1 ± 2.1 [b]	5.2 ± 0.5 [b]
Guillard F/2 Medium	1.22 ± 0.21 [c,d]	304 ± 15 [d]	16.6 ± 2.4 [d]	20.5 ± 1.8 [b]	7.8 ± 1.0 [c]
PM Medium	0.82 ± 0.16 [a,b,c]	421 ± 30 [b,c]	11.2 ± 1.2 [a,b]	19.1 ± 1.8 [b]	4.6 ± 0.5 [b]
Walne Medium	1.28 ± 0.22 [c,d]	285 ± 15 [e]	15.4 ± 1.8 [d]	23.6 ± 2.5 [b,c]	7.8 ± 0.8 [c]
Water Culture Medium	1.15 ± 0.20 [c,d]	413 ± 25 [b,c]	14.2 ± 1.6 [c,d]	19.4 ± 1.8 [b]	5.5 ± 0.5 [b]

Different letters in the same row indicate significant differences according to Tukey's test ($p < 0.05$).

Principal component analysis (PCA) was used to determine the correlations between the studied parameters. Figure 1 graphically illustrates all the correlations obtained, with 97.54% of the variance explained by the first two axes (F1 = 87.66% and F2 = 9.88%). This statistical analysis classified the culture media into three distinct groups:

First group: Comprising Walne medium and Guillard F/2 medium, which showed promising results in terms of biomass, antioxidant activity, and active biomolecule content. The IC50 of antioxidant activity ranged from 285 μg/mL to 304 μg/mL, and the fucoxanthin content reached 7.8 mg/g.

Second group: Including 3N-BBM+V, ASP-M, CHU-10, Conway, PM, WC aquatic culture, and Erdschreiber media, this group showed moderate results in terms of biomass, antioxidant activity, and active biomolecule content. The IC50 of antioxidant activity ranged from 405 μg/mL to 440 ± 2.5 μg/mL, and the fucoxanthin content ranged from 4.6 mg/g to 5.5 mg/g.

Third group: Containing only natural seawater, which served as a negative control with inferior results compared with enriched media. The measured antioxidant activity was 481 μg/mL, and the fucoxanthin content was 2.8 mg/g.

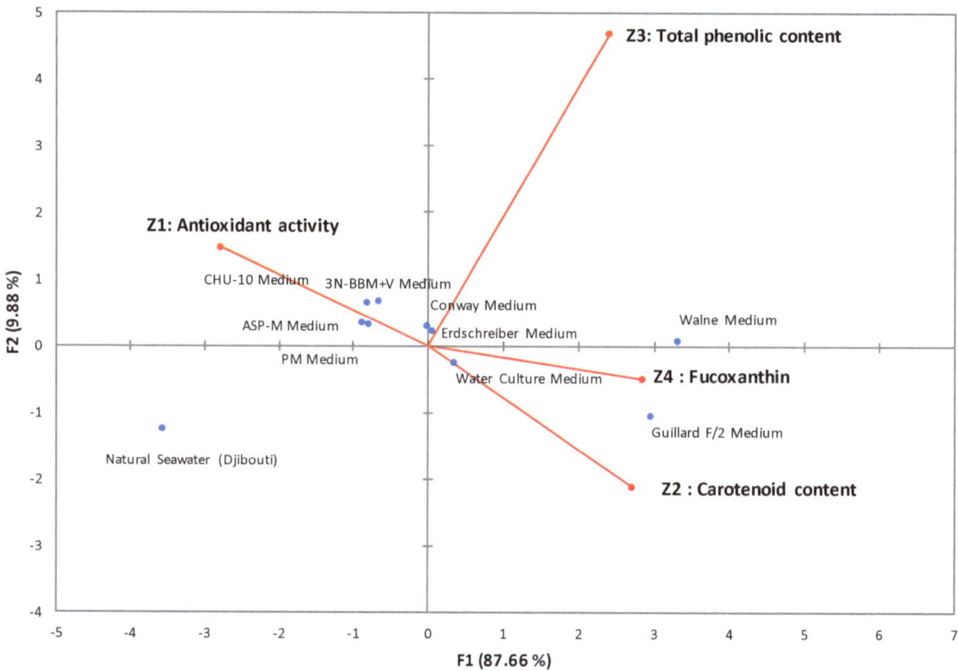

Figure 1. Correlations between antioxidant activity, carotenoid content, total phenolic content, and fucoxanthin content of *I. galbana* cultures. The various circles, represented by distinct colors, symbolize groups sharing common characteristics.

The experimental and statistical results led to the selection of Walne and Guillard F/2 media as the most effective due to their antioxidant power and high fucoxanthin content. These media share essential macronutrients, micronutrients, and vitamins, with Guillard F/2 medium distinguished by including Na_2EDTA, a chelating agent absent from Walne medium. The differences between these environments are mainly based on the specific proportions of the shared compounds, distinctly influencing the growth and development of microalgae [29,30]. Finally, Walne medium was favored for subsequent studies because of its high fucoxanthin content (7.8 mg/g).

2.2. Correlation between Antioxidant Activity and Compound Content

The correlation between antioxidant activity (Z1) and the contents of carotenoids (Z2), total phenolic compounds (Z3), and fucoxanthin (Z4) in cultures of the microalga *I. galbana* was studied using two statistical tools: the matrix of correlation and multiple regression modeling, including multiple linear regression (MLR) and nonlinear multiple regression (MNLR).

The correlation matrix in Table 2 shows the correlation coefficients between each pair of variables. It reveals a strong positive correlation between carotenoid content (Z2) and fucoxanthin (Z4), indicating that fucoxanthin is a significant component of carotenoids. In contrast, antioxidant activity (Z1) showed negative correlations with carotenoid and fucoxanthin content, suggesting that lower values correspond to higher antioxidant activity.

The results of multiple regression modeling (Table S1) show that the two models, MLR and MNLR, perform very well with high coefficients of determination (R^2) of 0.954 and 0.986, respectively. However, the nonlinear model (MNLR) shows greater explanatory capacity, indicating a more complex relationship between variables, which is better captured by this model. In terms of predictive accuracy, MNLR has a mean square error (MCE) of 156.573, lower than that of MLR (262.950), confirming better accuracy of the nonlinear model. Similarly, the root mean square error (RMCE) is lower for MNLR (12.513) than for MLR (16.216), further emphasizing the superior fit of the nonlinear model to the observed data.

Table 2. Correlation matrix.

Matrix	Z2	Z3	Z4	Z1
Z2	1			
Z3	0.660	1		
Z4	0.943	0.794	1	
Z1	−0.913	−0.706	−0.970	1

Z1: antioxidant activity; Z2: carotenoid content; Z3: total phenolic content; and Z4: fucoxanthin.

Figure S1 illustrates the linearity of the predictive results compared with the experimental results, demonstrating that MNLR aligns better with the experimental data than MLR. In conclusion, the results indicate that the relationships between antioxidant activity (Z1) and the contents of carotenoids (Z2), total phenolic compounds (Z3), and fucoxanthin (Z4) are well correlated, and nonlinear models are more appropriate for capturing the complexity of these relationships in *I. galbana* cultures.

2.3. Optimization of Experimental Conditions by Experimental Design

The experiments were planned to optimize the experimental conditions aimed at maximizing the production of fucoxanthin following the statistical tool of the experimental plan, taking into consideration five operational parameters: temperature, pH, light intensity, air flow rate, and CO_2 flow rate. Table S2 presents the results obtained for the fucoxanthin content for each of the 32 tests according to the polynomial model of full factorial designs at two levels (−1 and +1).

A first-order polynomial mathematical model was developed from the results obtained, including the main factors and higher-order interactions up to the fourth order between these five parameters. The coefficients of the fucoxanthin production model equation are listed in Table S3. The average yield achieved by the optimization of conditions was 7.93 mg/g with a coefficient of variation of 1.11%.

The statistical model derived from the experimental design is extremely accurate and well fitted to the experimental data. The statistical parameters presented in Table S4 (coefficient of determination: $R^2 = 0.99$, adjusted R^2: $R^2_{adj} = 0.99$, predicted R^2: $R^2_{pre} = 0.91$, and adequate precision: $P = 87.36\%$) indicate that the model explains a large part of the observed variation and has good predictive ability for new observations.

Table S5 presents the analysis of variance (ANOVA), which is used to assess the significance of the effects of different factors in the experimental design. The model is considered significant with an F value of 386.15, with only a 4.02% probability that such a high F value is due to chance. p-Values of less than 0.05 indicate that some model terms are substantial—D, E, AE, CE, DE, ACE, CDE, and BCDE—while others are not.

Finally, Figure S2 graphically illustrates the main interactions, which are significant contributors to the variation in fucoxanthin concentration, by showing interactions in one dimension (D and E), two dimensions (AE, CE, and DE), and three dimensions (ACE and CDE).

2.4. In Silico Study

The results of molecular docking of fucoxanthin with the three selected proteins (3FS1, 3L2C, and 8BBK) are shown in Figure 2 and Table 3.

The three proteins studied play an important role in antioxidant processes, as follows:

The 3FS1 protein, corresponding to HNF4a (hepatocyte nuclear factor 4 alpha), plays an essential role in the cellular antioxidant process by modulating the expression of genes associated with detoxification and the response to oxidative stress. As a transcription factor, HNF4a induces the expression of genes encoding antioxidant enzymes such as superoxide dismutase (SOD), catalase, and glutathione-S-transferases. These enzymes neutralize reactive oxygen species (ROS) and attenuate oxidative damage in liver cells and other tissues where HNF4a is expressed [31,32].

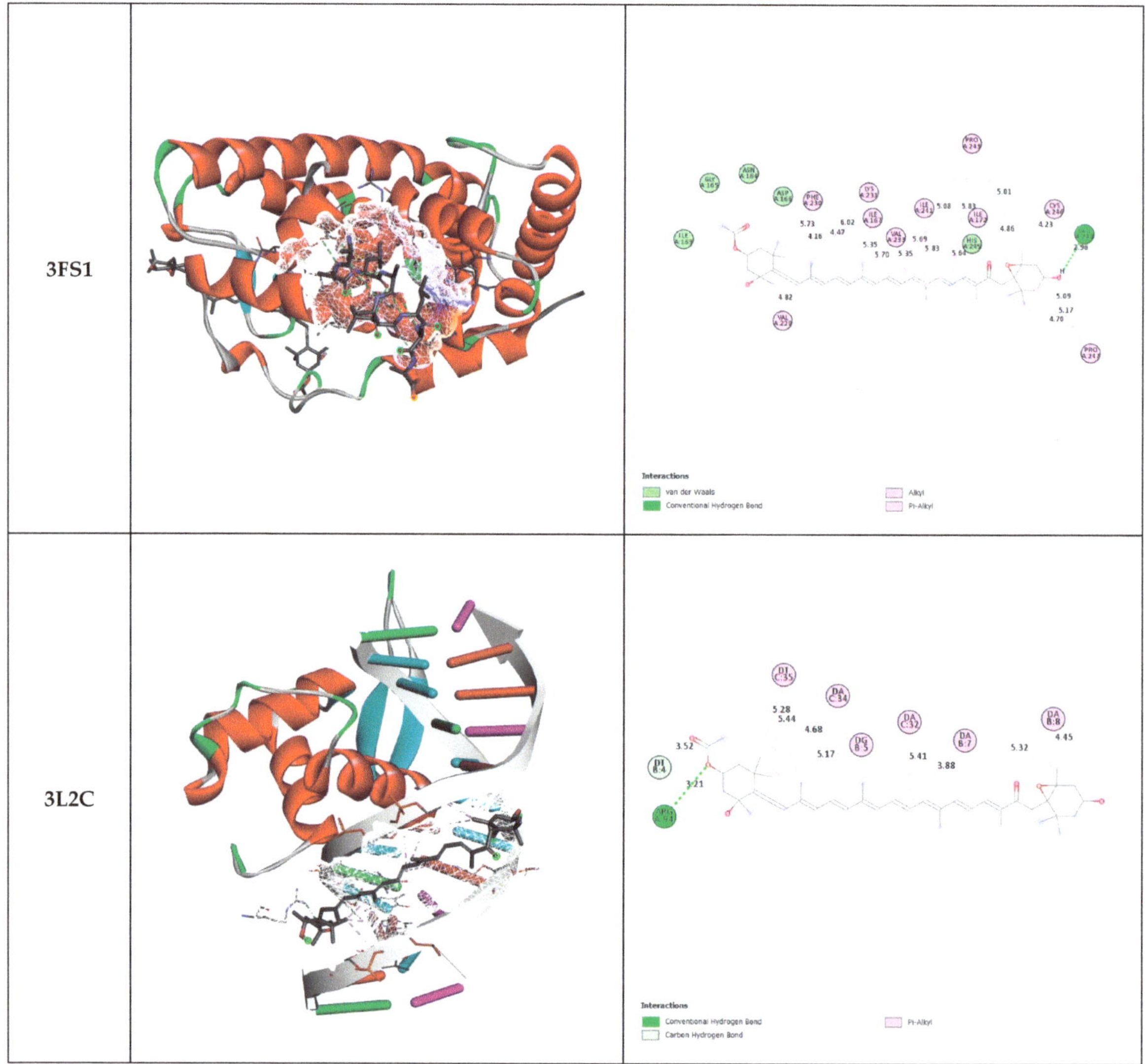

Figure 2. *Cont.*

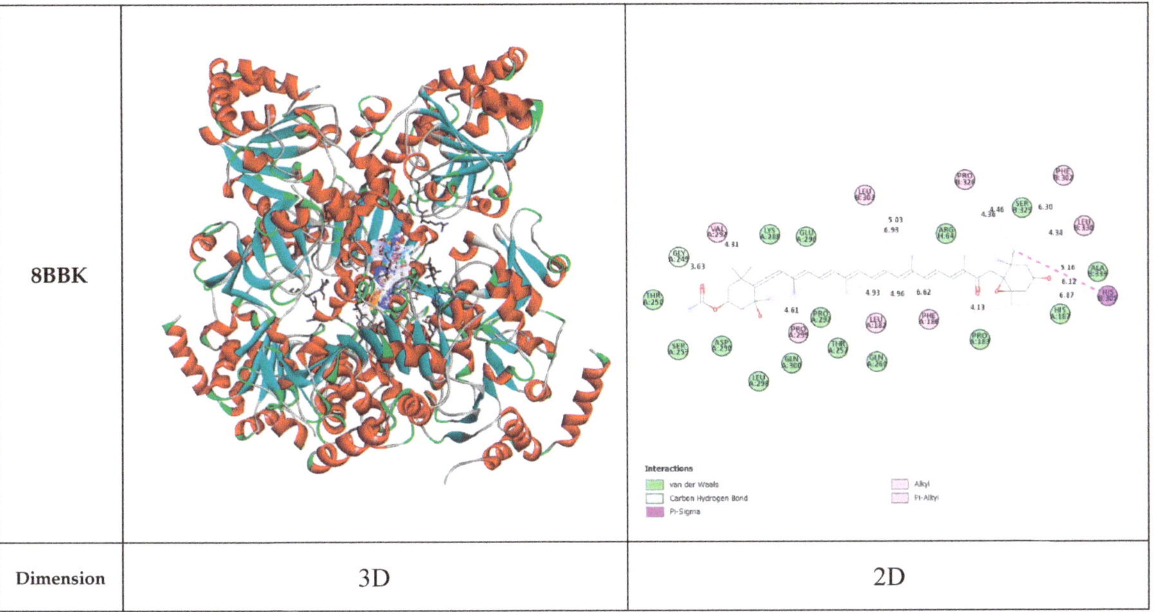

8BBK		
Dimension	3D	2D

Figure 2. Three-dimensional and two-dimensional docked views of fucoxanthin with 3FS1, 3L2C, and 8BBK proteins, respectively.

Table 3. Qualitative and energetic characteristics of molecular docking of fucoxanthin with antioxidant activity proteins.

Proteins	3FS1	3L2C	8BBK
Binding affinity (kcal/mol)	−7.3	−9.4	−9.3
pKi	5.35	6.89	6.82
Ligand efficiency (kcal/mol)	0.1521	0.1958	0.1938
Ligand–protein interactions	25	10	31
Number of π-σ bond	0	0	1
Number of alkyl bond	15	0	10
Number of π-alkyl bond	4	8	4
Number of conventional hydrogen bond	1	1	1
Number of carbon–hydrogen bond	0	1	0
Number of van der Waals bond	5	0	15

The 3L2C protein, identified as FOXO4 (forkhead box O4), plays a role in the antioxidant response by regulating the expression of genes encoding antioxidant enzymes and DNA repair proteins. In response to various cellular signals, including oxidative stress, FOXO4 migrates to the cell nucleus, where it binds to specific DNA sequences and activates the transcription of genes such as SOD, glutathione peroxidase, and other enzymes and antioxidants. This action helps neutralize ROS and restore redox balance in cells, thus reducing oxidative damage [33–35].

As for the 8BBK protein, corresponding to Sirt3 (Sirtuin 3), Sirt3 participates in the antioxidant process by deacetylating and activating key enzymes involved in defense against oxidative stress. Notably, Sirt3 deacetylates and activates SOD2 (MnSOD) localized in mitochondria, which increases the activity of this critical antioxidant enzyme. SOD2 converts superoxide to hydrogen peroxide, which is then broken down into water and oxygen by other enzymes. In addition, Sirt3 regulates other proteins involved in the mitochondrial antioxidant response, thereby preserving the functional integrity of mitochondria and reducing cellular oxidative stress [36,37].

Fucoxanthin exhibits a stronger interaction with 3L2C and 8BBK proteins compared with 3FS1, indicating a higher affinity of fucoxanthin for FOXO4 (3L2C) and Sirt3 (8BBK) compared with HNF4a (3FS1). This higher affinity suggests that fucoxanthin has an increased propensity to bind FOXO4 and Sirt3. Additionally, higher ligand efficiency implies that the ligand performs better per unit size in binding to the protein. In this context, fucoxanthin demonstrates slightly higher efficacy when interacting with FOXO4 (3L2C) and Sirt3 (8BBK) compared with HNF4a (3FS1).

The correlation with the principal component analysis (PCA) of all thermodynamic parameters and interaction bonds of the different types of molecular docking is presented in Figure 3. The analysis of this representation, based on the F1 and F2 axes (97.71%), proves that the specific interactions between fucoxanthin and the HNF4a (3FS1) and Sirt3 (8BBK) proteins are well correlated. On the other hand, the interactions of these last two proteins are not correlated with FOXO4, which suggests that fucoxanthin adopts two different mechanisms depending on the proteins studied.

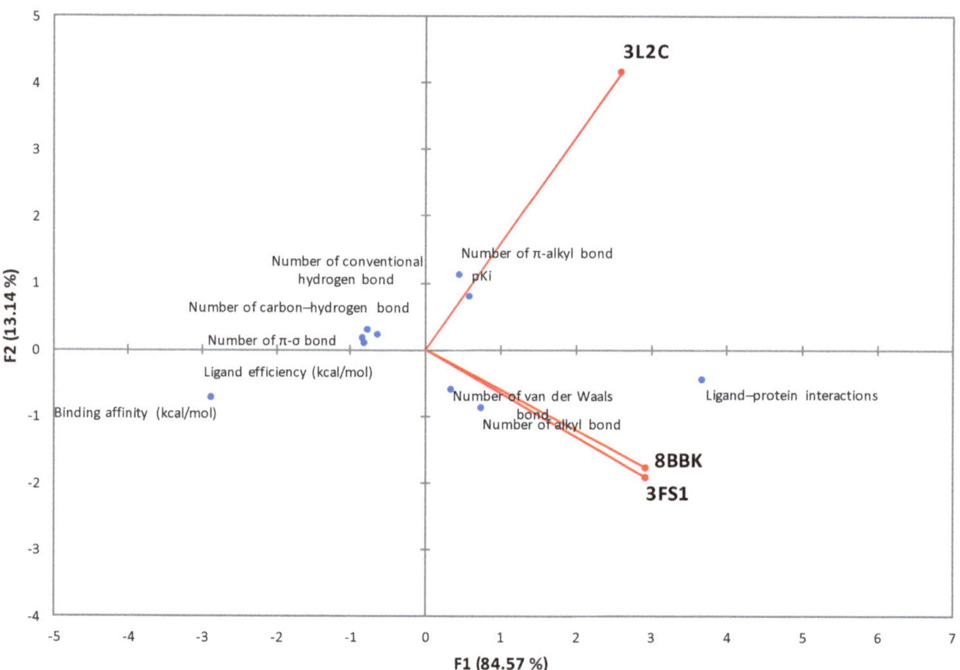

Figure 3. Correlation between the molecular docking parameters between fucoxanthin and the proteins studied.

3. Discussion

Fucoxanthin is widely recognized for its bioactive potential and substantial antioxidant activity, making it a compound with many industrial applications. These advantages are enhanced by continued advances in extraction and quantification techniques and a better understanding of its biological properties [38].

The quantification of fucoxanthin, an abundant carotenoid in microalgal biomass, particularly in *I. galbana*, is essential for its industrial and therapeutic applications [39]. In the pharmaceutical sector, it is used to develop food supplements and drugs due to its beneficial antioxidant, anticancer, anti-diabetes, anti-inflammatory effects, etc. [40–43]. Nutraceuticals incorporate it into functional food products for its health properties. In the cosmetic field, it is incorporated into skin care formulations for its anti-ageing effects and ability to protect against UV damage. In agriculture, its bioactive properties make it a potential candidate as a biofertilizer and plant protective agent [44].

The main objective of this study was to quantify fucoxanthin and valorize its antioxidant properties as an abundant carotenoid in *I. galbana* while maximizing its extraction and use in various industrial sectors. This included optimizing production and standardizing and establishing accurate quantification methods to ensure the quality and consistency of fucoxanthin products, facilitating their introduction to the market. Furthermore, the exploitation of the antioxidant properties of fucoxanthin has been explored for the development of new products while promoting the use of natural and renewable sources to support sustainable industrial practices.

The significant results of this study showed that Walne and Guillard F/2 culture media were the most effective in producing fucoxanthin in *I. galbana* culture due to their high antioxidant potential and increased fucoxanthin content. Correlation analysis revealed a strong association between antioxidant activity and fucoxanthin and carotenoid content and a moderate correlation with total phenols. Regression models, whether linear or nonlinear, have been used to model these relationships, with the nonlinear model showing a better ability to explain and predict antioxidant activity, suggesting a more accurate consideration of complex interactions. Using experimental designs, optimal cultivation conditions were identified, including temperature, pH, light intensity, air flow rate, and CO_2 flow rate, with particular attention paid to the last two parameters to influence the production of fucoxanthin significantly. The developed mathematical models demonstrated robust predictive ability and offer promising prospects for future optimization of fucoxanthin production in a sustainable industrial setting.

Recent works (Table 4) that have received good recognition in the literature concerning optimizing fucoxanthin production are numerous, among which the most important over the last 5 years are presented below.

In our study on fucoxanthin production from *I. galbana*, we explored various culture media and advanced analytical approaches to optimize the production of this bioactive pigment. Compared with previous studies, our analysis reveals several important and divergent aspects.

Medina et al. (2019) [45] demonstrated that methanol and ethanol are the most effective solvents to extract fucoxanthin from *I. galbana*, with significantly higher yields than petroleum ether and n-hexane. Their optimization of extraction time also highlighted the potential of *I. galbana* as a natural source of fucoxanthin. In contrast, our study shows that optimization of culture conditions and media, such as Walne and Guillard media, also plays a critical role in enhancing fucoxanthin production.

Gao et al. (2020) [46] optimized fucoxanthin production in *Tisochrysis lutea* by adjusting culture and irradiance parameters. They observed that higher dilution rates and specific irradiance maximized production. Our results support these observations to some extent, but our multidimensional approach allowed for further optimization by combining culture, statistical analysis, and bioinformatics validation techniques, which could provide additional insights for large-scale production.

Pereira et al. (2021) [47] illustrated seasonal variations in fucoxanthin production, showing that *Phaeodactylum tricornutum* performs better in autumn/winter, while *T. lutea* performs better in spring/summer. While their study highlights the feasibility of continuous production, our study highlights the impact of specific growth conditions and analytical approaches on maximizing production.

Table 4. Main works of optimization of fucoxanthin production.

Work	Reference	Year	Study Objective	Methodology	Key Results	Quantity of Fucoxanthin	Implications
1	Médine et al. [45]	2019	Examine the extraction of fucoxanthin from *I. galbana* for obesity prevention.	Comparison of solvents (methanol, ethanol, petroleum ether, n-hexane). Optimization of extraction time.	Best yields with methanol (6.282 mg/g DW) and ethanol (4.187 mg/g DW). Optimal extraction in 10 min with 100% ethanol.	6.282 mg/g DW (methanol), 4.187 mg/g DW (ethanol)	*I. galbana* is a promising source of fucoxanthin for the food industry.
2	Pereira et al. [46]	2021	Optimize industrial-scale fucoxanthin production.	Cultivation in 15 m^3 tubular photobioreactors. Seasonal comparison between *P. tricornutum* and *T. lutea*.	*P. tricornutum* achieved 2.87 g DW L^{-1} and 0.7% DW fucoxanthin (7 mg/g DW) in fall/winter. *T. lutea* was more productive in spring/summer.	7 mg/g DW	Feasibility of continuous fucoxanthin production year-round.
3	Gao et al. [47]	2020	Optimize fucoxanthin production in *T. lutea*.	Batch and continuous experiments, adjusting parameters like temperature, irradiation, and dilution rate.	Maximum productivity at 30 °C and 300 µmol m^{-2} s^{-1}. High dilution rates (0.53 and 0.80 d^{-1}) and light absorption of 2.23 mol m^{-2} d^{-1} favored high fucoxanthin content.	16.39 mg/g DW	Light absorption can predict fucoxanthin content.
4	Mohamadnia et al. [48]	2021	Optimize fucoxanthin production in *T. lutea* using response surface methodology.	Adjustment of culture conditions in 1 L batch photobioreactors with polynomial second-order modeling.	Optimal conditions: salinity 36.27 g L^{-1}, starch 3.90 g L^{-1}, nitrate 0.162 g L^{-1}.	79.4 mg/g DW	Optimization of culture parameters to maximize fucoxanthin production.
5	Mohamadnia et al. [49]	2022	Refine fucoxanthin production using response surface methodology (RSM).	Adjustment of concentrations of glutamic acid, trisodium citrate, succinic acid, sodium aspartate, and pyruvate.	Optimal concentrations: sodium aspartate 7.5 mM, sodium pyruvate 7.5 mM, glutamic acid 3.29 mM. Productivity of 22.4 mg L^{-1} day^{-1}.	22.4 mg L^{-1} day^{-1}	Metabolic optimization strategies to increase fucoxanthin production.
6	McElroy et al. [50]	2023	Integrate biorefining to valorize *Saccharina latissima* biomass.	Optimized extraction of fucoxanthin at 40 MPa. Integration with mannitol and alginate extraction. Life cycle analysis.	4.15% yield for fucoxanthin. Extraction of 67.27% to 69.38% of alginates. Reduction in environmental impact identified.	41.5 mg/g DW	Integrated biorefining processes to reduce environmental footprint.

Table 4. Cont.

Work	Reference	Year	Study Objective	Methodology	Key Results	Quantity of Fucoxanthin	Implications
7	Xia et al. [51]	2023	Assess the impact of CO_2 concentration and frequency on fucoxanthin production.	Comparison of continuous and intermittent CO_2 supply at different concentrations.	Continuous CO_2 at 5% achieved maximum biomass productivity (0.33 g L^{-1} day^{-1}). Intermittent CO_2 at 5% optimized fucoxanthin accumulation.	0.56 mg/g DW	Improved fucoxanthin accumulation with intermittent CO_2 supply.
8	Bo et al. [52]	2023	Study the effect of spermidine on fucoxanthin biosynthesis in *Isochrysis* sp.	Addition of spermidine under different light intensities and assessment of cell proliferation and pigment synthesis.	Optimal cell density of 5.40×10^6 cells/mL after 11 days. Maximum diadinoxanthin (1.09 mg/g DW) and fucoxanthin under low light intensity.	6.11 mg/g DW	Spermidine enhances fucoxanthin production and mitigates photosystem damage under high light intensity.
9	Garcia-García et al. [53]	2024	Explore extraction of fucoxanthin and DHA from *T. lutea* using green solvents.	Use of green solvents and advanced extraction techniques, such as ultrasonic-assisted extraction with 2-methyltetrahydrofuran and ethanol.	High extraction yields of fucoxanthin and DHA. 2-Methyl-tetrahydrofuran-enriched extracts showed better composition.	High (exact quantity not provided)	Advanced extraction techniques to preserve bioactivity of extracts.
10	Manochkumar et al. [54]	2024	Optimize fucoxanthin production using machine learning.	Development of a machine learning model to predict fucoxanthin yield based on phytohormone supplementation.	Random Forest and ANN models showed improved accuracy with hormone descriptors.	-	Combining UV spectrometry and ML algorithms for precise fucoxanthin predictions and production optimization.
11	This work	-	Optimize fucoxanthin production from *I. galbana* and validate antioxidant potential.	Test various culture media. Evaluate antioxidant potential. Use PCA and regression models. Optimize growth conditions (air flow rate and CO_2 flow). Perform molecular docking analysis.	Walne and Guillard media most effective. Strong correlation between antioxidant activity and fucoxanthin. Air flow rate and CO_2 flow are key factors. Fucoxanthin interacts with antioxidant proteins.	13.4 mg/g DW	Improved production methods. Validated antioxidant benefits. Insights for future research and applications.

The works of Mohamadnia et al. (2021) [48] and Mohamadnia et al. (2022) [49] on optimizing fucoxanthin production using response surface methodology confirm the importance of precise culture conditions. Their approach identified optimal parameters for *T. lutea*, which is consistent with our observation that growth conditions play an essential role. However, our use of in silico analyses to study the molecular interactions of fucoxanthin added an additional dimension to understanding its bioactive properties.

McElroy et al. (2023) [50] implemented an integrated biorefinery approach to extract fucoxanthin and other compounds from *Saccharina latissima*. Their process integration to reduce the ecological footprint is particularly relevant, although our study mainly focused on optimizing culture conditions and validating the results using bioinformatics approaches.

A study conducted by Xia et al. (2023) [51] showed that intermittent CO_2 supply promotes better fucoxanthin accumulation in *I. galbana*, a result in agreement with our observation that optimization of culture conditions is crucial for increased production. Furthermore, a study conducted by Bo et al. (2023) [52] revealed that spermidine positively influences fucoxanthin biosynthesis, highlighting the importance of biological factors in the production of this pigment.

Finally, Garcia-García et al. (2024) [53] explored the use of green solvents for fucoxanthin extraction, highlighting the advantages of modern techniques over traditional methods. This reinforces our conclusion that the use of advanced methods and solvents is essential to optimize extraction while preserving the bioactivity of the extracts.

In conclusion of this discussion, our study, through its multidimensional approach integrating the optimization of growth conditions, advanced statistical analysis, and bioinformatics validation, provides a more comprehensive and innovative overview for fucoxanthin production. These results contribute to a better understanding of the parameters influencing the production of this pigment and offer promising prospects for its industrial application.

The mentioned research on the production of fucoxanthin from microalgae has focused its attention on three key aspects: optimization of environmental variations; precision in the adjustment of culture parameters such as temperature, lighting, pH, and salinity and the sources of nutrients, O_2 and CO_2; and the improvement of extraction processes using specific solvents and optimized conditions (Figure 4) [54–58]. These studies highlight the remarkable importance of adapting culture conditions to maximize productivity and fucoxanthin concentration in microalgae cultures, considering seasonal variations, while judiciously choosing extraction methods to ensure high yields of this bioactive compound.

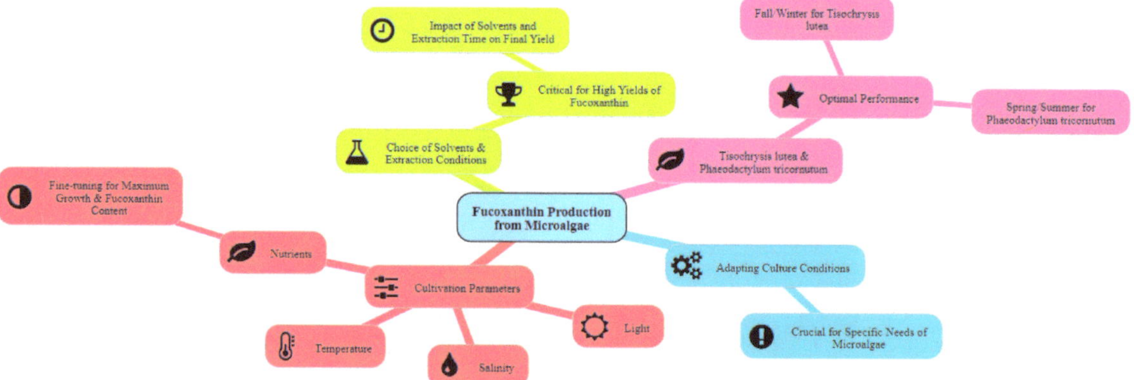

Figure 4. The main factors in the production of fucoxanthin from microalgae (created with www.map-this.com, accessed on 3 June 2024).

Molecular docking results of fucoxanthin with 3FS1 (HNF4a), 3L2C (FOXO4), and 8BBK (Sirt3) proteins revealed distinct interactions, highlighting a marked preference of fucoxanthin for FOXO4 and Sirt3 over HNF4a. These proteins have remarkable biochemical processes in the fine regulation of antioxidant responses and defense against oxidative stress, as evidenced by their ability to influence the expression of enzymes such as superoxide dismutase and glutathione peroxidase [59–61]. The analysis highlighted the specificity of these interactions, revealing that fucoxanthin adopts diverse mechanisms depending on the target proteins. This functional diversity reinforces the therapeutic potential of fucoxanthin as an antioxidant. By optimizing the production of fucoxanthin from *I. galbana* under specified conditions, this study laid a solid foundation for future biotechnological applications aimed at exploiting the promising antioxidant properties of this natural compound.

4. Material and Methods

4.1. Cultivation of the Microalgae I. galbana

The marine microalgae strain *I. galbana* was isolated from the coast of Tadjoura in Djibouti (1°46′58.084″ N, 42°53′1.667″ E). Initial isolation was conducted using Falcon conical tubes and cell isolation through micropipettes under a microscope after successive dilutions. The microalgae were then cultured in a liquid medium and on agar plates. Cultivation was carried out in seawater collected from the same location, which was sterilized and enriched with various culture media, including 3N-BBM+V medium, ASP-M medium, CHU-10 medium, Conway medium, Erdschreiber medium, Guillard F/2 medium, PM medium, Walne medium, and WC aquatic culture medium. Some of these media were supplemented with silicate (sodium metasilicate, Sigma-Aldrich 307815, Missouri, United States), particularly CHU-10 medium, Conway medium, and PM medium.

Initially, *I. galbana* was cultured at 25 °C in shakers using natural seawater under continuous illumination of 150 μmol m^{-2} s^{-1}. Growth was monitored by measuring the optical density at 680 nm (Shimadzu UV-1601 spectrophotometer, Kyoto, Japan) every 2 days over a growth period of approximately 10 days. In the subsequent phase, cultures were sub-cultured according to the specific experimental conditions (selected medium, temperature, pH, lighting, flow rate, and CO_2 flow rate) until the biomass reached the stationary phase (15th day of culture). Biomass was then harvested by centrifugation at 2000 rpm (Thermo Scientific™ Megafuge™ 8, Waltham, MA, USA) for 10 min, freeze-dried (Christ Alpha-2–4 LDPlus, Osterode am Harz, Germany), and stored at −20 °C.

4.2. Procedure for Obtaining Extracts for Analyses

The extraction process was performed using the maceration method. For this, 1 g of *I. galbana* biomass was extracted with 100 mL of a methanol/chloroform mixture (1:1, *v/v*) for 12 h at room temperature in complete darkness. This extraction was repeated three times, and all extracts were combined. The pooled extracts were then filtered through Whatman No. 4 filter paper and concentrated under reduced pressure using a rotary evaporator. The final extracts were stored in amber glass vials at −20 °C until further use [62].

4.3. Determination of Antioxidant Activity by the DPPH Method

The antioxidant activity by scavenging DPPH free radicals was evaluated following the method of Flieger et al. (2020) [63]. A stock solution of each extract was prepared in methanol at a concentration of 20 mg/mL. Then, 20 μL of each extract was added to 180 μL of DPPH radical solution (60 μM) in 96-well plates, resulting in final concentrations of 50, 100, 200, 500, and 1000 μg/mL. The samples were shaken to ensure thorough mixing. Since some colored extracts can absorb at 520 nm, a control (blank sample) was prepared by adding 20 μL of each sample solution to 180 μL of methanol. The absorbance of the samples was measured at 30, 60, and 120 min at 520 nm using methanol as a blank in a Shimadzu UV-1601 spectrophotometer. The IC$_{50}$ (50% inhibitory concentration) was calculated to compare the free radical scavenging efficiency. DPPH radical scavenging activity was calculated using the following equation:

$$\text{Scavenging effect}(\%) = 100 \times \left[1 - \left(\frac{AS - AB}{AC}\right)\right]$$

AS: Sample absorbance is the absorbance of the methanolic solution of DPPH in the presence of all the extracts and the standard.

AB: Blank absorbance is the absorbance of the sample of extracts in methanol (without DPPH to subtract the absorbance of the colored extracts).

AC: Control absorbance is the absorbance of the methanolic solution of DPPH.

4.4. Estimation of Carotenoid Content

A method described by Zhou et al. (2020) [64] was employed to determine the carotenoid content of *I. galbana* samples. Approximately 2 g of each sample was mixed with 25 mL of methanol, vortexed for 10 min, and filtered through Whatman No. 1 filter paper. The filtrate was fractionated with 20 mL of petroleum ether and subsequently washed with 100 mL of distilled water. Any residual water was removed using Whatman No. 1 filter paper coated with 5 g of anhydrous sodium sulfate. The extract volume was adjusted to 25 mL with ethanol, and the absorbance was measured at 450 nm using a spectrophotometer.

4.5. Total Phenolic Content

Total phenolic content was determined using the Folin–Ciocalteu reagent, according to the procedure described by Pauliuc (2020) [65]. Each sample (1 mg/mL) was mixed with 5 mL of reagent, diluted (1:10 v/v) with water, and mixed with 4 mL of 7.5% sodium carbonate. The total phenolic content was measured at 750 nm using a spectrophotometer. Gallic acid was used as a standard (0 to 200 mg/mL). The result was expressed as mg of gallic acid equivalent (GAE) per 100 g of sample.

4.6. Quantification of Fucoxanthin

Fucoxanthin quantification was carried out according to the method described by Li et al. (2021) [66] using high performance liquid chromatography (HPLC) coupled with UV detection. The analyses were carried out on an Agilent 1200 HPLC System (Agilent Technologies), including a quaternary pump and a diode array detector (DAD). Chromatographic separation was carried out using a YMC-Pack ODS-A, C18 column (250 mm × 4.6 mm, 5 μm). The column temperature was regulated at 30 °C to optimize the separation of the analytes. The mobile phase, consisting of a methanol–water mixture (80:20, v/v), was supplied at a constant flow rate of 1 mL/min. Samples were injected at a volume of 20 μL. The detection of fucoxanthin was performed at a wavelength of 450 nm, specifically chosen for this analysis. Standard solutions of fucoxanthin were prepared by dissolving the compound in methanol to obtain a stock solution, which was then diluted to prepare standard solutions with concentrations ranging from 0.5 to 12 ppb; hence, Figure S3 presents the calibration curve made for this quantification. The microalgae extracts were also diluted in methanol and injected to a volume of 20 μL under the same operating conditions to quantify fucoxanthin. Before being injected into the HPLC system, all solutions were degassed and filtered through a 0.22 μm membrane to remove particles that could damage the column or disrupt the analysis.

4.7. Statistical Studies

Numerical data were collected from three replicates for each test to ensure the accuracy of the results. Type A uncertainty assessment was performed for the statistical analysis of numerical data. The tests were then subjected to Student's t-test ($p < 0.05$) to assess their significance. To detect significant differences between groups of samples, analysis of variance (ANOVA) was performed, followed by Tukey's test for multiple comparisons.

This test was used to identify significant variations between groups and provide a detailed analysis of the observed differences.

For correlation studies, data modeling was conducted using XLSTAT version 2016, employing several statistical correlation methods to determine relationships between parameters. Principal component analysis (PCA) was utilized to transform a set of correlated variables into a reduced number of uncorrelated principal components [67]. This analysis evaluated the correlation between the enrichment of culture media of the microalga *I. galbana* and antioxidant activity, as well as the contents of carotenoids, total phenols, and fucoxanthin, classifying the environments according to these studied parameters. Additionally, the correlation between antioxidant activity, carotenoids, total phenols, and fucoxanthin was analyzed using simple linear regressions between each pair of parameters and multiple linear regression (MLR) and multiple nonlinear regression (MNLR) [68]. These statistical approaches aimed to establish the relationships between antioxidant activity and the contents of carotenoids, total phenols, and fucoxanthin. Design-Expert 13 was used for all calculations and graphical representations for optimization studies utilizing experimental designs. The methodology aimed to minimize the experimental conditions for producing fucoxanthin from the microalga *I. galbana*.

The experimental design used was a full factorial design based on the following five X_i factors:

Factor 1 = T: temperature (25 °C and 30 °C);
Factor 2 = pH: pH (6.5 and 7.5);
Factor 3 = light intensity: LI (100 μmol/m^2/s and 500 μmol/m^2/s);
Factor 4 = air flow rate: AFR (0.5 L/min and 1.0 L/min);
Factor 5 = CO$_2$ flow rate: CO$_2$FR (0.1–0.2 L/min).

Each factor was studied at two levels—high (+1) and low (−1)—resulting in 2^5 = 32 trials in total. The polynomial model included the following:

$$Fucoxanthin\ (mg/g) = a_0 + \sum_{i=1}^{n} a_i X_i + \sum_{i=1}^{n}\sum_{j=1}^{n-1} a_{ij} X_i X_j + \sum_{i=1}^{n}\sum_{j=1}^{n-1}\sum_{k=1}^{n-2} a_{ijk} X_i X_j X_k + \sum_{i=1}^{n}\sum_{j=1}^{n-1}\sum_{k=1}^{n-2}\sum_{l=1}^{n-3} a_{ijkl} X_i X_j X_k X_L + a_{ijklm} X_i X_j X_k X_l X_m$$

A mean: a_0;
5 main effects for each factor: a_i;
10 interactions of order 2: a_{ij};
10 interactions of order 3: a_{ijk};
5 interactions of order 4: a_{ijkl};
1 interaction of order 5: a_{ijklm}.

All values obtained are presented as mean ± uncertainty, with a significance level of 5% for each experiment, as determined by statistical analysis using the Student's *t*-test. Each experimental variant was tested in triplicate.

4.8. In Silico Study

This in silico study used molecular docking methodology to perform a virtual screening of the crystal structures of a few antioxidant proteins available in the RCSB database, including PDB IDs (3FS1, 3L2C, and 8BBK). The choice of these crystal structures for molecular docking was guided by their relevance to the study objectives aimed at evaluating the antioxidant activity of fucoxanthin. These proteins were selected based on their potential involvement in antioxidant processes, as confirmed by previous works, as well as their structural availability in the RCSB database, ensuring their validity and feasibility for comparative computational analysis [69–71]. This approach allowed the investigation of specific interactions with proteins relevant to this study, thereby contributing to a cross-sectional understanding of the potential effects of fucoxanthin. To carry out this analysis, several software tools were employed, including MGLtools 1.4.6, Autodock 4.0, Autogrid

4.0, BIOVIA Discovery Studio Visualizer 2.5, ChemBiodraw Ultra 12.0, and Chemdraw 3D 10.0. Initially, protein structures were prepared using the BIOVIA Discovery Studio Visualizer to remove heteroatoms, co-crystallized ligands, and solvents to optimize conditions for docking. Autodock tools were then used to assign appropriate polar charges and generate optimized pdbqt files for each protein structure. For fucoxanthin, the structure was drawn with ChemDraw Ultra, then energetically minimized with Chem 3D Pro before being converted to pdbqt format via OpenBabel GUI. Structure-based virtual screening was performed with Autodock 4, docking fucoxanthin independently into the active site of each target protein. Ligand–protein interactions were visualized and analyzed with the BIOVIA Discovery Studio Visualizer. To validate the results, root mean square deviation (RMSD) values were calculated, ensuring that accepted poses had RMSD values of less than 2.0 for ligands redocked by co-crystallization.

5. Conclusions

This study successfully optimized the production of fucoxanthin from cultures of *I. galbana*. F/2 media were identified as the most effective, promoting high fucoxanthin content and improving the overall antioxidant potential of the alga. Analyses demonstrated close correlations between fucoxanthin and total carotenoids, linked to antioxidant activity. A specific statistical model (MNLR) effectively captured the complex interactions between various factors influencing fucoxanthin production. Furthermore, the experiments identified air flow rate and CO_2 flow rate as crucial factors for maximizing fucoxanthin yields. Computational molecular docking analyses showed that fucoxanthin binds most efficiently to FOXO4 and Sirt3, two proteins playing critical roles in the antioxidant response and regulation of key enzymes against oxidative stress. Furthermore, the results confirmed these specific interactions, highlighting the propensity of fucoxanthin to adopt different mechanisms depending on the targeted proteins. These findings suggest the significant potential of fucoxanthin in antioxidant and therapeutic applications. In conclusion, all the data from this work validate the robustness of the developed model, which is very promising for the future optimization of fucoxanthin production in biotechnological applications.

Supplementary Materials: The following supporting information can be downloaded at: https://www.mdpi.com/article/10.3390/md22080358/s1, Figure S1. displays a validation graph depicting the predicted antioxidant activity of the marine microalga *I. galbana*; Figure S2. Illustrates the representations of fucoxanthin across one-dimensional, two-dimensional, and three-dimensional formats; Figure S3. Calibration curve of fucoxanthin by HPLC-UV; Table S1. Statistical parameters of the multiple regression of antioxidant activity as a function of the content of carotenoids, total phenolic compounds and fucoxanthin in cultures of the microalga *I. galbana*; Table S2. Fucoxanthin is used for each test in all experiments according to the design of the experiments; Table S3. Coefficients of optimized models for fucoxanthin production; Table S4. Statistical studies of the model; Table S5. ANOVA for the selected factorial model.

Author Contributions: Conceptualization: F.M.A.-L. and T.A.; Data curation: A.A. (Ayoub Ainane) and F.M.A.-L.; Formal analysis: H.M. and T.A.; Funding acquisition: F.M.A.-L.; Investigation: F.M.A.-L. and L.A.; Methodology: A.A. (Ayoub Ainane), F.M.A.-L. and L.A.; Project administration: F.M.A.-L.; Resources: P.P.J., A.M.A. and A.A. (Ahmad Ali); Software: H.M., A.M.A. and A.A. (Ayoub Ainane); Supervision: F.M.A.-L.; Validation: F.M.A.-L. and T.A.; Visualization: A.A. (Ayoub Ainane), F.M.A.-L. and T.A.; Roles/Writing—original draft: F.M.A.-L. and T.A.; Writing—review & editing: F.M.A.-L., A.A. (Ahmad Ali), P.P.J. and T.A. All authors have read and agreed to the published version of the manuscript.

Funding: This research received no external funding.

Data Availability Statement: Data are contained within the article.

Conflicts of Interest: The authors declare no conflicts of interest.

References

1. Bürck, M.; Ramos, S.D.P.; Braga, A.R.C. Enhancing the Biological Effects of Bioactive Compounds from Microalgae through Advanced Processing Techniques: Pioneering Ingredients for Next-Generation Food Production. *Foods* **2024**, *13*, 1811. [CrossRef] [PubMed]
2. Maghzian, A.; Aslani, A.; Zahedi, R. A comprehensive review on effective parameters on microalgae productivity and carbon capture rate. *J. Environ. Manag.* **2024**, *355*, 120539. [CrossRef] [PubMed]
3. Chen, Y.; Liang, H.; Du, H.; Jesumani, V.; He, W.; Cheong, K.-L.; Li, T.; Hong, T. Industry chain and challenges of microalgal food industry—A review. *Crit. Rev. Food Sci. Nutr.* **2024**, *64*, 4789–4816. [CrossRef]
4. Ainane, T.; Abdoul-Latif, F.M.; El Yaagoubi, B.; Boujaber, N.; Oumaskour, K.; Ainane, A. Nannochloropsis oculata Microalga for Fertilization: Bioguided Fractionation of N-Hexane Extract by Stimulating Growth Activity for Cucurbita moschata. *Pharmacologyonline* **2021**, *3*, 1803–1809.
5. Dhandwal, A.; Bashir, O.; Malik, T.; Salve, R.V.; Dash, K.K.; Amin, T.; Shams, R.; Wani, A.W.; Shah, Y.A. Sustainable microalgal biomass as a potential functional food and its applications in food industry: A comprehensive review. *Environ. Sci. Pollut. Res.* **2024**, 1–19. [CrossRef]
6. Guehaz, K.; Boual, Z.; Abdou, I.; Telli, A.; Belkhalfa, H. Microalgae's polysaccharides, are they potent antioxidants? Critical review. *Arch. Microbiol.* **2023**, *206*, 14. [CrossRef]
7. Duan, X.; Xie, C.; Hill, D.R.A.; Barrow, C.J.; Dunshea, F.R.; Martin, G.J.O.; Suleria, H.A. Bioaccessibility, bioavailability and bioactivities of carotenoids in microalgae: A review. *Food Rev. Int.* **2023**, *40*, 230–259. [CrossRef]
8. Chandrika, U.G. Carotenoid Dyes—Properties and Production. *Handb. Nat. Color.* **2023**, 351–369.
9. Meléndez-Martínez, A.J.; Esquivel, P.; Rodriguez-Amaya, D.B. Comprehensive review on carotenoid composition: Transformations during processing and storage of foods. *Food Res. Int.* **2023**, *169*, 112773. [CrossRef] [PubMed]
10. Gómez-Sagasti, M.T.; López-Pozo, M.; Artetxe, U.; Becerril, J.M.; Hernández, A.; García-Plazaola, J.I.; Esteban, R. Carotenoids and their derivatives: A "Swiss Army knife-like" multifunctional tool for fine-tuning plant-environment interactions. *Environ. Exp. Bot.* **2023**, *207*, 105229. [CrossRef]
11. Xu, J.; Lin, J.; Peng, S.; Zhao, H.; Wang, Y.; Rao, L.; Liao, X.; Zhao, L. Development of an HPLC-PDA Method for the Determination of Capsanthin, Zeaxanthin, Lutein, β-Cryptoxanthin and β-Carotene Simultaneously in Chili Peppers and Products. *Molecules* **2023**, *28*, 2362. [CrossRef]
12. González-Peña, M.A.; Ortega-Regules, A.E.; Anaya de Parrodi, C.; Lozada-Ramírez, J.D. Chemistry, occurrence, properties, applications, and encapsulation of carotenoids—A review. *Plants* **2023**, *12*, 313. [CrossRef]
13. Bohn, T.; de Lera, A.R.; Landrier, J.-F.; Rühl, R. Carotenoid metabolites, their tissue and blood concentrations in humans and further bioactivity via retinoid receptor-mediated signalling. *Nutr. Res. Rev.* **2023**, *36*, 498–511. [CrossRef]
14. Jiří, B.; Lenka, V.; Josef, S.; Věra, K. Exploring carotenoids: Metabolism, antioxidants, and impacts on human health. *J. Funct. Foods* **2024**, *118*, 106284.
15. Zhang, Z.; Wei, Z.; Xue, C. Delivery systems for fucoxanthin: Research progress, applications and future prospects. *Crit. Rev. Food Sci. Nutr.* **2024**, *64*, 4643–4659. [CrossRef]
16. Abdoul-Latif, F.M.; Ainane, A.; Aboubaker, I.H.; Ali, A.M.; Mohamed, H.; Jutur, P.P.; Ainane, T. Unlocking the Green Gold: Exploring the Cancer Treatment and the Other Therapeutic Potential of Fucoxanthin Derivatives from Microalgae. *Pharmaceuticals* **2024**, *17*, 960. [CrossRef]
17. Sun, H.; Yang, S.; Zhao, W.; Kong, Q.; Zhu, C.; Fu, X.; Zhang, F.; Liu, Z.; Zhan, Y.; Mou, H.; et al. Fucoxanthin from marine microalgae: A promising bioactive compound for industrial production and food application. *Crit. Rev. Food Sci. Nutr.* **2023**, *63*, 7996–8012. [CrossRef] [PubMed]
18. Mapelli-Brahm, P.; Gómez-Villegas, P.; Gonda, M.L.; León-Vaz, A.; León, R.; Mildenberger, J.; Rebours, C.; Saravia, V.; Vero, S.; Vila, E.; et al. Microalgae, seaweeds and aquatic bacteria, archaea, and yeasts: Sources of carotenoids with potential antioxidant and anti-inflammatory health-promoting actions in the sustainability era. *Mar. Drugs* **2023**, *21*, 340. [CrossRef]
19. Gonçalves, C.d.A.; Figueredo, C.C. What we really know about the composition and function of microalgae cell coverings? — An overview. *Acta Bot. Bras.* **2020**, *34*, 599–614. [CrossRef]
20. Zarrinmehr, M.J.; Farhadian, O.; Heyrati, F.P.; Keramat, J.; Koutra, E.; Kornaros, M.; Daneshvar, E. Effect of ni-trogen concentration on the growth rate and biochemical composition of the microalga, Isochrysis galbana. *Egypt. J. Aquat. Res.* **2020**, *46*, 153–158. [CrossRef]
21. Matos, J.; Afonso, C.; Cardoso, C.; Serralheiro, M.L.; Bandarra, N.M. Yogurt Enriched with *Isochrysis galbana*: An Innovative Functional Food. *Foods* **2021**, *10*, 1458. [CrossRef]
22. Solomonova, E.S.; Shoman, N.Y.; Akimov, A.I. Hormesis effect of the herbicide glyphosate on growth and lipid synthesis in the microalga *Isochrysis galbana*, an object of industrial cultivation. *Aquac. Int.* **2024**, 1–9. [CrossRef]
23. Wang, Y.-Y.; Xu, S.-M.; Cao, J.-Y.; Wu, M.-N.; Lin, J.-H.; Zhou, C.-X.; Zhang, L.; Zhou, H.-B.; Li, Y.-R.; Xu, J.-L.; et al. Co-cultivation of *Isochrysis galbana* and Marinobacter sp. can enhance algal growth and docosahexaenoic acid production. *Aquaculture* **2022**, *556*, 738248. [CrossRef]
24. Abdoul-Latif, F.M.; Oumaskour, K.; Boujaber, N.; Ainane, A.; Mohamed, J.; Ainane, T. Formulations of a cosmetic product for hair care based on extract of the microalga *Isochrysis galbana*: In vivo and in vitro activities. *J. Anal. Sci. Appl. Biotechnol.* **2021**, *3*, 15–19.

25. Pocha, C.K.R.; Chia, W.Y.; Chew, K.W.; Munawaroh, H.S.H.; Show, P.L. Current advances in recovery and biorefinery of fucoxanthin from Phaeodactylum tricornutum. *Algal Res.* **2022**, *65*, 102735. [CrossRef]

26. Kim, S.M.; Kang, S.-W.; Kwon, O.-N.; Chung, D.; Pan, C.-H. Fucoxanthin as a major carotenoid in Isochrysis aff. galbana: Characterization of extraction for commercial application. *J. Korean Soc. Appl. Biol. Chem.* **2012**, *55*, 477–483. [CrossRef]

27. Stock, A.; Subramaniam, A. Accuracy of empirical satellite algorithms for mapping phytoplankton diagnostic pigments in the open ocean: A supervised learning perspective. *Front. Mar. Sci.* **2020**, *7*, 599. [CrossRef]

28. Shannon, E.; Abu-Ghannam, N. Optimisation of fucoxanthin extraction from Irish seaweeds by response surface methodology. *J. Appl. Phycol.* **2017**, *29*, 1027–1036. [CrossRef]

29. Grubišić, M.; Šantek, B.; Kuzmić, M.; Čož-Rakovac, R.; Šantek, M.I. Enhancement of Biomass Production of Diatom *Nitzschia* sp. S5 through Optimisation of Growth Medium Composition and Fed-Batch Cultivation. *Mar. Drugs* **2024**, *22*, 46. [CrossRef]

30. Trinh, K.S. Comparison of Growth of Chlorella vulgaris in Flat-Plate Photobioreactor Using Batch, Fed-Batch, and Repeated Fed-Batch Techniques with Various Concentrations of Walne Medium. *J. Tech. Educ. Sci.* **2023**, *18*, 9–15. [CrossRef]

31. Somade, O.T.; Ajayi, B.O.; Adeyi, O.E.; Dada, T.A.; Ayofe, M.A.; Inalu, D.C.; Ajiboye, O.I.; Shonoiki, O.M.; Adelabu, A.O.; Onikola, R.T.; et al. Ferulic acid inter-ventions ameliorate NDEA-CCl4-induced hepatocellular carcinoma via Nrf2 and p53 upregulation and Akt/PKB-NF-κB-TNF-α pathway downregulation in male Wistar rats. *Toxicol. Rep.* **2024**, *12*, 119–127. [CrossRef] [PubMed]

32. Alqrad, M.A.; El-Agamy, D.S.; Ibrahim, S.R.; Sirwi, A.; Abdallah, H.M.; Abdel-Sattar, E.; El-Halawany, A.M.; Elsaed, W.M.; Mohamed, G.A. SIRT1/Nrf2/NF-κB Signaling Mediates Anti-Inflammatory and Anti-Apoptotic Activities of Oleanolic Acid in a Mouse Model of Acute Hepatorenal Damage. *Medicina* **2023**, *59*, 1351. [CrossRef] [PubMed]

33. Jeung, D.; Lee, G.-E.; Chen, W.; Byun, J.; Nam, S.-B.; Park, Y.-M.; Lee, H.S.; Kang, H.C.; Lee, J.Y.; Kim, K.D.; et al. Ribosomal S6 kinase 2-forkhead box protein O4 signaling pathway plays an essential role in melanogenesis. *Sci. Rep.* **2024**, *14*, 9440. [CrossRef] [PubMed]

34. Salama, A.; Elgohary, R.; Amin, M.M.; Elwahab, S.A. Impact of protocatechuic acid on alleviation of pulmonary damage induced by cyclophosphamide targeting peroxisome proliferator activator receptor, silent information regulator type-1, and fork head box protein in rats. *Inflammopharmacology* **2023**, *31*, 1361–1372. [CrossRef] [PubMed]

35. Omorou, M.; Huang, Y.; Gao, M.; Mu, C.; Xu, W.; Han, Y.; Xu, H. The forkhead box O3 (FOXO3): A key player in the regulation of ischemia and reperfusion injury. *Cell. Mol. Life Sci.* **2023**, *80*, 102. [CrossRef] [PubMed]

36. Dhillon, V.S.; Shahid, M.; Deo, P.; Fenech, M. Reduced SIRT1 and SIRT3 and Lower Antioxidant Capacity of Seminal Plasma Is Associated with Shorter Sperm Telomere Length in Oligospermic Men. *Int. J. Mol. Sci.* **2024**, *25*, 718. [CrossRef] [PubMed]

37. Mishra, Y.; Kaundal, R.K. Role of SIRT3 in mitochondrial biology and its therapeutic implications in neuro-degenerative disorders. *Drug Discov. Today* **2023**, *28*, 103583. [CrossRef]

38. Miyashita, K.; Beppu, F.; Hosokawa, M.; Liu, X.; Wang, S. Bioactive significance of fucoxanthin and its effective extraction. *Biocatal. Agric. Biotechnol.* **2020**, *26*, 101639. [CrossRef]

39. Wu, S.; Xie, X.; Huan, L.; Zheng, Z.; Zhao, P.; Kuang, J.; Liu, X.; Wang, G. Selection of optimal flocculant for effective harvesting of the fucoxanthin-rich marine microalga *Isochrysis galbana*. *J. Appl. Phycol.* **2016**, *28*, 1579–1588. [CrossRef]

40. Bigagli, E.; D'ambrosio, M.; Cinci, L.; Niccolai, A.; Biondi, N.; Rodolfi, L.; Nascimiento, L.B.D.S.; Tredici, M.R.; Luceri, C. A comparative in vitro evaluation of the anti-inflammatory effects of a *Tisochrysis lutea* extract and fucoxanthin. *Mar. Drugs* **2021**, *19*, 334. [CrossRef]

41. Méresse, S.; Fodil, M.; Fleury, F.; Chénais, B. Fucoxanthin, a marine-derived carotenoid from brown seaweeds and microalgae: A promising bioactive compound for cancer therapy. *Int. J. Mol. Sci.* **2020**, *21*, 9273. [CrossRef] [PubMed]

42. Kawee-Ai, A.; Kim, A.T.; Kim, S.M. Inhibitory activities of microalgal fucoxanthin against α-amylase, α-glucosidase, and glucose oxidase in 3T3-L1 cells linked to type 2 diabetes. *J. Oceanol. Limnol.* **2019**, *37*, 928–937. [CrossRef]

43. Foo, S.C.; Yusoff, F.M.; Ismail, M.; Basri, M.; Yau, S.K.; Khong, N.M.; Chan, K.W.; Ebrahimi, M. Antioxidant capacities of fucoxanthin-producing algae as influenced by their carotenoid and phenolic contents. *J. Biotechnol.* **2017**, *241*, 175–183. [CrossRef] [PubMed]

44. Sandhya, S.V.; Vijayan, K.K. Symbiotic association among marine microalgae and bacterial flora: A study with special reference to commercially important *Isochrysis galbana* culture. *J. Appl. Phycol.* **2019**, *31*, 2259–2266. [CrossRef]

45. Medina, E.; Cerezal, P.; Morales, J.; Ruiz-Domínguez, M.C. Fucoxanthin from marine microalga *Isochrysis galbana*: Optimization of extraction methods with organic solvents. *Dyna* **2019**, *86*, 174–178. [CrossRef]

46. Gao, F.; (Cabanelas, I.I.T.; Wijffels, R.H.; Barbosa, M.J. Process optimization of fucoxanthin production with Tisochrysis lutea. *Bioresour. Technol.* **2020**, *315*, 123894. [CrossRef]

47. Pereira, H.; Sá, M.; Maia, I.; Rodrigues, A.; Teles, I.; Wijffels, R.H.; Navalho, J.; Barbosa, M. Fucoxanthin production from Tisochrysis lutea and Phaeodactylum tricornutum at industrial scale. *Algal Res.* **2021**, *56*, 102322. [CrossRef]

48. Mohamadnia, S.; Tavakoli, O.; Faramarzi, M.A. Enhancing production of fucoxanthin by the optimization of culture media of the microalga Tisochrysis lutea. *Aquaculture* **2021**, *533*, 736074. [CrossRef]

49. Mohamadnia, S.; Tavakoli, O.; Faramarzi, M.A. Optimization of metabolic intermediates to enhance the production of fucoxanthin from Tisochrysis lutea. *J. Appl. Phycol.* **2022**, *34*, 1269–1279. [CrossRef]

50. McElroy, C.R.; Kopanitsa, L.; Helmes, R.; Fan, J.; Attard, T.M.; Simister, R.; van den Burg, S.; Ladds, G.; Bailey, D.S.; Gomez, L.D. Integrated biorefinery approach to valorise Saccharina latissima biomass: Combined sustainable processing to produce biologically active fuco-xanthin, mannitol, fucoidans and alginates. *Environ. Technol. Innov.* **2023**, *29*, 103014. [CrossRef]

51. Xia, Y.; Sekine, M.; Hirahara, M.; Kishinami, H.; Yusoff, F.M.; Toda, T. Effects of concentration and frequency of CO2 supply on productivity of marine microalga *Isochrysis galbana*. *Algal Res.* **2023**, *70*, 102985. [CrossRef]

52. Bo, Y.; Wang, S.; Ma, F.; Manyakhin, A.Y.; Zhang, G.; Li, X.; Zhou, C.; Ge, B.; Yan, X.; Ruan, R.; et al. The influence of spermidine on the build-up of fucoxanthin in Isochrysis sp. Acclimated to varying light intensities. *Bioresour. Technol.* **2023**, *387*, 129688. [CrossRef] [PubMed]

53. García-García, P.; Ospina, M.; Señoráns, F.J. Tisochrysis lutea as a source of omega-3 polar lipids and fucoxanthin: Extraction and characterization using green solvents and advanced techniques. *J. Appl. Phycol.* **2024**, *36*, 1697–1708. [CrossRef]

54. Manochkumar, J.; Jonnalagadda, A.; Cherukuri, A.K.; Vannier, B.; Janjaroen, D.; Ramamoorthy, S. Enhancing cellular production of fucoxanthin through machine learning assisted predictive approach in *Isochrysis galbana*. *bioRxiv* **2024**, *26*, 591373.

55. Khaw, Y.S.; Yusoff, F.M.; Tan, H.T.; Mazli, N.A.I.N.; Nazarudin, M.F.; Shaharuddin, N.A.; Omar, A.R.; Takahashi, K. Fucoxanthin production of microalgae under different culture factors: A systematic review. *Mar. Drugs* **2022**, *20*, 592. [CrossRef] [PubMed]

56. Sun, J.; Zhou, C.; Cheng, P.; Zhu, J.; Hou, Y.; Li, Y.; Zhang, J.; Yan, X. A simple and efficient strategy for fucoxanthin extraction from the microalga Phaeodactylum tricornutum. *Algal Res.* **2022**, *61*, 102610. [CrossRef]

57. Mohamadnia, S.; Tavakoli, O.; Faramarzi, M.A.; Shamsollahi, Z. Production of fucoxanthin by the microalga Ti-sochrysis lutea: A review of recent developments. *Aquaculture* **2020**, *516*, 734637. [CrossRef]

58. Sun, Z.; Wang, X.; Liu, J. Screening of Isochrysis strains for simultaneous production of docosahexaenoic acid and fucoxanthin. *Algal Res.* **2019**, *41*, 101545. [CrossRef]

59. Ranjith, H.V.; Sagar, D.; Kalia, V.K.; Dahuja, A.; Subramanian, S. Differential activities of antioxidant enzymes, superoxide dismutase, peroxidase, and catalase vis-à-vis phosphine resistance in field populations of lesser grain borer (*Rhyzopertha dominica*) from India. *Antioxidants* **2023**, *12*, 270. [CrossRef]

60. Vašková, J.; Kočan, L.; Vaško, L.; Perjési, P. Glutathione-related enzymes and proteins: A review. *Molecules* **2023**, *28*, 1447. [CrossRef]

61. Halliwell, B. Understanding mechanisms of antioxidant action in health and disease. *Nat. Rev. Mol. Cell Biol.* **2024**, *25*, 13–33. [CrossRef] [PubMed]

62. Lfitat, A.; Ed-dahmani, I.; Bousraf, F.Z.; Belhaj, A.; Ainane, T.; Taleb, M.; Gourch, A.; Abdellaoui, A. Antifungal Activity of the Argan Kernels Extracts and the Heat Treatment Effect on the Activity Variation. *Indones. Food Nutr. Prog.* **2023**, *20*, 10–23. [CrossRef]

63. Flieger, J.; Flieger, M. The [DPPH•/DPPH-H]-HPLC-DAD method on tracking the antioxidant activity of pure antioxidants and goutweed (*Aegopodium podagraria* L.) hydroalcoholic extracts. *Molecules* **2020**, *25*, 6005. [CrossRef] [PubMed]

64. Zhou, W.; Niu, Y.; Ding, X.; Zhao, S.; Li, Y.; Fan, G.; Zhang, S.; Liao, K. Analysis of carotenoid content and diversity in apricots (Prunus armeniaca L.) grown in China. *Food Chem.* **2020**, *330*, 127223. [CrossRef]

65. Pauliuc, D.; Dranca, F.; Oroian, M. Antioxidant activity, total phenolic content, individual phenolics and physico-chemical parameters suitability for Romanian honey authentication. *Foods* **2020**, *9*, 306. [CrossRef] [PubMed]

66. Li, M.; Feng, H.; Ouyang, X.; Ling, J. Determination of Fucoxanthin in Bloom-Forming Macroalgae by HPLC–UV. *J. Chromatogr. Sci.* **2021**, *59*, 978–982. [CrossRef] [PubMed]

67. Abdoul-Latif, F.M.; Elmi, A.; Merito, A.; Nour, M.; Risler, A.; Ainane, A.; Bignon, J.; Ainane, T. Essential oil of Ruta chalepensis L. from Djibouti: Chemical Analysis and Modeling of In Vitro Anticancer Profiling. *Separations* **2022**, *9*, 387. [CrossRef]

68. Aamir, H.; Aamir, K.; Javed, M.F. Linear and Non-Linear Regression Analysis on the Prediction of Compressive Strength of Sodium Hydroxide Pre-Treated Crumb Rubber Concrete. *Eng. Proc.* **2023**, *44*, 5. [CrossRef]

69. Muhammed, M.T.; Aki-Yalcin, E. Molecular docking: Principles, advances, and its applications in drug discovery. *Lett. Drug Des. Discov.* **2024**, *21*, 480–495. [CrossRef]

70. Mohamed Abdoul-Latif, F.; Ainane, A.; Merito, A.; Aboubaker, I.H.; Mohamed, H.; Cherroud, S.; Ainane, T. The Effects of Khat Chewing among Djiboutians: Dental Chemical Studies, Gingival Histopathological Analyses and Bioinformatics Approaches. *Bioengineering* **2024**, *11*, 716. [CrossRef]

71. Chen, G.; Seukep, A.J.; Guo, M. Recent advances in molecular docking for the research and discovery of potential marine drugs. *Mar. Drugs* **2020**, *18*, 545. [CrossRef] [PubMed]

marine drugs

Article

Light Intensity Enhances the Lutein Production in *Chromochloris zofingiensis* Mutant LUT-4

Qiaohong Chen [1], Mingmeng Liu [2], Wujuan Mi [1], Dong Wan [1], Gaofei Song [1], Weichao Huang [1] and Yonghong Bi [1,*]

[1] State Key Laboratory of Freshwater Ecology and Biotechnology, Institute of Hydrobiology, Chinese Academy of Sciences, Wuhan 430072, China; qiaohongchen@ihb.ac.cn (Q.C.); miwj@ihb.ac.cn (W.M.); wangdong@ihb.ac.cn (D.W.); song@ihb.ac.cn (G.S.); huangwc@ihb.ac.cn (W.H.)
[2] School of Civil Engineering, Hubei Engineering University, Xiaogan 432000, China; lmm007@hbeu.edu.cn
* Correspondence: biyh@ihb.ac.cn

Abstract: *Chromochloris zofingiensis*, a unicellular green alga, is a potential source of natural carotenoids. In this study, the mutant LUT-4 was acquired from the chemical mutagenesis pool of *C. zofingiensis* strain. The biomass yield and lutein content of LUT-4 reached 9.23 g·L^{-1}, and 0.209% of dry weight (DW) on Day 3, which was 49.4%, and 33% higher than that of wild-type (WT), respectively. The biomass yields of LUT-4 under 100, 300, and 500 µmol/m^2/s reached 8.4 g·L^{-1}, 7.75 g·L^{-1}, and 6.6 g·L^{-1}, which was 10.4%, 21%, and 29.6% lower compared with the control, respectively. Under mixotrophic conditions, the lutein yields were significantly higher than that obtained in the control. The light intensity of 300 µmol/m^2/s was optimal for lutein biosynthesis and the content of lutein reached 0.294% of DW on Day 3, which was 40.7% more than that of the control. When LUT-4 was grown under 300 µmol/m^2/s, a significant increase in expression of genes implicated in lutein biosynthesis, including phytoene synthase (*PSY*), phytoene desaturase (*PDS*), and lycopene epsilon cyclase (*LCYe*) was observed. The changes in biochemical composition, Ace-CoA, pyruvate, isopentenyl pyrophosphate (IPP), and geranylgeranyl diphosphate (GGPP) contents during lutein biosynthesis were caused by utilization of organic carbon. It was thereby concluded that 300 µmol/m^2/s was the optimal culture light intensity for the mutant LUT-4 to synthesize lutein. The results would be helpful for the large-scale production of lutein.

Keywords: *Chromochloris zofingiensis*; biomass; lutein; light intensity; organic carbon availability

Citation: Chen, Q.; Liu, M.; Mi, W.; Wan, D.; Song, G.; Huang, W.; Bi, Y. Light Intensity Enhances the Lutein Production in *Chromochloris zofingiensis* Mutant LUT-4. *Mar. Drugs* **2024**, *22*, 306. https://doi.org/10.3390/md22070306

Academic Editors: Cecilia Faraloni and Eleftherios Touloupakis

Received: 30 May 2024
Revised: 27 June 2024
Accepted: 27 June 2024
Published: 29 June 2024

1. Introduction

Lutein, a naturally occurring carotenoid, has garnered significant attention due to its potential health benefits, including the scavenging of free radicals, prevention or amelioration of cardiovascular diseases, age-related macular degeneration (AMD), Alzheimer's Disease (AD), and certain forms of cancer, as well as its advantageous effects on skin health [1,2]. Currently, the primary source for the commercial production of lutein is marigold petals [3]. In addition to requiring considerable effort to separate the petals and extract lutein from the marigold flowers, which comprise a meager 0.03% of dry weight (DW), the cultivation of the marigold plant is a labor-intensive undertaking [4]. Although lutein is frequently present in vegetables, not all populations receive enough of it on a daily basis. Thus, it is important to look for high-quality lutein sources for dietary supplements.

Microalgae are abundant sources of carotenoids, which can operate as primary carotenoids during photosynthesis or as secondary carotenoids in reaction to unfavorable conditions [5]. As an essential pigment for photosynthetic processes, microalgae lutein production is correlated with photosynthetic activity. In contrast to terrestrial plants, microalgae exhibit superior rates of growth and photosynthetic efficiency [6]. Some microalgal species, including *Chlorella protothecoides* [7], *Dunaliella salina* [8], *Muriellopsis* sp. [4],

Parachlorella sp. JD-076 [9], *Scenedesmus obliquus* FSP-3 [4], and *Chlorella vulgaris* UTEX 265 [10], have been studied for lutein production, with limited progress.

Since photosynthetic pigment synthesis is a physiological response of microalgae cells to high light stress, photoautotrophic cultivation is frequently used in the production of lutein [11]. However, the dry weight in photoautotrophic microalgal cells was lower than that of those cultured heterotrophically [12]. Currently, mixotrophic culture integrates the benefits of autotrophy and heterotrophy [13]. Microalgae engaged in mixotrophic culture take up both organic and inorganic nutrients in the presence of light under conditions of aerobic respiration and photosynthesis [14]. Mixotrophic cultivation of microalgae has been shown to have numerous advantages, including a reduction in the photo-inhibitory effect of photosynthetic capacity, biomass loss at night, and photo-oxidative damage during the cultivation period [15]. In addition, light intensity is also an important parameter that has influenced the production of lutein in microalgae. When *Scenedesmus obliquus* FSP-3 was exposed to white light instead of blue, green, or red light, high lutein production was seen at a light intensity of 300 μmol/m^2/s [4]. The growth of *Coccomyxa onubensis* under white light with an intensity of 400 μmol/m^2/s led to a notable lutein production [16]. Thus, the generation of lutein from microalgae with high biomass and lutein content is now possible via mixotrophic culturing.

Chromochloris zofingiensis is a Chlorophyceae class green microalga that exhibits rapid growth under three different trophic modes (i.e., autotrophy, heterotrophy, and mixotrophy) [17]. Chlorophyll degradation and the accumulation of secondary carotenoids occurred when *C. zofingiensis* ceased photosynthesis in the presence of glucose [18]. The mutant CZ-LZM3 of *C. zofingiensis* strain has described a deficiency in astaxanthin accumulation but, on the other hand, accumulated significant quantities of three distinct carotenoids (namely lutein, zeaxanthin, and β-carotene) during heterotrophic cultivation [8]. Nevertheless, the contents of lutein in this mutant (i.e., CZ-LZM3) decreased significantly during heterotrophic growth. Therefore, it is essential to exert significant efforts to augment lutein production in *C. zofingiensis*.

This study investigated the biological profiles of *C. zofingiensis* mutant LUT-4 under different light intensities by linking the physiological properties and molecular characteristics to evaluate the potential of LUT-4 to produce lutein. The growth characteristics of mutant LUT-4 under different light intensities were determined. The main organic composition (i.e., lipid, protein, and carbohydrate) of mutant LUT-4 under optimal light intensity was measured. The metabolites and carotenogenesis genes involved in lutein biosynthesis were detected. This study would have substantial implications for natural lutein production by *C. zofingiensis* mutant LUT-4.

2. Results and Discussion

2.1. Isolation and Pigment of C. zofingiensis LUT-4 Strain

From the ethyl methyl sulfonate (EMS) mutagenesis pool, the colonies that are differentiated from wild-type (WT) by color were chosen to screen the mutants that can synthesize lutein, other than astaxanthin. A total of 28 mutants with a yellow or yellow-green appearance in color were identified from 30,000 mutants. Through pigment analysis by using HPLC, four stable mutants (LUT-1, LUT-2, LUT-3, and LUT-4) were selected due to their capacity to accumulate lutein (Figure 1a). In contrast to the wild-type, these mutants accumulated a large quantity of lutein but lacked astaxanthin. When cultivated as colonies on agar under heterotrophic conditions, they exhibited a yellow or yellow–green color, while the wild type was an orange color (Figure 1b).

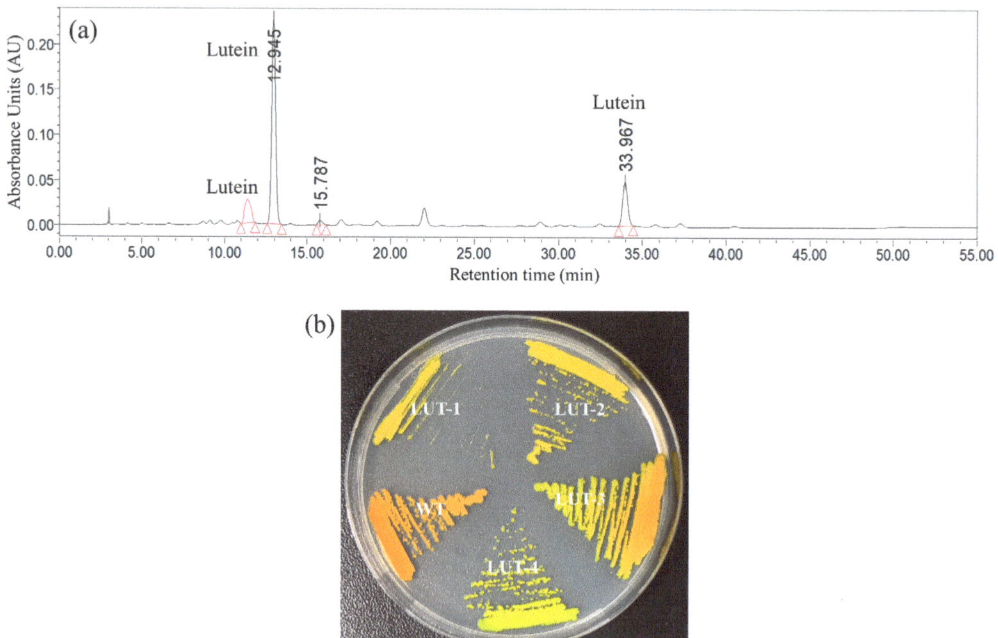

Figure 1. The comparison between wild-type (WT) and LUT-4. (**a**) The pigment profiles of lutein colonies were analyzed by using HPLC. (**b**) Color appearance of *C. zofingiensis* wild-type and lutein colonies on agar plates under heterotrophic conditions.

When microalgal cells were cultivated under heterotrophic conditions, the biomass concentration and cell density of lutein mutants exhibited significant differences from that of WT (Figure 2a,b). As compared to WT and other mutants, LUT-4 showed significantly increased biomass accumulation, up to the highest yield of 11.55 g·L^{-1} and 13.7 g·L^{-1} on Day 4 under N-replete and N-deprived conditions, demonstrating 29.3% and 19.8% higher yield than that of WT (8.93 g·L^{-1}, 11.44 g·L^{-1}, $p < 0.001$), respectively. Furthermore, the biomass yields of LUT-2 and LUT-3 were also higher compared with WT under heterotrophic conditions (Figure 2a,b). Conversely, the biomass yield of LUT-1 was significantly lower compared to WT ($p < 0.001$, Figure 2a,b). LUT-2, LUT-3, and LUT-4 cells were dividing faster than those of WT under heterotrophic conditions. On Day 4, under N-replete conditions, the cell density of LUT-2, LUT-3, and LUT-4 was 6.8×10^7 cell·mL^{-1}, 6.05×10^7 cell·mL^{-1}, and 7.68×10^7 cell·mL^{-1}, which was 1.6-, 1.4-, and 1.8-fold higher than that of WT (4.23×10^7 cell·mL^{-1}), respectively. Under N-deprived conditions, the cell density of LUT-2, LUT-3, and LUT-4 was 7.41×10^7 cell·mL^{-1}, 7.62×10^7 cell·mL^{-1}, and 7.89×10^7 cell·mL^{-1}, respectively, on Day 4, which was higher compared to WT (7.01×10^7 cell·mL^{-1}, Figure 2d).

A prevalent organic carbon source utilized in the heterotrophic cultivation of microalgae is glucose. The glucose consumption of lutein mutants under N-replete and N-deprived conditions is presented in Figure 2c,d. The results showed that the glucose consumption of lutein mutants differed from WT. After four days of cultivation, the residual glucose concentration of LUT 1, LUT2, LUT-3, and LUT-4 under N-replete conditions was 9.45 g·L^{-1}, 4.95 g·L^{-1}, 4.22 g·L^{-1}, and 3.75 g·L^{-1}, respectively (Figure 2e). The utilization of glucose in LUT-2, LUT-3, and LUT-4 was quicker compared to WT under N-replete conditions. Similarly, the residual glucose concentration of LUT-2 and LUT-4 in the culture medium under N-deprived conditions was utilized quicker compared with WT (Figure 2f). Conversely, the glucose consumption of LUT-1 and LUT-3 was slower than that of WT.

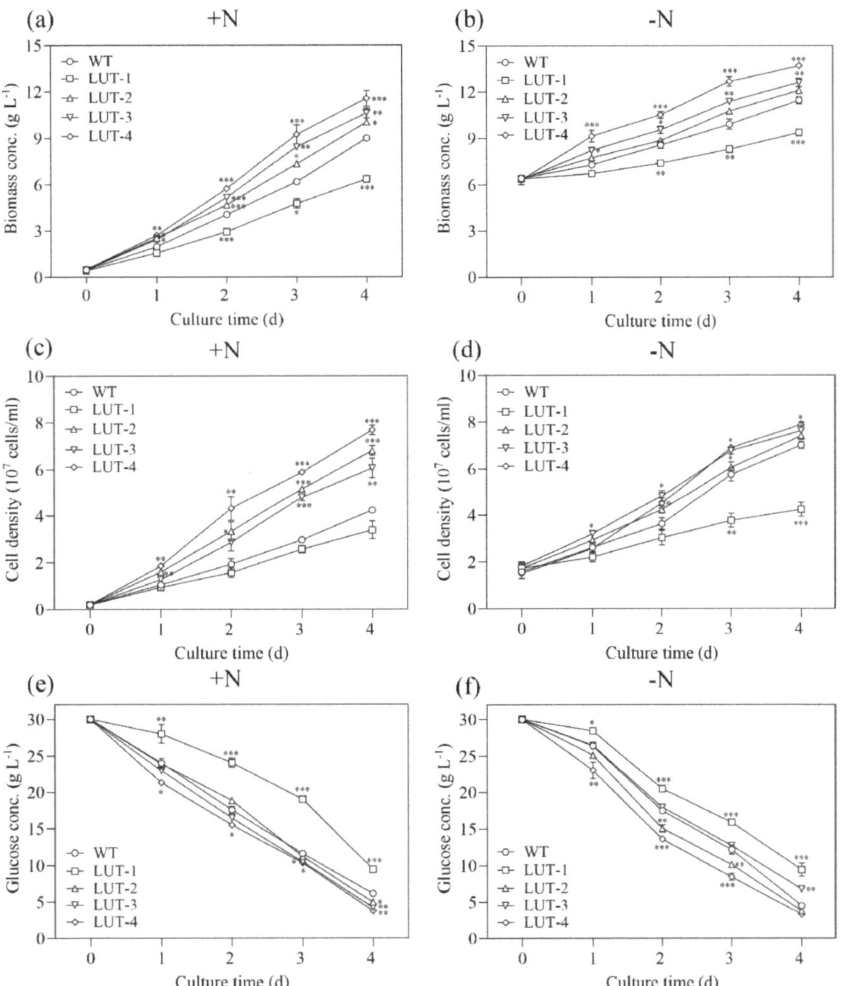

Figure 2. The growth and glucose consumption of wild-type (WT) and lutein mutants of *C. zofingiensis* under N-replete and N-deprived conditions. (**a,b**) Biomass concentration. (**c,d**) Cell density. (**e,f**) Glucose concentration. Values represent mean ± SD (n = 3). *, **, and *** are statistically significant at $p < 0.05$, $p < 0.01$, and $p < 0.001$, while lutein mutants are compared, respectively, with WT, at a given time.

HPLC analysis demonstrated that LUT-1, LUT-2, LUT-3, and LUT-4 mainly accumulated lutein. The content of lutein in LUT-1, LUT-2, LUT-3, and LUT-4 under N-replete conditions was 0.167%, 0.173%, 0.183%, and 0.195% of DW on Day 4, which was 14.4%, 18.5%, 25.3%, and 33.6% higher compared with WT (0.146% of DW), respectively (Figure 3a). Under N-deprived conditions, the lutein content of LUT-1, LUT-2, LUT-3, and LUT-4 reached 0.127%, 0.134%, 0.144%, and 0.154% of DW on Day 4, respectively, which was significantly higher compared with WT (0.072% of DW, $p < 0.001$, Figure 3b).

Collectively, LUT-4 exhibited increased cellular mass, quicker cell division, and high lutein content under heterotrophic conditions in comparison to WT. The biomass yield of LUT-4 was significantly higher compared with WT under heterotrophic conditions, suggesting that the specific mutations did not impact the growth potential. In general, mutagenesis is defined as the process by which heritable alterations arise in the genetic information of an organism [19]. EMS was classified as a chemical mutagen due to its ability to induce insertion and site-direction mutagenesis in DNA sequences [20]. The mutations could occur not only in genes related to carotenoid synthesis but also in other genes. In addition, the mutation sites need to be detected in further studies.

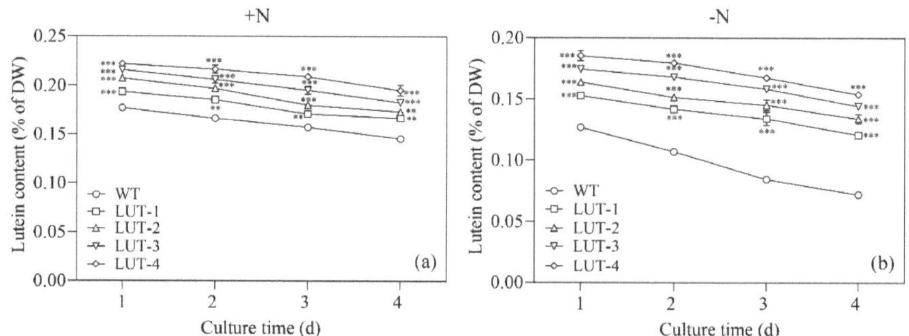

Figure 3. Changes in lutein contents in lutein mutants of *C. zofingiensis* strain under N-replete (**a**) and N-deprived (**b**) conditions. Values represent mean ± SD (*n* = 3). ** and *** are statistically significant at $p < 0.01$ and $p < 0.001$, while lutein mutants are compared with WT, respectively, at a given time.

2.2. Effect of Light Intensity on Growth and Lutein Accumulation of LUT-4

When LUT-4 was cultivated heterotrophically, the lutein content declined to 0.154% of DW. In contrast to the outcomes observed in comparable research, this performance remains unsatisfactory [21,22]. Light intensity is frequently the most influential factor in cell growth and lutein accumulation [23]. An increase in light intensity from 100 μmol/m^2/s to 500 μmol/m^2/s resulted in a substantial decrease in biomass concentration ($p < 0.05$, Figure 4a). After four days of cultivation, the highest biomass obtained under 100 μmol/m^2/s, 300 μmol/m^2/s, and 500 μmol/m^2/s was 10.8 g·L^{-1}, 9.95 g·L^{-1}, and 8 g·L^{-1}, which was 7.3%, 14.6%, and 31.3% lower, respectively, compared with the control (11.65 g·L^{-1}). When the light intensity was increased from 100 μmol/m^2/s to 500 μmol/m^2/s, the glucose consumption was enhanced (Figure 4b).

Furthermore, the composition of microalgae cells may also alter in response to variations in ambient light intensity, particularly for light-related chemicals like carotenoids and chlorophyll [11]. The research discovered that under high light intensity, the quality of main xanthophylls in microalgae, such as lutein, tended to decrease [24]. The reduction in size of light-harvesting cells, where lutein is located and generated, may be the reason for the drop in lutein content under high-intensity light [4]. As shown in Figure 4c, the highest lutein content reached 0.294% of dry weight (DW) when a light intensity at 300 μmol/m^2/s was used. When the light intensity was increased, the lutein content decreased. Under 100 μmol/m^2/s, the lutein content reached 0.26% of DW on Day 4, which was slightly lower than that under 300 μmol/m^2/s (0.277% of DW). Among the results above, the light intensity of 300 μmol/m^2/s was optimal for lutein production, and the highest yield of lutein was 0.028 g·L^{-1}, on Day 4.

The results found above suggest that light intensity has inverse effects on biomass and lutein accumulation. For instance, the microalgae cultivated under 300 μmol/m^2/s had a high lutein concentration, but their biomass yield was slightly lower compared to the control. Notably, the lutein content achieved in this study was better than most of those reported in the literatures [22,25]. LUT-4 appeared to be an excellent microalgal feedstock for the commercial production of lutein, as demonstrated by the present study, which attributes its lutein content to a comparatively high level of accumulation at a light intensity of 300 μmol/m^2/s. To further increase the lutein yield of this strain on a large scale and thereby render it more economically viable, additional engineering research is required.

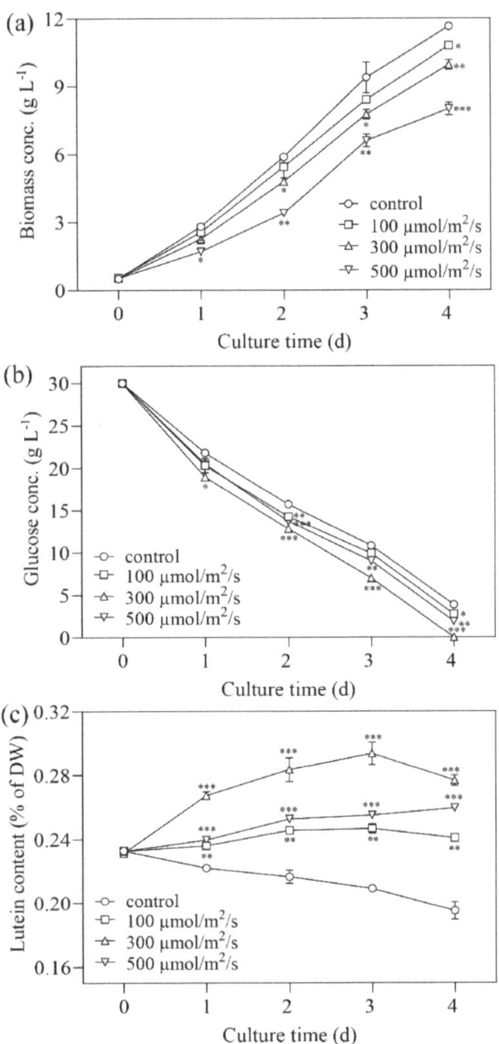

Figure 4. Effects of light intensities on the growth of *C. zofingiensis* mutant LUT-4 and lutein accumulation. (**a**) Biomass concentration. (**b**) Glucose concentration. (**c**) Lutein content. Values represent mean $\pm$ SD (n = 3). *, **, and *** are statistically significant at $p < 0.05$, $p < 0.01$, and $p < 0.001$, while 100, 300, 500 μmol/m^2/s light intensity are compared, respectively, with the control, at a given time.

2.3. Effects of Light Intensity on Biochemical Composition

Based on the results above, 300 μmol/m^2/s was the optimal light intensity for lutein accumulation in LUT-4. To make better use of this cultivation mode, the biological profiles were investigated in the following experiment. Under 300 μmol/m^2/s, the TFA content in LUT-4 rose from 16.32% to 24.89% of DW, which was 15.7% higher compared with the control (21.51% of DW, Figure 5a). On Day 4, when LUT-4 was grown at a light intensity of 300 μmol/m^2/s, the main fatty acids were C16:0, C18:1, C18:2, and C18:3 (Figure 5b). The abundances of C16:0 and C18:3 declined under 300 μmol/m^2/s. In contrast, the abundance of C18:1, which comprised as much as 32.55% of TFA, increased significantly, while C18:3 decreased drastically (Figure 5b). The results confirmed the previous research, which discovered that *C. zofingiensis* increased the abundance of C18:1 and reduced the abundances of C18:3 in response to stressful conditions [26].

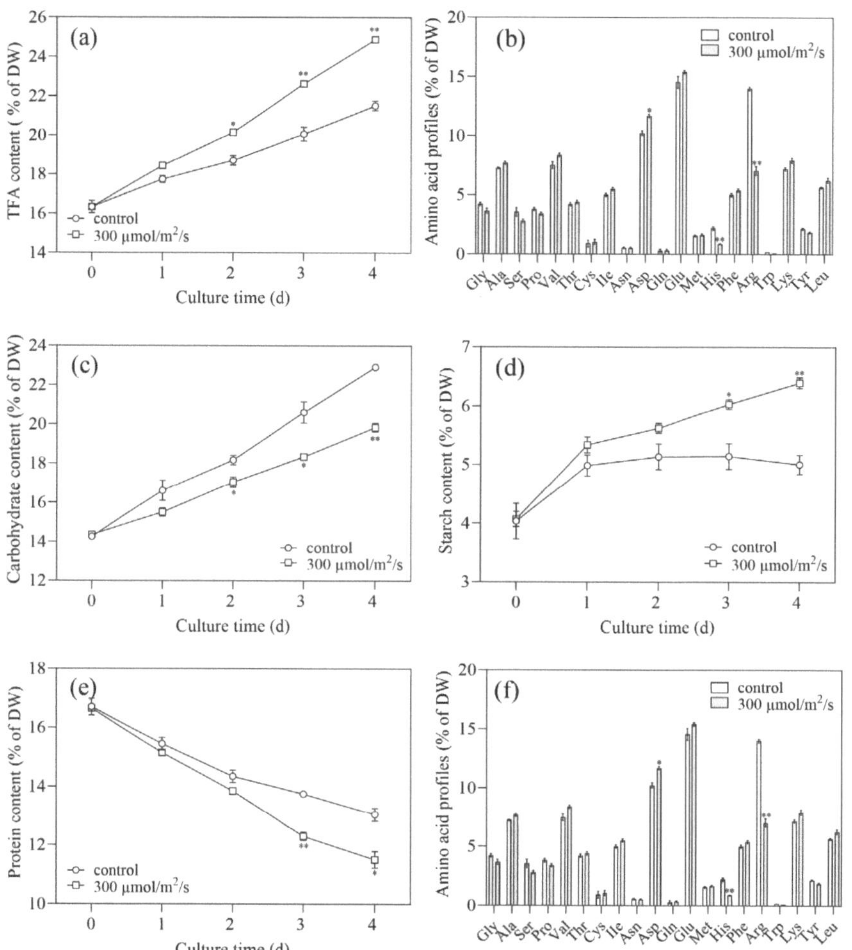

Figure 5. The biochemical changes in *C. zofingiensis* mutant LUT-4 were affected by optimal light intensity. (**a**) TFA contents. (**b**) Fatty acid profiles. (**c**) Carbohydrate content. (**d**) Starch content. (**e**) protein content. (**f**) Amino acid profiles. Values represent mean ± SD (n = 3). * and ** are statistically significant at $p < 0.05$ and $p < 0.01$, while 300 μmol/m^2/s light intensity is compared with the control, at a given time.

As shown in Figure 5e, the light intensity of 300 μmol/m^2/s significantly decreased protein content, by 11.5% of DW. In several reported microalgae, intracellular protein tended to degrade under stress conditions, providing the carbon skeleton and energy for lipid biosynthesis [27]. The degraded protein was suggested to first guide carbohydrate biosynthesis and then lipid. The results showed that the content of carbohydrates was significantly lower compared to the control ($p < 0.05$). However, the starch content in LUT-4 under 300 μmol/m^2/s increased from 4.07% to 6.4% of DW.

The results showed that the amino acid composition of LUT-4 was the same as the wild-type [28]. Interestingly, light intensity altered the composition of essential amino acids including Gly, Ala, Ser, Pro, Val, Thr, Ile, Asp, Glu, His, Phe, Arg, Lys, Tyr, and Leu (Figure 5f). Notably, the contents of His and Arg dramatically reduced ($p < 0.01$). Moreover, the contents of Ala, Val, Thr, Ile, Asp, and Glu increased in LUT-4. Thus, it could be concluded that light intensity altered the protein composition of LUT-4. Taken together, the alterations in biochemical compositions under 300 μmol/m^2/s were mainly caused by the utilization of organic carbon.

2.4. Carbon Availability Comparison

The activity of the synthetic pathway and the availability of carbon molecules are both critical for lipid biosynthesis [29]. As shown in Figure 6a, Ace-CoA content increased significantly at a light intensity of 300 μmol/m^2/s, which was primarily via the central carbon metabolism ($p < 0.05$). Ace-CoA serves as the primary precursor for lipid biosynthesis and may undergo conversion into C16:0 and C18:0 as part of fatty acid metabolism [30]. The lipid content is determined by the activity of the synthetic pathway when the precursor is adequate (i.e., Ace-CoA). Generally, Ace-CoA could join the biosynthesis of fatty acids continuously and rapidly. The results showed that the light intensity of 300 μmol/m^2/s could increase Ace-CoA content and activate the pathway for fatty acid biosynthesis. These findings were consistent with the results above. Briefly, the C16:0 content was reduced and the C18:1 content was increased at a light intensity of 300 μmol/m^2/s (Figure 5b).

Glucose can be directed to participate in carotenoid metabolism once it has been assimilated by cells [31]. In this study, the lutein content in LUT-4 could be increased to 0.294% of DW at a light intensity of 300 μmol/m^2/s. The synthesis of carotenoids in green alga commences with IPP, which is generated through the non-mevalonate pathway by 3-phosphoglyceraldehyde (G3P) and pyruvate [32]. The light intensity of 300 μmol/m^2/s significantly improved pyruvate content compared to that of the control, which suggested this cultivation method supplied more available carbon molecules for carotenoid synthesis ($p < 0.05$, Figure 6b). However, the content of IPP and GGPP (i.e., the downstream metabolite of IPP) declined at a light intensity of 300 μmol/m^2/s (Figure 6c,d). In comparison to the control, the lutein content was greater, even though the GGPP content was lower at a light intensity of 300 μmol/m^2/s. The findings indicated that, with the aid of suitable light intensity, carbon molecules could be converted to lutein, resulting in a reduction in its precursor metabolites. Thus, lutein accumulation was dependent on the availability of an abundance of carbon molecules, which was demonstrated by the carbon-use nature of lutein synthesis. In addition, the light intensity of 300 μmol/m^2/s resulted in the highest lutein content of 0.294% of DW, on Day 3 (Figure 4c). The implemented strategy (i.e., at a light intensity of 300 μmol/m^2/s) increased carbon availability relative to the control by increasing the rate of glucose uptake and pyruvate content. The consumption rate of GGPP content was accelerated by the strategy. As previous investigations have unveiled, the conversion of GGPP involves a limited number of enzymes, which are likewise examined in the subsequent section [27,33].

Several essential enzymes, including *PSY*, *PDS*, *LCYb*, *LCYe*, and *BKT*, are typically involved in sequential chain transformations that generate the diverse carotenoid family [34]. The carotenoid metabolism gene *PSY* limits the pace at which two GGPP contents may be condensed to produce lycopene [32]. The *PDS* gene, which is involved in carotenoid biosynthesis, enables microalgae to convert phytoene into ζ-carotene [35]. Compared with the control, the light intensity of 300 μmol/m^2/s significantly improved the relative expression level of *PSY* and *PDS*, respectively ($p < 0.05$, Figure 6e,f). *LCYe* is the gene responsible for lutein synthesis, whereas *LCYb* and *BKT* are genes intimately associated with astaxanthin synthesis [27,36]. Alterations in these genes may direct the flow of carbon into carotenoids, either as primary or secondary metabolites [36]. LUT-4 cells cultivated under 300 μmol/m^2/s increased the relative expression level of *LCYe*, which suggested the strategies had the potential to enhance the synthetic ability of Lutein ($p < 0.05$, Figure 6h). Moreover, the light intensity (i.e.,300 μmol/m^2/s) significantly decreased the expression level of *LCYb* and *BKT* compared to that under the control ($p < 0.05$). Thus, the light intensity of 300 μmol/m^2/s had the most pronounced impact on promoting lutein synthesis, aligning with the observed lutein content values. The current study revealed that LUT-4 seems to be a great microalgal feedstock for commercial lutein production because of its comparatively high lutein content at a light intensity of 300 μmol/m^2/s.

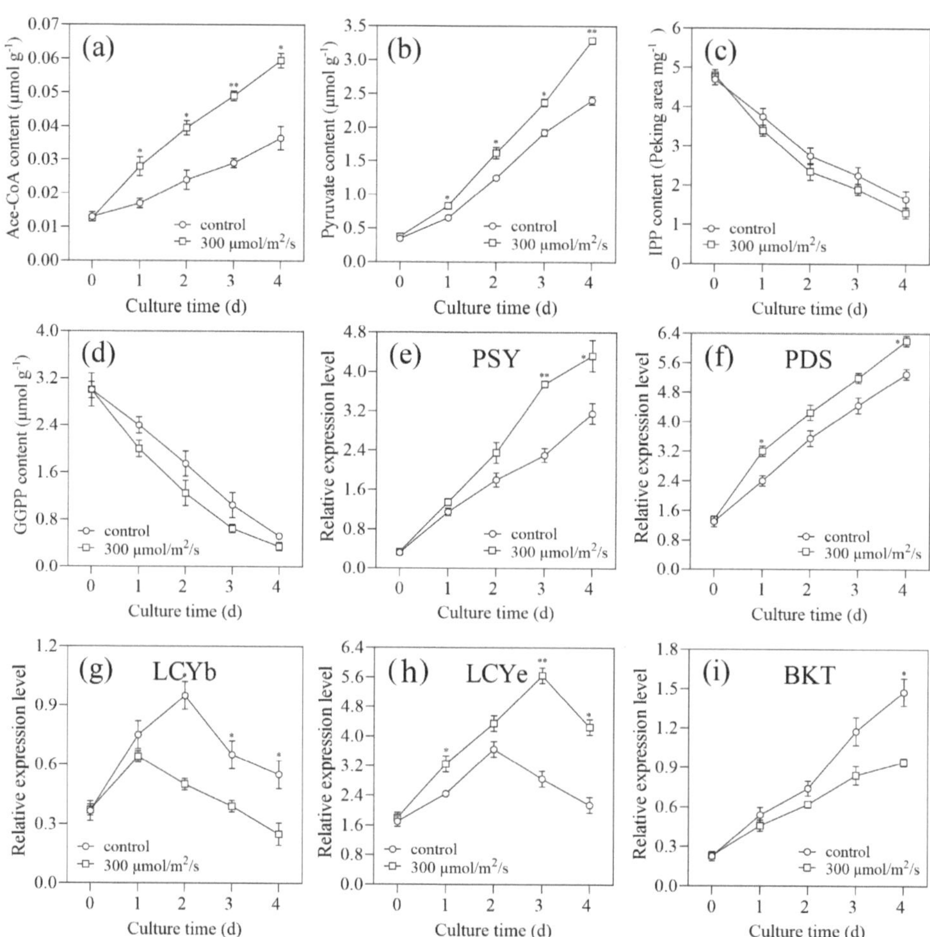

Figure 6. Effects of 300 μmol/m^2/s light intensity on metabolism and gene expression of *C. zofingiensis* mutant LUT-4. Variations in (**a**) Ace-CoA, (**b**) pyruvate, (**c**) isopentenyl pyrophosphate (IPP), and (**d**) geranylgeranyl diphosphate (GGPP) contents during cultivation periods. Variation in gene expression in (**e**) phytoene synthase (*PSY*), (**f**) phytoene desaturase (*PDS*), (**g**) Lycopene beta cyclase (*LCYb*), (**h**) lycopene epsilon cyclase (*LCYe*), and (**i**) Beta-carotenoid ketolase (*BKT*) during cultivation periods. Values represent mean ± SD (*n* = 3). * and ** are statistically significant at *p* < 0.05 and *p* < 0.01, while 300 μmol/m^2/s light intensity is compared with the control, at a given time.

3. Materials and Methods

3.1. Microalgal Strains and Growth Conditions

The American Type Culture Collection (ATCC, Rockville, MD, USA) provided the wild-type *Chromochloris zofingiensis* (ATCC30412), which was cultivated in the modified Endo medium. The pH of the modified Endo medium was first set to 6.5 by using the 3 M NaOH solution [37]. After transferring a single colony of *C. zofingiensis* into 250 mL Erlenmeyer flasks with 100 mL sterilized medium, the flasks were orbitally shaken at 180 rpm for seven days at 26 °C.

Mutagenesis was performed by harvesting WT cells at the early logarithmic phase by using centrifugation (1000, 5 min) and washing twice with phosphate-buffered saline (PBS, pH = 6.5). After treating microalgal cells for one hour in the dark with 2% (w/v) ethyl methyl sulfonate (EMS, Sigma-Aldrich, St. Louis, MO, USA), 10% (w/v) Na_2SO_3 was added to stop the mutagenesis process. The treated cells were resuspended in the refreshed modified Endo medium for 24 h in the absence of light after being rinsed and washed with PBS. Agar plates with modified Endo medium were used to cultivate approximately 20,000 colonies after mutagenesis. In contrast to the orange WT colonies, those exhibiting a green–yellow hue were chosen for cultivation in liquid-modified Endo medium supplemented with 30 g·L^{-1} glucose.

Log phase microalgal cells were inoculated into N-replete-modified Endo medium at an initial cell density of 2×10^6 cell·mL^{-1} for 4 days of heterotrophic cultivation. The gathered cells were then reintroduced into N-deprived-modified Endo medium, while the initial biomass concentration was kept at 6 g·L^{-1}. Furthermore, various light intensities (i.e., 100, 300, and 500 $\mu mol/m^2/s$) were applied to optimize the light supply conditions to increase lutein production by LUT-4. The same sample of microalgal cells was inoculated for the following experiment.

3.2. Determination of Dry Weight and Residual Glucose Concentration

Samples were taken at each time point, cleaned twice with 0.5 M NH_4HCO_3, and passed through a 1.2 μm pore-size dry GF/C filter paper (Whatman, Life Sciences, Maidstone, UK), which had been pre-weighed and dried at 105 °C in an oven for a whole night. Before measuring the dry weight, the Whatman GF/C filter paper was put in a desiccator for 20 min to enable the temperature to drop. The measurement of glucose concentration was performed by utilizing a Gold-Accu Glucose Monitoring System (Model BGMS-1; Sinocare Inc., Changsha, China).

3.3. Lutein Content Determination

Samples were extracted by utilizing previously published procedures [8]. A YMC Carotenoid (250 × 4.6 mm, 5 μm) column separated the lutein at 30 °C. A sample aliquot of ten microliters was introduced into the Waters Associates, Milford, MA, USA HPLC system outfitted with a 2998 photodiode array detector (Waters, Milford, MA, USA). Eluent A (methanol/methyl tert-butyl ether/water = 81:15:4, v/v) and eluent B (methanol/methyl tert-butyl ether/water = 43.5:52.5:4, v/v) made up the mobile phase. A gradient procedure was employed to separate the lutein: 0% B for 45 min, followed by a two-minute increase in gradient to 100% of B, and an eight-minute hold at 0% B. The rate of flow was 1.0 mL·min^{-1}. For quantification, the lutein standard (Sigma-Aldrich, St. Louis, MO, USA) was utilized as the calibrant. Lutein was detected by comparing the retention durations and absorption spectra of specific peaks in the chromatogram to the standards, and peak areas were extracted for quantification using calibrant curves produced with the standards.

3.4. Determination of Organic Composition

Methods previously documented were employed to extract the samples [12]. The quantification of total fatty acid content was accomplished through the combination of Agilent 7890B gas chromatography and 5977A mass spectrometry (GC-MS) (Agilent, Santa Clara, CA, USA).

The supernatant was discarded after the samples were ground and incubated for 30 min at 80 °C with an 80% ethanol solution. Amylase was introduced into the granules to aid in the hydrolysis of starch, which was accomplished by heating the pellets. Anthrone and sulfuric acid were added to the released glucose and incubated for 10 min at 95 °C. The starch content was determined by translating the optical density at 562 nm to units of glucose.

Samples were incubated with 0.5 mL of acetic acid for 20 min at 80 °C before being treated with 10 mL of acetone. After the supernatant was removed, 2.5 mL of 4 M trifluoroacetic acid was added and the samples were then heated for 4 h. The optical density at 490 nm (OD_{490}) was determined by boiling samples in a solution containing chromogenic reagent for twenty minutes. A standard curve was constructed by utilizing glucose to determine the total content of carbohydrates.

The protein content was determined in strict adherence to the established approach [38]. The protein content was calculated by using a conversion factor of 6.25 to the total nitrogen content of the samples, as measured by an automated Kjeldahl analyzer (UDK 159-VELP, Usmate Velate, Italy). The amino acid composition was analyzed by using the A300 auto amino acid analyzer (Membra Pure, Bodenheim, Germany) equipped with a TS263 column.

3.5. Isopentenyl Pyrophosphate (IPP) and Geranylgeranyl Diphosphate (GGPP) Quantification

To ascertain IPP and GGPP, microalgal cells were extracted and subsequently subjected to liquid nitrogen pulverizing to disrupt them. A vacuum freeze-drying machine (LP 20, Ilshinbiobase Co., Ltd., Dongducheon, Korea) was then utilized to dehydrate the metabolites that had been extracted with methanol. The samples were purified by a solid-phase extraction (SPE) column (Waters, Milford, MA, USA) and analyzed by UPLC-MS/MS (Waters, Milford, MA, USA), according to the earlier study [39]. The standards of IPP and GGPP were acquired from Sigma-Aldrich (St. Louis, MO, USA).

3.6. Measurement of Ace-CoA and Pyruvate

For measurement of Ace-CoA contents, the microalgal cells were extracted and analyzed by using the Ace-CoA assay reagent (Sigma, MAK039). Pyruvate content was determined by extracting pyruvate and quantifying it using the pyruvate assay reagent (Sigma, MAK071), in accordance with the provided instructions.

3.7. Determination of Expression Levels of mRNA

The RNA Plant Plus Reagent (Tiangen, Beijing, China) was utilized to extract the total RNA of *C. zofingiensis*, in accordance with the manufacturer's instructions. Reverse transcription of the RNA to cDNA was then performed according to the instructions, using the QuantScript RT Kit reagent (Tiangen, Beijing, China). One Step SYBR PrimeScript PLUS RT-PCR Kit reagent (TaKaRa, Tokyo, Japan) was utilized to conduct real-time PCR. Table 1 lists the available Primers. The mRNA expression level was stabilized by using the *C. zofingiensis* actin (ACT) gene as the internal control.

Table 1. PCR primers used for RT-PCR to quantify expression level of carotenogenesis genes.

Gene	Forward (5′-3′)	Reverse (5′-3′)
PSY	CACCAGGTTGTCAGAGTCCA	ACTAGTGTGTTGCTGACTCT
PDS	GATGAATGTATTTGCTGAACT	GGCCAGTGCCTTAGCCATAG
LCYe	TCAAAGCACAGGCGAACAAACA	AACGTCGGGACCTATAAGTCCG
LCYb	CGCAGGCGAAAAATTCCTGT	TAAGGAATGTCACACCGCTGG
BKT	GGTGCTCAAAGTGGGGTGGT	CCATTTCCCACATATTGCACCT
ACT	TGCCGAGCGTGAAATTGTGA	CGTGAATGCCAGCAGCCTCCA

3.8. Statistical Analysis

The experiments in this research were carried out in triplicate. The statistical differences between the two data sets were determined by using the unpaired Student's *t*-test. A combination of one-way ANOVA and Dunnett's post hoc test was utilized to assess multiple comparisons. The statistical analysis was conducted by using GraphPad Prism version 10 (GraphPad Software, La Jolla, CA, USA). The software allowed for approval at a level below $p < 0.05$.

4. Conclusions

In this study, four lutein mutants were isolated from chemical mutagenesis. By comparing the growth and lutein content, LUT-4 was selected for further study. Results showed that 300 μmol/m^2/s was the most suitable light intensity for lutein accumulation. The biochemical composition, Ace-CoA, pyruvate, IPP, and GGPP content alterations demonstrated that the light intensity could enhance the use of organic carbon for lutein biosynthesis. Moreover, the elevated expression of *PSY*, *PDS*, and *LCYe* genes could facilitate the generation of lutein. Overall, this study supplied a feasible technique for producing natural lutein by LUT-4.

Author Contributions: Conceptualization, Q.C.; data curation, Q.C. and M.L.; formal analysis, Q.C. and M.L.; funding acquisition, Q.C. and Y.B.; investigation, G.S., W.H. and Y.B.; methodology, W.M.; project administration, D.W.; writing—original draft, Q.C.; writing—review and editing, Q.C. and Y.B. All authors have read and agreed to the published version of the manuscript.

Funding: This investigation was supported by the Postdoctoral Fellowship Program of CPSF (No. GZC20232917) and Knowledge Innovation Program of Wuhan-Shuguang Project (No. 2022 020801020144).

Institutional Review Board Statement: Not applicable.

Data Availability Statement: Data are contained within the article.

Conflicts of Interest: The authors declare no conflicts of interest.

References

1. Nwachukwu, I.D.; Udenigwe, C.C.; Aluko, R.E. Lutein and Zeaxanthin: Production Technology, Bioavailability, Mechanisms of Action, Visual Function, and Health Claim Status. *Trends Food Sci. Technol.* **2016**, *49*, 74–84. [CrossRef]
2. Lim, L.S.; Mitchell, P.; Seddon, J.M.; Holz, F.G.; Wong, T.Y. Age-Related Macular Degeneration. *Lancet* **2012**, *379*, 1728–1738. [CrossRef]
3. Lin, J.H.; Lee, D.J.; Chang, J.S. Lutein Production from Biomass: Marigold Flowers versus Microalgae. *Bioresour. Technol.* **2015**, *184*, 421–428. [CrossRef]
4. Ho, S.H.; Chan, M.C.; Liu, C.C.; Chen, C.Y.; Lee, W.L.; Lee, D.J.; Chang, J.S. Enhancing Lutein Productivity of an Indigenous Microalga *Scenedesmus Obliquus* FSP-3 Using Light-Related Strategies. *Bioresour. Technol.* **2014**, *152*, 275–282. [CrossRef]
5. Liu, C.; Hu, B.; Cheng, Y.; Guo, Y.; Yao, W.; Qian, H. Carotenoids from Fungi and Microalgae: A Review on Their Recent Production, Extraction, and Developments. *Bioresour. Technol.* **2021**, *337*, 125398. [CrossRef]
6. Li, D.; Yuan, Y.; Cheng, D.; Zhao, Q. Effect of Light Quality on Growth Rate, Carbohydrate Accumulation, Fatty Acid Profile and Lutein Biosynthesis of *Chlorella* sp. AE10. *Bioresour. Technol.* **2019**, *291*, 121783. [CrossRef]
7. Shi, X.-M.; Zhang, X.-W.; Chen, F. Heterotrophic Production of Biomass and Lutein by *Chlorella protothecoides* on Various Nitrogen Sources. *Enzyme Microb. Technol.* **2000**, *27*, 312–318. [CrossRef]
8. Chen, Q.; Chen, Y.; Xiao, L.; Li, Y.; Zou, S.; Han, D. Co-Production of Lutein, Zeaxanthin, and β-Carotene by Utilization of a Mutant of the Green Alga *Chromochloris zofingiensis*. *Algal Res.* **2022**, *68*, 102882. [CrossRef]
9. Heo, J.; Shin, D.S.; Cho, K.; Cho, D.H.; Lee, Y.J.; Kim, H.S. Indigenous Microalga *Parachlorella* sp. JD-076 as a Potential Source for Lutein Production: Optimization of Lutein Productivity via Regulation of Light Intensity and Carbon Source. *Algal Res.* **2018**, *33*, 1–7.
10. Gong, M.; Bassi, A. Investigation of *Chlorella vulgaris* UTEX 265 Cultivation under Light and Low Temperature Stressed Conditions for Lutein Production in Flasks and the Coiled Tree Photo-Bioreactor (CTPBR). *Appl. Biochem. Biotechnol.* **2017**, *183*, 652–671. [CrossRef]
11. Kona, R.; Pallerla, P.; Addipilli, R.; Sripadi, P.; Venkata Mohan, S. Lutein and β-Carotene Biosynthesis in *Scenedesmus* sp. SVMIICT1 through Differential Light Intensities. *Bioresour. Technol.* **2021**, *341*, 125814. [CrossRef]
12. Chen, Q.; Chen, Y.; Xu, Q.; Jin, H.; Hu, Q.; Han, D. Effective Two-Stage Heterotrophic Cultivation of the Unicellular Green Microalga *Chromochloris zofingiensis* Enabled Ultrahigh Biomass and Astaxanthin Production. *Front. Bioeng. Biotechnol.* **2022**, *10*, 834230. [CrossRef]
13. Chen, W.C.; Hsu, Y.C.; Chang, J.S.; Ho, S.H.; Wang, L.F.; Wei, Y.H. Enhancing Production of Lutein by a Mixotrophic Cultivation System Using Microalga *Scenedesmus obliquus* CWL-1. *Bioresour. Technol.* **2019**, *291*, 121891. [CrossRef]
14. Yen, H.W.; Sun, C.H.; Ma, T.W. The Comparison of Lutein Production by *Scenesdesmus* Sp. in the Autotrophic and the Mixotrophic Cultivation. *Appl. Biochem. Biotechnol.* **2011**, *164*, 353–361. [CrossRef]
15. Chen, C.Y.; Liu, C.C. Optimization of Lutein Production with a Two-Stage Mixotrophic Cultivation System with *Chlorella sorokiniana* MB-1. *Bioresour. Technol.* **2018**, *262*, 74–79. [CrossRef]
16. Zheng, H.; Wang, Y.; Li, S.; Nagarajan, D.; Varjani, S.; Lee, D.J.; Chang, J.S. Recent Advances in Lutein Production from Microalgae. *Renew. Sustain. Energy Rev.* **2022**, *153*, 111795. [CrossRef]
17. Mao, X.; Zhang, Y.; Wang, X.; Liu, J. Novel Insights into Salinity-Induced Lipogenesis and Carotenogenesis in the Oleaginous Astaxanthin-Producing Alga *Chromochloris zofingiensis*: A Multi-Omics Study. *Biotechnol. Biofuels* **2020**, *13*, 73. [CrossRef]
18. Roth, M.S.; Westcott, D.J.; Iwai, M.; Niyogi, K.K. Hexokinase Is Necessary for Glucose-Mediated Photosynthesis Repression and Lipid Accumulation in a Green Alga. *Commun. Biol.* **2019**, *2*, 347. [CrossRef]

19. Wu, M.; Zhang, H.; Sun, W.; Li, Y.; Hu, Q.; Zhou, H.; Han, D. Metabolic Plasticity of the Starchless Mutant of *Chlorella sorokiniana* and Mechanisms Underlying Its Enhanced Lipid Production Revealed by Comparative Metabolomics Analysis. *Algal Res.* **2019**, *42*, 101587. [CrossRef]

20. Oladosu, Y.; Rafii, M.Y.; Abdullah, N.; Hussin, G.; Ramli, A.; Rahim, H.A.; Miah, G.; Usman, M. Principle and Application of Plant Mutagenesis in Crop Improvement: A Review. *Biotechnol. Biotechnol. Equip.* **2016**, *30*, 1–16. [CrossRef]

21. Sánchez, J.F.; Fernández, J.M.; Acién, F.G.; Rueda, A.; Pérez-Parra, J.; Molina, E. Influence of Culture Conditions on the Productivity and Lutein Content of the New Strain *Scenedesmus almeriensis*. *Process Biochem.* **2008**, *43*, 398–405. [CrossRef]

22. Wei, D.; Chen, F.; Chen, G.; Zhang, X.W.; Liu, L.J.; Zhang, H. Enhanced Production of Lutein in Heterotrophic *Chlorella protothecoides* by Oxidative Stress. *Sci. China C Life Sci.* **2008**, *51*, 1088–1093. [CrossRef]

23. Ma, R.; Zhang, Z.; Ho, S.H.; Ruan, C.; Li, J.; Xie, Y.; Shi, X.; Liu, L.; Chen, J. Two-Stage Bioprocess for Hyper-Production of Lutein from Microalga *Chlorella sorokiniana* FZU60: Effects of Temperature, Light Intensity, and Operation Strategies. *Algal Res.* **2020**, *52*, 102119. [CrossRef]

24. Vaquero, I.; Mogedas, B.; Ruiz-Domínguez, M.C.; Vega, J.M.; Vílchez, C. Light-Mediated Lutein Enrichment of an Acid Environment Microalga. *Algal Res.* **2014**, *6*, 70–77. [CrossRef]

25. Del Campo, J.A.; Moreno, J.; Rodríguez, H.; Vargas, M.A.; Rivas, J.; Guerrero, M.G. Carotenoid Content of Chlorophycean Microalgae: Factors Determining Lutein Accumulation in *Muriellopsis* sp. (Chlorophyta). *J. Biotechnol.* **2000**, *76*, 51–59. [CrossRef]

26. Wu, T.; Yu, L.; Zhang, Y.; Liu, J. Characterization of Fatty Acid Desaturases Reveals Stress-Induced Synthesis of C18 Unsaturated Fatty Acids Enriched in Triacylglycerol in the Oleaginous Alga *Chromochloris zofingiensis*. *Biotechnol. Biofuels* **2021**, *14*, 184. [CrossRef]

27. Sun, H.; Ren, Y.; Fan, Y.; Lu, X.; Zhao, W.; Chen, F. Systematic Metabolic Tools Reveal Underlying Mechanism of Product Biosynthesis in *Chromochloris zofingiensis*. *Bioresour. Technol.* **2021**, *337*, 125406. [CrossRef]

28. Sun, H.; Ren, Y.; Lao, Y.; Li, X.; Chen, F. A Novel Fed-Batch Strategy Enhances Lipid and Astaxanthin Productivity without Compromising Biomass of *Chromochloris zofingiensis*. *Bioresour. Technol.* **2020**, *308*, 123306. [CrossRef]

29. Liang, M.H.; Wang, L.; Wang, Q.; Zhu, J.; Jiang, J.G. High-Value Bioproducts from Microalgae: Strategies and Progress. *Crit. Rev. Food Sci. Nutr.* **2019**, *59*, 2423–2441. [CrossRef]

30. Chen, Z.; Chen, J.; Liu, J.; Li, L.; Qin, S.; Huang, Q. Transcriptomic and Metabolic Analysis of an Astaxanthin-Hyperproducing *Haematococcus pluvialis* Mutant Obtained by Low-Temperature Plasma (LTP) Mutagenesis under High Light Irradiation. *Algal Res.* **2020**, *45*, 101746. [CrossRef]

31. Sun, H.; Li, X.; Ren, Y.; Zhang, H.; Mao, X.; Lao, Y.; Wang, X.; Chen, F. Boost Carbon Availability and Value in Algal Cell for Economic Deployment of Biomass. *Bioresour. Technol.* **2020**, *300*, 122640. [CrossRef]

32. Jin, H.; Lao, Y.M.; Zhou, J.; Zhang, H.J.; Cai, Z.H. Simultaneous Determination of 13 Carotenoids by a Simple C18 Column-Based Ultra-High-Pressure Liquid Chromatography Method for Carotenoid Profiling in the Astaxanthin-Accumulating *Haematococcus pluvialis*. *J. Chromatogr. A* **2017**, *1488*, 93–103. [CrossRef]

33. Zhang, Z.; Huang, J.J.; Sun, D.; Lee, Y.; Chen, F. Two-Step Cultivation for Production of Astaxanthin in *Chlorella Zofingiensis* Using a Patented Energy-Free Rotating Floating Photobioreactor (RFP). *Bioresour. Technol.* **2017**, *224*, 515–522. [CrossRef]

34. Sun, H.; Zhao, W.; Mao, X.; Li, Y.; Wu, T.; Chen, F. High-Value Biomass from Microalgae Production Platforms: Strategies and Progress Based on Carbon Metabolism and Energy Conversion. *Biotechnol. Biofuels* **2018**, *11*, 227.

35. Zhang, H.; Chen, A.; Huang, L.; Zhang, C.; Gao, B. Transcriptomic Analysis Unravels the Modulating Mechanisms of the Biomass and Value-Added Bioproducts Accumulation by Light Spectrum in *Eustigmatos* Cf. Polyphem (Eustigmatophyceae). *Bioresour. Technol.* **2021**, *338*, 125523. [CrossRef]

36. Varela, J.C.; Pereira, H.; Vila, M.; León, R. Production of Carotenoids by Microalgae: Achievements and Challenges. *Photosynth. Res.* **2015**, *125*, 423–436. [CrossRef]

37. Jin, H.; Chuai, W.; Li, K.; Hou, G.; Wu, M.; Chen, J.; Wang, H.; Jia, J.; Han, D.; Hu, Q. Ultrahigh-Cell-Density Heterotrophic Cultivation of the Unicellular Green Alga *Chlorella sorokiniana* for Biomass Production. *Biotechnol. Bioeng.* **2021**, *118*, 4138–4151. [CrossRef]

38. Wang, L.; Chen, C.; Tang, Y.; Liu, B. A Novel Hypothermic Strain, *Pseudomonas reactans* WL20-3 with High Nitrate Removal from Actual Sewage, and Its Synergistic Resistance Mechanism for Efficient Nitrate Removal at 4 °C. *Bioresour. Technol.* **2023**, *385*, 129389. [CrossRef]

39. Lao, Y.M.; Lin, Y.M.; Wang, X.S.; Xu, X.J.; Jin, H. An Improved Method for Sensitive Quantification of Isoprenoi Diphosphates in the Astaxanthin-Accumulating *Haematococcus pluvialis*. *Food Chem.* **2022**, *375*, 131911.

Article

Maximizing Polysaccharides and Phycoerythrin in *Porphyridium purpureum* via the Addition of Exogenous Compounds: A Response-Surface-Methodology Approach

Sanjiong Yi [1], Ai-Hua Zhang [1,*], Jianke Huang [1,*], Ting Yao [1], Bo Feng [1], Xinghu Zhou [2], Yadong Hu [2,*] and Mingxuan Pan [2]

[1] Jiangsu Province Engineering Research Center for Marine Bio-Resources Sustainable Utilization, College of Oceanography, Hohai University, Nanjing 210098, China; 18136562630@163.com (S.Y.); 15850576061@163.com (T.Y.); hhu335338350@163.com (B.F.)

[2] Jiangsu Innovation Center of Marine Bioresource, Jiangsu Coast Development Investment Co., Ltd., Jiangsu Coast Development Group Co., Ltd., Nanjing 210019, China; zhouxinghu@jsyhkf.com (X.Z.); panmingxuan@jsyhkf.com (M.P.)

* Correspondence: ahz@hhu.edu.cn (A.-H.Z.); jkhuang@hhu.edu.cn (J.H.); huyadong@jsyhkf.com (Y.H.)

Citation: Yi, S.; Zhang, A.-H.; Huang, J.; Yao, T.; Feng, B.; Zhou, X.; Hu, Y.; Pan, M. Maximizing Polysaccharides and Phycoerythrin in *Porphyridium purpureum* via the Addition of Exogenous Compounds: A Response-Surface-Methodology Approach. *Mar. Drugs* **2024**, *22*, 138. https://doi.org/10.3390/md22030138

Academic Editors: Cecilia Faraloni and Eleftherios Touloupakis

Received: 28 February 2024
Revised: 18 March 2024
Accepted: 19 March 2024
Published: 21 March 2024

Abstract: Phycoerythrin and polysaccharides have significant commercial value in medicine, cosmetics, and food industries due to their excellent bioactive functions. To maximize the production of biomass, phycoerythrin, and polysaccharides in *Porphyridium purpureum*, culture media were supplemented with calcium gluconate (CG), magnesium gluconate (MG) and polypeptides (BT), and their optimal amounts were determined using the response surface methodology (RSM) based on three single-factor experiments. The optimal concentrations of CG, MG, and BT were determined to be 4, 12, and 2 g L^{-1}, respectively. The RSM-based models indicated that biomass and phycoerythrin production were significantly affected only by MG and BT, respectively. However, polysaccharide production was significantly affected by the interactions between CG and BT and those between MG and BT, with no significant effect from BT alone. Using the optimized culture conditions, the maximum biomass (5.97 g L^{-1}), phycoerythrin (102.95 mg L^{-1}), and polysaccharide (1.42 g L^{-1}) concentrations met and even surpassed the model-predicted maximums. After optimization, biomass, phycoerythrin, and polysaccharides concentrations increased by 132.3%, 27.97%, and 136.67%, respectively, compared to the control. Overall, this study establishes a strong foundation for the highly efficient production of phycoerythrin and polysaccharides using *P. purpureum*.

Keywords: *Porphyridium purpureum*; polysaccharide; phycoerythrin; response surface methodology; polypeptides; gluconate

1. Introduction

Porphyridium purpureum, a member of the Rhodophyta, has attracted significant attention as a source of high-value bioactive substances, such as phycobiliproteins [1,2], long-chain polyunsaturated fatty acids [3,4], and sulfated polysaccharides [5]. Phycobiliproteins are crucial components of light-harvesting pigments in Cyanobacteria, Rhodophyta, and Cryptophyta [6]. According to their spectroscopic properties, phycobiliproteins with a pink/red coloration are classified as phycoerythrin (PE, 540–570 nm); those with blue coloration are phycocyanin (PC, 610–620 nm); and those with bluish-green coloration are allophycocyanin (APC, 650–655 nm) [1]. B-phycoerythrin and R-phycocyanin are present in *P. purpureum* [7]. Studies have demonstrated that phycoerythrin has significant antioxidant [8], immune-regulating [9], and anticancer effects [10]. Phycoerythrin extracted from *P. purpureum* is predominantly used in the food industry and has been approved as a food colorant due to its health benefits, intense fluorescence, and vivid color [11]. Furthermore, the polysaccharides derived from *P. purpureum* have seen a wide range of applications in the medical, cosmetics, and food industries [12,13] due to their excellent anti-inflammatory, anti-viral, anti-oxidant, and immunomodulating properties [14–16]. The commercial price of phycoerythrin varies significantly depending on its purity, but the market price of purified phycoerythrin was most recently reported to be USD 200 per milligram [17]. Moreover, there has been a continuous increase in the market demand for phycoerythrin in recent years owing to it being a natural product and having various functional properties [18]. However, the instability of phycoerythrin under adverse conditions such as high temperatures, low pH, and exposure to light remains a critical issue that limits its widespread application [19].

To date, much research has been conducted regarding how to enhance the concentrations of phycoerythrin and polysaccharides in *P. purpureum* to better meet market demands. The main strategies have focused on improving the culture media [1], supplying exogenous substances [20], optimizing environmental factors [21], and changing the culture method [22]. The optimization of the culture medium is a primary method for enhancing microalgal growth and increasing the production of bioactive substances. Nitrogen is a crucial nutrient for microalgal growth, and both nitrogen concentration and nitrogen type can impact the production of biomass and bioactive substances in microalgae [23]. This connection was exemplified by the work of Sánchez-Saavedra et al. [24], who found that the biomass productivity (173.2 mg L^{-1} d^{-1}) of *P. cruentum* was higher when the alga was cultured at a $NaNO_3$ concentration of 0.075 g L^{-1} compared to higher $NaNO_3$ concentrations (i.e., 0.45 and 0.225 g L^{-1}). Additionally, the presence of an organic carbon source can influence the yields of biomass and bioactive substances in *P. purpureum* [23]. For instance, the maximum biomass of *P. purpureum* CoE1 was achieved with a 0.5% (*w/v*) glucose dosage, while the maximum arachidonic acid (ARA) concentration was obtained with a 0.38% (*w/v*) glycerol dosage [22]. Furthermore, adding exogenous substances is essential for maximizing microalgal biomass and the concentrations of bioactive substances. Numerous studies have investigated how the biomass and the production of bioactive substances are enhanced by supplementing with phytohormones, metal ions, and vitamins [25–27]. One such study showed that the ARA concentration of *P. purpureum* was enhanced by stimulation with 20 mg L^{-1} of 5-aminolevulinic acid, with a peak yield of 170.32 mg L^{-1}, which represented a 70.82% increase compared to the control [20]. Hence, it is essential to add the suitable substances to culture media at concentrations within the appropriate ranges to maximize the production efficiency of active substances.

Response surface methodology (RSM) is an important statistical optimization tool that has been widely used for experimental modeling. This method reduces the number of experiments required and optimizes the interactions among the experimental process parameters in various processes [28]. Previously, RSM has been utilized to optimize the microalgae-culture process and significantly improve microalgal productivity. The optimal concentrations of sodium chloride, magnesium sulfate, sodium nitrate, and dipotassium hydrogen phosphate have been determined using the RSM, with the highest PE content in *P. purpureum* reaching 3.3% under optimized conditions [1]. To maximize the PB content of *P. cruentum*, RSM was used to determine the optimal conditions of temperature (10 °C) and light intensity (30 µmol m^{-2} s^{-1}), resulting in a maximum phycobiliprotein (PB) content of 2.9% [29]. Clearly, RSM has been proven to be an efficient and effective method for medium optimization.

A commercially produced combination of peptide complexes, commonly referred to as Bainengtai (BT) in China, is composed of enzymatic hydrolysates of high-quality plant proteins. This product is extensively used in the agriculture and feed industries to promote the growth of both plants and animals. Our previous study showed that BT enhanced phycocyanin production in *Arthrospira maxima* [30]. Calcium gluconate (CG) and magnesium gluconate (MG) dissociate into gluconic acid and cations in the medium solutions, so they can be considered to be a combination of a glucose and an ion under appropriate conditions [31]. Similarly, gluconate is primarily used as an additive in the food, pharmaceutical, health, and construction industries. Pang et al. [32] indicated that gluconate, the metabolic product of glucose, significantly increased the biomass of *Haematococcus pluvialis* compared to sodium acetate and ribose, making it a suitable candidate for use as an organic carbon source.

Therefore, in the present study, we investigated the effects of calcium gluconate (CG), magnesium gluconate (MG), and BT as additional supplements in *P. purpureum* culture. We aimed to determine the independent and interactive effects of the three factors (CG, MG, and BT concentrations) on biomass, phycoerythrin, and polysaccharide production by *P. purpureum*. Additionally, using RSM, this study focused on determining the optimal amounts of these substances to maximize biomass yield and the production of phycoerythrin and polysaccharides.

2. Results

2.1. Effects of Single Factors (CG, MG, and BT) on Microalgal Growth and the Accumulation of Bioactive Substances

It was found that the CG supplementation promotes microalgal growth and the accumulation of bioactive substances. As shown in Figure 1A,C, among all CG concentration levels, microalgal biomass and polysaccharide concentrations peaked when CG was added at 4 g L^{-1}. After 24 days of culturing, the maximum concentrations of biomass and polysaccharides reached 4.78 ± 0.03 and $0.7 \pm 0.01 \text{ g L}^{-1}$, respectively, which were 1.54 and 1.75 times higher than the concentrations in the control group. However, the maximum phycoerythrin concentration, which was 1.45 times higher than that in the control group (Figure 1B; $146.9 \pm 10.77 \text{ mg L}^{-1}$, day 20), was observed with a CG concentration of 2 g L^{-1}. These results indicated that the optimal CG concentration for *P. purpureum* growth and polysaccharide accumulation was 4 g L^{-1}, whereas a CG concentration of 2 g L^{-1} was optimal for phycoerythrin production.

MG supplementation also enhanced the production of biomass and bioactive substances in *P. purpureum*. As shown in Figure 1F, the addition of MG markedly increased the polysaccharide yield, which reached $0.76 \pm 0.02 \text{ g L}^{-1}$ on the 24th day. Compared to the control group, the maximum concentration of polysaccharide increased by 375% at an MG concentration of 11 g L^{-1}. In addition, the biomass and phycoerythrin concentrations (Figure 1D,E) peaked at $4.12 \pm 0.19 \text{ g L}^{-1}$ and $89.53 \pm 2.77 \text{ mg L}^{-1}$ on days 24 and 16, respectively. Compared to the control group, the maximum biomass and phycoerythrin concentrations increased by 74.6% and 23.3%, respectively, at an MG concentration of 12 g L^{-1}. As a consequence, an MG concentration of 12 g L^{-1} was identified as the optimal concentration of MG for *P. purpureum* growth and phycoerythrin accumulation. However, for polysaccharide production, an MG concentration of 11 g L^{-1} was optimal.

In terms of the impact of BT on microalgal growth and the accumulation of bioactive substances, the maximum biomass and phycoerythrin and polysaccharide concentrations in the BT treatment groups were generally higher than those in the control group, except at a BT concentration of 0.5 g L^{-1}. On the 24th day, the biomass and polysaccharide concentrations reached their maximum values of $2.35 \pm 0.17 \text{ g L}^{-1}$ and $0.267 \pm 0.002 \text{ g L}^{-1}$, respectively, at a BT concentration of 2 g L^{-1} (Figure 1G,I). However, when the BT concentration was 1.5 g L^{-1}, the concentration of phycoerythrin (Figure 1H) reached its maximum ($83.02 \pm 0.59 \text{ mg L}^{-1}$) on the 12th day. Therefore, the optimal BT concentrations for *P. purpureum* growth and polysaccharide accumulation were both 2 g L^{-1}, while that for phycoerythrin production was 1.5 g L^{-1}.

2.2. Model Fitting of RSM

The quadratic regression equations for the biomass, phycoerythrin, and polysaccharide concentrations were obtained using RSM (Table 1). The p-values for all the investigated responses were $p < 0.05$, showing the significance of the applied model [33]. The p-values for biomass concentration (0.0123), phycoerythrin concentration (0.0054) and polysaccharide concentration (0.0054) were all less than 0.05, demonstrating that the models for all responses were significant. At the same time, the high R^2 value (>0.8893) suggested that all the models fit the data well (Table 1). Furthermore, the relatively low variation coefficients (9.93–11.07%) and the lack of fit ($p > 0.05$) implied high experimental reliability and a strong correlation between the responses and the independent variables.

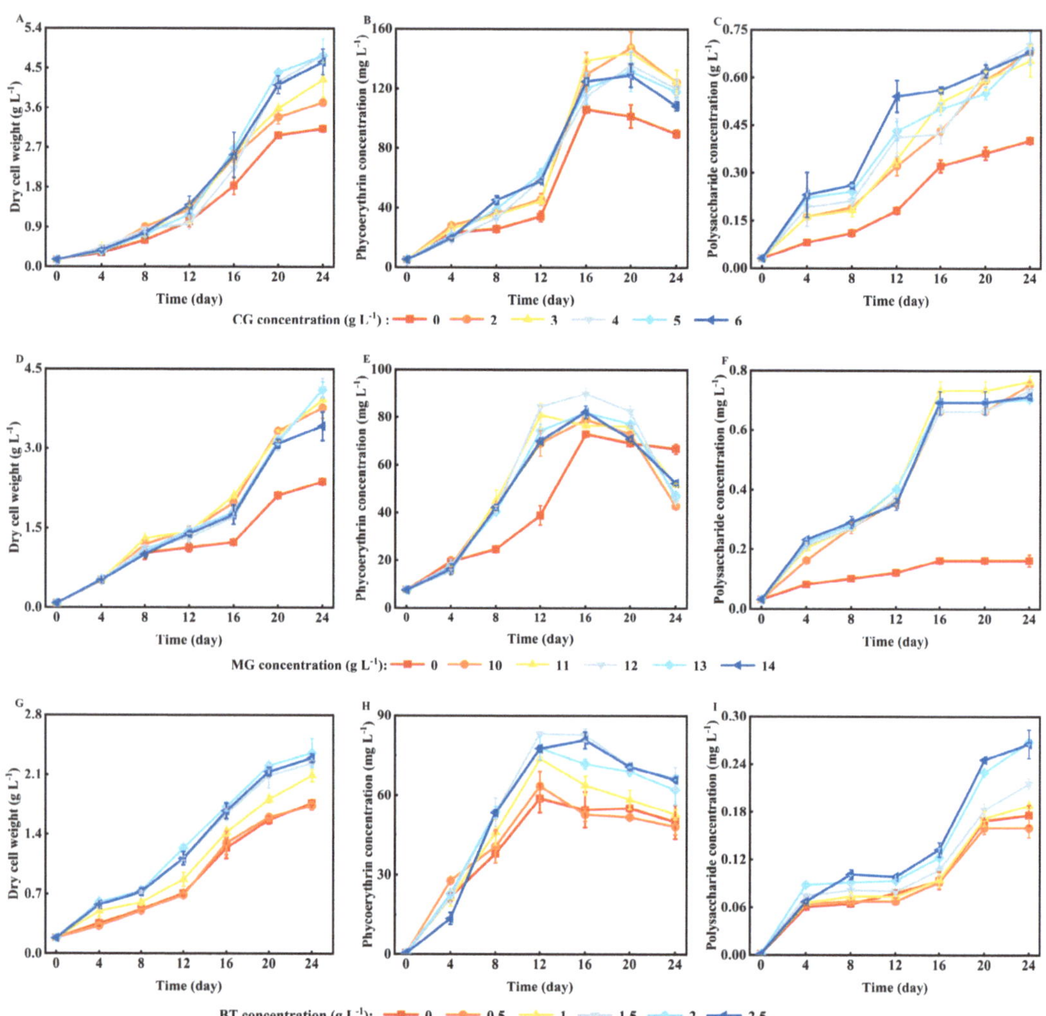

Figure 1. Effects of CG (**A–C**), MG (**D–F**), and BT (**G–I**) on the production of biomass, phycoerythrin and polysaccharides by *P. purpureum*. The data represent the average ± standard deviation ($n = 3$). CG, calcium gluconate; MG, magnesium gluconate; BT, polypeptides.

2.3. Combined Effects of Variables on Biomass, Phycoerythrin, and Polysaccharide Concentrations

The biomass, phycoerythrin, and polysaccharide concentrations under various experimental conditions are shown in Table 2, and the relationships between the three variables and the responses are depicted in 3D response surfaces and contour plots in Figures 2–4. The biomass concentration increased initially and then decreased with increasing BT concentration when the concentrations of CG and MG were fixed at 6 and 16 g L^{-1}, respectively (Figure 2B,C). Higher biomasses were observed at lower MG concentrations when CG and BT were fixed at 6 CG and 0.5 g L^{-1}, respectively (Figure 2A,C). In addition, the ANOVA results of the model indicated that single factors (A, C), interaction terms (AB, AC, BC), and quadratic terms (A^2, B^2) had non-significant effects on biomass concentration, with p-values of these factors all exceeding 0.05. The microalgal biomass was more sensitive to MG than to CG and BT (Table S1). In general, a lower MG concentration (8 g L^{-1}) contributed to increased biomass production.

Table 1. Analysis of variance for the response-surface models.

Source	Modified Equations with Significant Terms	p-Value	R^2	Adj.R^2	SD	Lack of Fit	C.V.%
Biomass con-centration	$5.21 + 0.28A - 0.705B + 0.0925C - 0.1475AB + 0.4875AC - 0.0225BC - 0.1187A^2 - 0.4137B^2 - 1.01C^2$	0.0123	0.8893	0.7469	0.4449	0.3384	9.93
Phycoerythrin concentra-tion	$74.11 - 1.6A - 0.85B - 8.84C - 5.89AB - 1.42AC + 0.165BC + 1.99A^2 - 4.96B^2 - 24.28C^2$	0.0054	0.9145	0.8046	6.78	0.6649	11.07
Polysaccharide concentra-tion	$1.02 + 0.1588A + 0.1338B + 0.0525C + 0.0225AB + 0.11AC + 0.125BC - 0.0647A^2 - 0.0748B^2 - 0.1472C^2$	0.0054	0.9142	0.8039	0.0915	2.45	10.32

A, calcium gluconate; B, magnesium gluconate; C, polypeptides; R^2, coefficient of determination; Adj.R^2, adjusted R^2; SD, standard deviation; CV, coefficient of variation.

Table 2. Experimental data and predicted values based on established models of biomass and phycoerythrin and polysaccharide concentrations.

		Variables						Responses		
		CG		MG		BT		Biomass	Phycoerythrin Concentration	Polysaccharide Concentration
Std	Run	(g L^{-1})		(g L^{-1})		(g L^{-1})		(g L^{-1})	(mg L^{-1})	(g L^{-1})
		Coded	Actural	Coded	Actural	Coded	Actural	Actural	Actural	Actural
1	5	−1	2	−1	8	0	2	5.77	84.23	1
2	13	1	6	−1	8	0	2	4.87	70.01	1.13
3	16	−1	2	1	16	0	2	4.02	56.78	0.76
4	8	1	6	1	16	0	2	4.36	36.91	0.64
5	6	−1	2	0	12	−1	0.5	5.28	69.73	0.52
6	17	1	6	0	12	−1	0.5	4.22	56.78	0.78
7	9	−1	2	0	12	1	3.5	4.91	44.05	1.06
8	7	1	6	0	12	1	3.5	3.78	60.77	1.29
9	3	0	4	−1	8	−1	0.5	3.32	43.79	0.64
10	14	0	4	1	16	−1	0.5	3.5	33.3	1.09
11	4	0	4	−1	8	1	3.5	5.31	65.09	1.03
12	10	0	4	1	16	1	3.5	5.33	72.4	1.02
13	2	0	4	0	12	0	2	6.08	72.02	0.91
14	15	0	4	0	12	0	2	3.25	52.51	0.71
15	12	0	4	0	12	0	2	4.77	78.83	0.93
16	11	0	4	0	12	0	2	3.57	82.06	0.81
17	1	0	4	0	12	0	2	3.86	62.7	0.76

CG, calcium gluconate; MG, magnesium gluconate; BT, polypeptide.

The relationships between the phycoerythrin concentration and the three independent variables were analyzed. The concentration of phycoerythrin declined with increasing BT concentration when the concentrations of CG and MG were fixed at 6 and 16 g L^{-1}, respectively (Figure 3B,C). Meanwhile, the concentration of phycoerythrin first increased and then decreased with increasing MG concentration, but this response was dependent on CG concentration (Figure 3A). However, the ANOVA results for the model indicated that the single factors (A, B), interaction terms (AB, AC, BC), and quadratic terms (A^2, B^2) had non-significant effects on phycoerythrin production ($p > 0.05$). Compared to CG and MG, BT had a stronger influence on phycoerythrin production (Table S2). Therefore, the BT concentration range 1.5–2.5 g L^{-1} was identified as optimal for phycoerythrin accumulation.

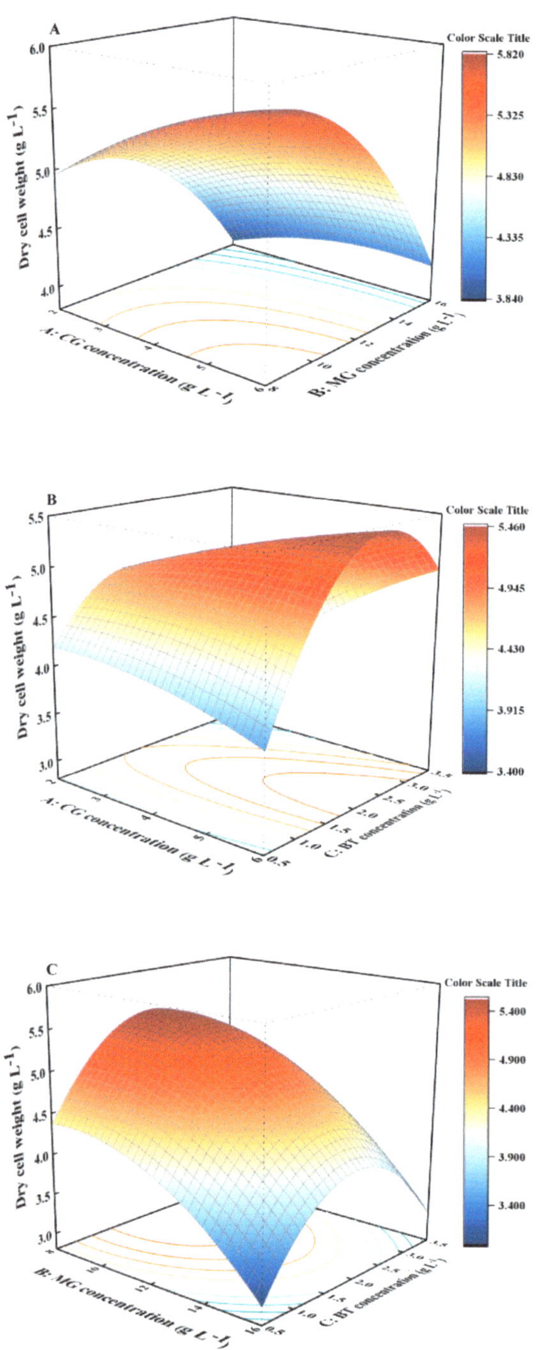

Figure 2. Response-surface plots showing the combined effect of the CG, MG, and BT concentration for responses in terms of biomass concentration. (**A**): interaction between CG and MG; (**B**): interaction between CG and BT; (**C**): interaction between MG and BT; CG, calcium gluconate; MG, magnesium gluconate; BT, polypeptide.

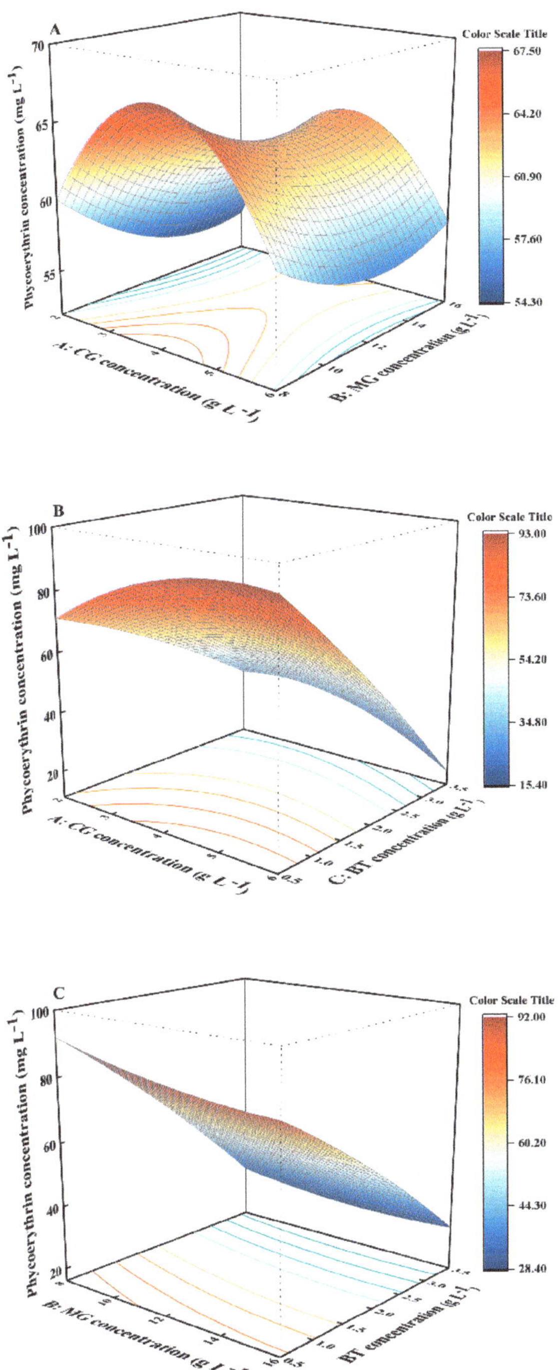

Figure 3. Response-surface plots showing the combined effect of the CG, MG, and BT concentration for responses in terms of phycoerythrin concentration. (**A**): interaction between CG and MG; (**B**): interaction between CG and BT; (**C**): interaction between MG and BT; CG, calcium gluconate; MG, magnesium gluconate; BT, polypeptide.

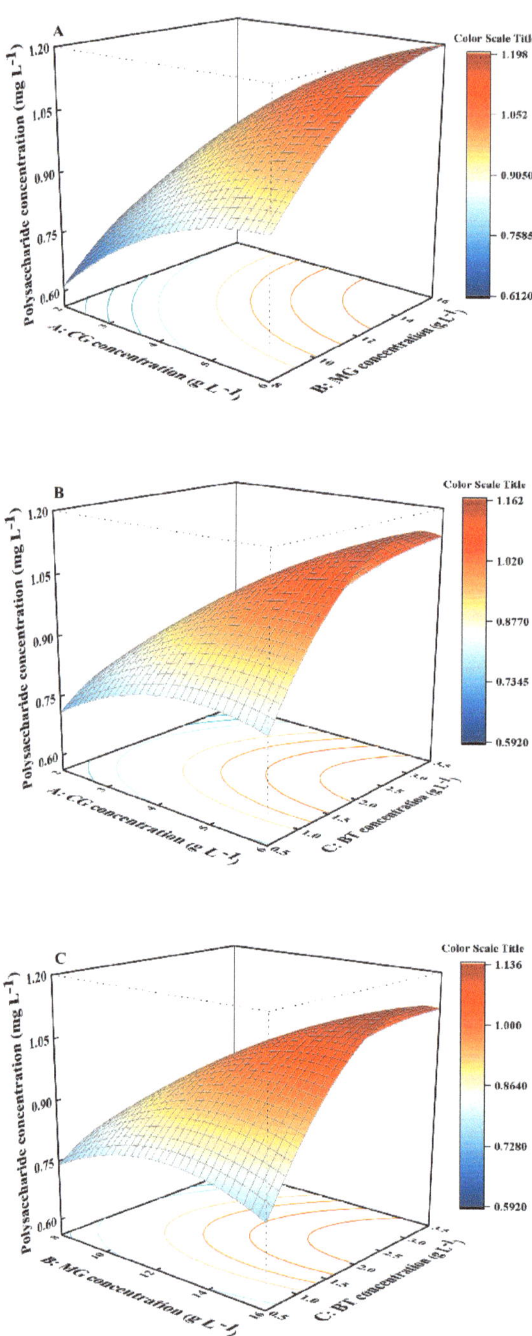

Figure 4. Response-surface plots showing the combined effect of the CG, MG, and BT concentration for responses of polysaccharide concentration. (**A**): interaction between CG and MG; (**B**): interaction between CG and BT; (**C**): interaction between MG and BT; CG, calcium gluconate; MG, magnesium gluconate; BT, polypeptide.

In terms of polysaccharide concentrations, an initially increasing and then decreasing trend was observed as the BT concentration increased from 1.0 to 3.5 g L^{-1} when CG and MG concentrations were fixed at 6 and 16 g L^{-1}, respectively (Figure 4B,C). The highest polysaccharide concentration occurred over the range of BT concentrations from 2.0 to 3.5 g L^{-1}. However, polysaccharide concentration was less sensitive to BT ($p > 0.05$) than to CG and MG ($p < 0.05$). In addition, the ANOVA results of the model revealed that single factors (A, B), interaction terms (AB, AC), and quadratic terms (C^2) had significant effects on polysaccharide concentration ($p < 0.05$) (Table S3). A strong interactive effect was observed between the concentrations of CG and BT, as well as between the concentrations of MG and BT, with significant p-values of 0.0472 and 0.0293, respectively. Overall, when the concentrations of CG and MG were held constant, relatively high BT concentrations (2.0–3.5 g L^{-1}) were found to be more conducive to polysaccharide accumulation.

2.4. Optimization and Experimental Validation

A comparison between the actual experimental data and the predicted data from the quadratic regression model is illustrated in Figure 5. The strong linear relationship between the two data sets indicated that the predictions aligned well with the experimental results, suggesting that the model is suitable for prediction and optimization. According to the model prediction, the maximum concentrations of biomass, phycoerythrin, and polysaccharides can reach as high as 5.90 g L^{-1}, 98.17 mg L^{-1}, and 1.32 g L^{-1}, respectively, when the concentrations of CG, MG, and BT are at optimal levels (Figure 6). Indeed, these levels were achieved in the verification experiments, where the measured biomass, phycoerythrin, and polysaccharide concentrations all slightly exceeded their respective predicted concentrations, at 5.97 g L^{-1}, 102.95 mg L^{-1}, and 1.42 g L^{-1}, respectively.

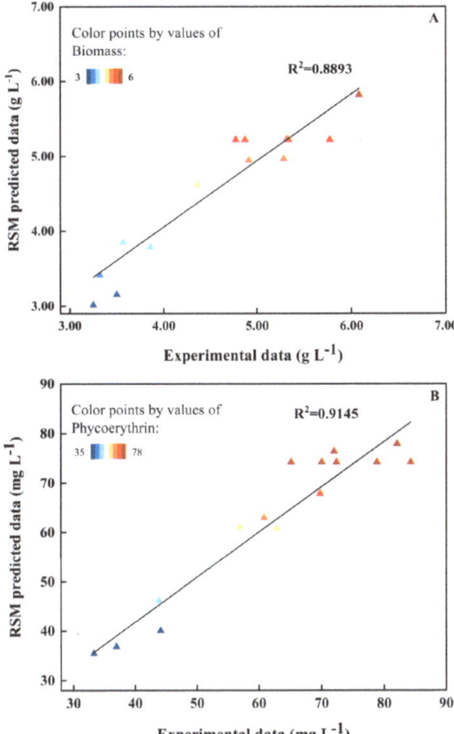

Figure 5. *Cont.*

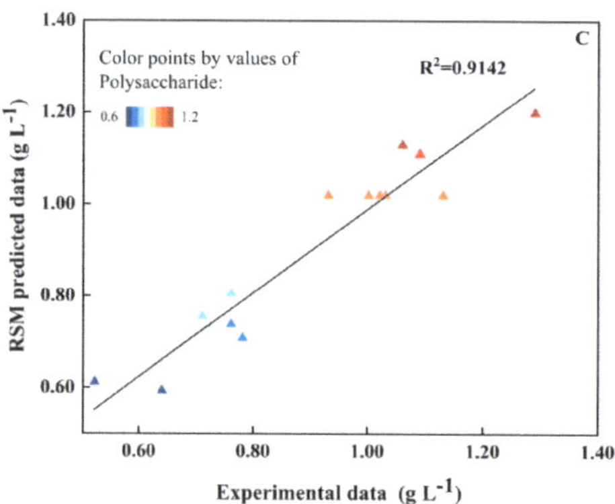

Figure 5. Comparison between experimental and predicted values of (**A**) biomass concentration, (**B**) phycoerythrin concentration, and (**C**) polysaccharide concentration.

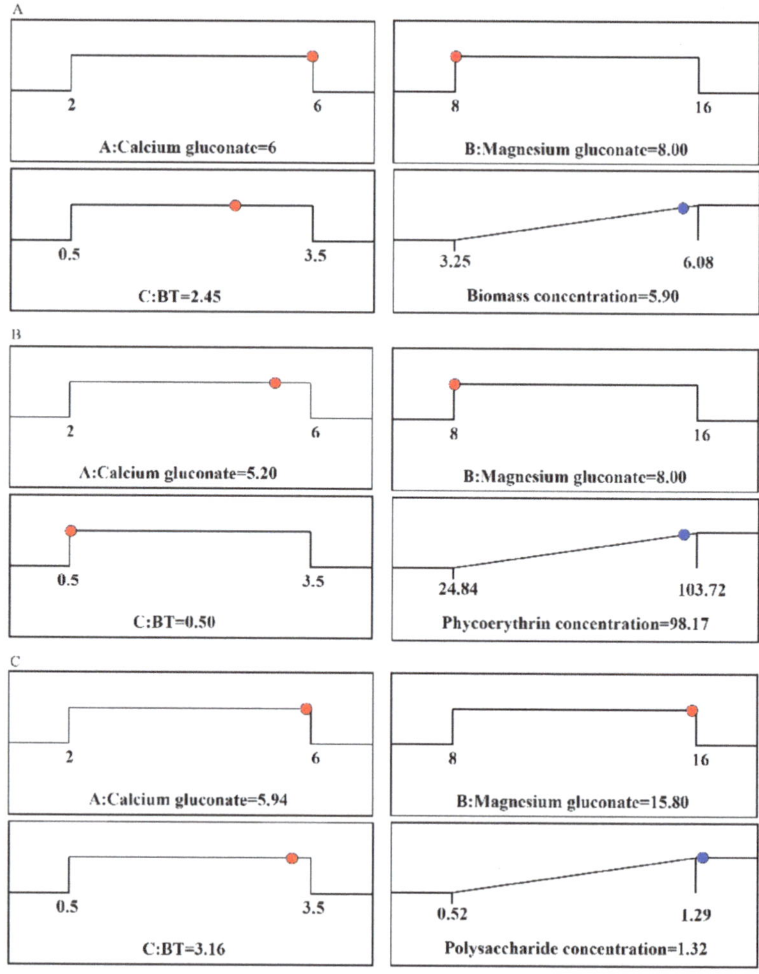

Figure 6. Optimal conditions predicted by the models for (**A**) biomass concentration, (**B**) phycoerythrin concentration, and (**C**) polysaccharide concentration. Red dot: the optimal addition amount predicted by the model; Blue dot: the optimal response concentration predicted by the model.

3. Discussion

In this study, the growth and accumulation of bioactive substances in *P. purpureum* were evaluated under supplementation with three exogenous substances (CG, MG, and BT). CG and MG dissociate into gluconic acid and cations in the medium solutions, so they can be considered a combination of a carbon resource (glucose) and an ion (calcium or magnesium) under appropriate conditions. The experimental results showed that the biomass and concentrations of phycoerythrin and polysaccharides increased with the addition of CG or MG compared to the control. This effect was attributed to the presence of gluconate, which acted as an organic carbon source. A similar trend was observed in *H. pluvialis*, where the addition of gluconate increased the biomass productivity and photosynthetic efficiency [32]. Furthermore, as for gluconate, the concentrations of calcium and magnesium can also affect biomass and the accumulation of bioactive substances. A study of *Chlorella vulgaris* and *Scenedesmus obliquus* found that an increasing magnesium concentration positively affected the biomass and lipid content [26]. In their study, compared to the control group, the biomass yield of *C. vulgaris* and *S. obliquus* increased by 33% and 36%, respectively, at 150 mg L^{-1} MgSO$_4$. Furthermore, the lipid content increased to a maximum of 27% and 26% of dry cell weight in *C. vulgaris* and *S. obliquus*, respectively, at 100 mg L^{-1} MgSO$_4$. However, the increased calcium concentrations had a little impact on the growth of the microalgae. Interestingly, the lipid content of *C. vulgaris* and *S. obliquus* peaked at 331 mg L^{-1} and 224 mg L^{-1}, respectively, under calcium-starved conditions. Therefore, the combination of gluconate and metal ions can promote microalgal growth and the production of bioactive substances.

BT can be considered a nitrogen source and is primarily composed of polypeptides. Nitrogen is an essential nutrient for cell growth and is used to synthesize photosynthetic pigments, amino acids, coenzymes, and other compounds [24]. Therefore, the concentration of nitrogen can influence microalgal growth and the accumulation of bioactive substances. Over a gradient of NaNO$_3$ in nitrogen-free Zarrouk medium, the highest cell density of *Arthrospira platensis* was observed at a concentration of 40 mM NaNO$_3$, in the middle of the range [34]. Furthermore, the amounts of proteins and pigments in *A. platensis* decreased when the alga was cultured under conditions of nitrogen limitation [35]. In addition to nitrogen concentration, the type of nitrogen source can significantly affect the growth of microalgae and the accumulation of bioactive substances [36,37]. In the present study, as the BT concentration increased, the polysaccharides concentration gradually increased due to the increasing biomass concentration, but the change had little effect on the polysaccharide content (Table S4). However, after the initial increase, the biomass and phycoerythrin concentrations decreased as the BT concentration increased further, likely due to the high nitrogen concentration. Similar patterns were observed in *Neochloris oleoabundans*, whose growth was not enhanced at higher nitrate concentrations (15 and 20 mM) [38].

To date, most studies on *P. purpureum* have focused on the effects of individual exogenous substances and single or interactive environmental conditions on the production of biomass and high-value compounds [3,20,29]. To the best of our knowledge, this is the first study to investigate the interactive effects of exogenous substances on the growth of and accumulation of high-value compounds in *P. purpureum*. We evaluated the combined effects of exogenous substances on biomass, phycoerythrin production, and polysaccharide production in *P. purpureum* using the response surface methodology, applying a second-degree polynomial (i.e., quadratic model). This approach allowed us to determine the optimal supplement levels for maximizing the concentrations of biomass, phycoerythrin, and polysaccharides. MG was the only factor that exhibited a significant effect on biomass yield ($p < 0.01$). The MG concentration was negatively correlated with the biomass of *P. purpureum*, with the highest biomass occurring at a low MG concentration (8 g L^{-1}). This result agreed with a previous report that the biomass of *P. purpureum* was 1.82 times higher at a glucose concentration of 5 g L^{-1} compared to a concentration of 10 g L^{-1} [22]. It was also reported that the biomass yields of *C. vulgaris* and *S. obliquus* peaked when the magnesium concentration increased to double that of the control [26]. BT was the only factor that had a significant effect on phycoerythrin production ($p < 0.01$). Similarly, the highest phycoerythrin concentration was achieved at an extremely low concentration of BT (0.5 g L^{-1}). As a nitrogen source, BT enhanced the growth of microalgae and the accumulation of bioactive substances. In addition to influencing the growth of *P. purpureum*, the nitrogen source can also affect the synthesis of phycoerythrin. Nitrogen deficiency has been previously observed to decrease the content and stability of phycobilisomes associated with photosynthetic activity in *P. purpureum* [39]. It has also been reported that the maximum phycoerythrin concentration in *P. purpureum* UTEX LB 2757 occurred at a low nitrogen concentration of 0.075 g L^{-1} [17]. In this study, the production of polysaccharides was significantly affected by the interactions between CG and BT and between MG and BT ($p < 0.05$) but not by changes in BT alone ($p > 0.05$). The method of culture has also been observed to affect production, with increased biomass and production of bioactive substances by microalgae in mixotrophic culture compared to photoautotrophic and heterotrophic culture [40]. Moreover, there was an observed increase in oxidative phosphorylation and a weakening of photosynthesis in microalgal cells when an organic carbon source was added. Furthermore, the enhanced phosphorylation not only compensated for the loss of photosynthesis, but also substantially increased the biomass [41]. Compared to the microalgae in photoautotrophic culture, *Chlorella sorokiniana* showed increased biomass and production of bioactive substances due to changes in the metabolic genes involved, which were more closely related to carbon flux than to photosynthesis [42]. Therefore, it was speculated that the decreases in phycoerythrin and increases in polysaccharides induced by the organic carbon source were due to gene regulation in the related metabolic pathways.

In order to optimize the *P. purpureum* growth medium, we established predictor models for algal biomass, phycoerythrin, and polysaccharide concentrations using the response surface method. The maximum biomass (5.97 g L^{-1}), phycoerythrin (102.95 mg L^{-1}), and polysaccharide concentrations (1.42 g L^{-1}) were successfully achieved using the ideal conditions predicted by this model. These values were 132.3%, 27.97%, and 140.33% higher than those achieved in the initial ASW medium, respectively.

To date, various studies have investigated approaches for improving biomass, phycoerythrin production, and polysaccharide production in *P. purpureum*; a summary of their results is provided in Table 3. As can be seen, there have been fewer studies on phycoerythrin production. The resulting polysaccharide concentrations varied greatly, ranging from 0.23 to 4.62 g L^{-1}, with the majority concentrated in the range from 0.23 to 2.14 g L^{-1}. The significant variation in polysaccharide concentration was likely caused by the diversity of culture conditions among studies, which differed in terms of algal strains, culture medium, light intensity, and other factors. As shown in Table 3, a glass flask and a bioreactor were the primary devices used for culturing *P. purpureum*. It is worth noting that higher polysaccharide production (>2 g L^{-1}) was achieved in the small photobioreactors (<5 L) (Table 3) [21,43–45]. These high polysaccharide concentrations were attributed to differences in lighting conditions and were attained in photobioreactors rather than flasks. The polysaccharide concentrations in algae cultured in glass flasks (<1 g L^{-1}) or relatively large-scale photobioreactors (<1.4 g L^{-1}) were lower than that in our study (1.42 g L^{-1}).

It should be noted that the addition of CG, MG, and BT will increase production costs. Therefore, a brief economic analysis was conducted as follows. Under optimal conditions, the phycoerythrin concentration reached the maximum levels, with a 22.5 mg L^{-1} increase in phycoerythrin yield compared to the control. The extra cost of the supplemental substances required for 1 L optimized culture medium was calculated based on the commercial prices of the substances (CG, MG, and BT) and came to ~USD 0.4 L^{-1}. Considering the increased production of phycoerythrin (an additional 0.0225 g L^{-1}) and the market price of purified phycoerythrin (USD 200 g^{-1}), the extra production was valued at approximately USD 4.5 L^{-1}, which far exceeded the additional input cost (USD 0.4). Therefore, this approach to maximizing phycoerythrin production would be cost-effective and economically feasible.

Table 3. Summary of algal growth, phycoerythrin production and polysaccharide production in *Porphyridium* sp. reported in the literature and in this study.

Number	Medium	Special Culture Conditions	Culturing Scale	Biomass Concentration or Cell Number	Polysaccharide Concentration (g L^{-1})	PB and PE Concentrations (mg L^{-1})	Refs
1	F/2	N: P ratio	250 mL flask	5.94×10^9 cell L^{-1}	0.23	NA	[46]
2	F/2-RSE	Light, temperature, and nitrogen	250 mL glass flask	3.4 g L^{-1}	0.92	PB: 47.20 PE: 38.80	[29]
3	OMII	Consumption of N and P	30 L flat-plate photobioreactor	1.71×10^{10} cell L^{-1}	0.73	NA	[47]
4	OMI	Light regime	15 L plate photobioreactor	1.38×10^{10} cell L^{-1}	0.95	NA	[48]
5	ASW	Outdoor mass culture	72 L flat plate glass reactor	1.37×10^{11} cell L^{-1}	1.32	NA	[49]
6	ASW	Different nitrogen concentrations	6×60 cm photobioreactor	5.53 g L^{-1}	2.14	PB: 1010	[43]
7	Pm	Light, temperature and nitrogen	5 L photobioreactor	6.12×10^{10} cell L^{-1}	4.10	NA	[44]
8	ASW	Optimization of light and sodium bicarbonate	3 L batch culture of photobioreactor	15.2 g L^{-1}	4.5	PB 12.17 g/100 g	[45]
9	ASW	light intensities	BioIII fermenter	4.44×10^9 cell L^{-1}	4.63	NA	[21]
10	ASW	Addition of CG, MG, and BT	500 mL flask	5.97 g L^{-1}	1.42	102.95	

Overall, high productivity was achieved for both phycoerythrin and polysaccharide by adding CG, MG, and BT in quantities based on RSM to optimize the culture medium. However, the cultures in this study were limited to 500 mL flasks, so further testing would be required in scaled-up photobioreactors to confirm that these results are scalable. Therefore, the next step may involve using a larger-scale photobioreactor, investigating the influence of light intensity, and combining the photobioreactor with optimized light intensity to further maximize the production of phycoerythrin and polysaccharides.

4. Materials and Methods

4.1. Microalgal Strain

The marine microalgal strain *Porphyridium purpureum (Bory) K.M.Drew & R.Ross 1965* was obtained from the Freshwater Algae Culture Center at the Institute of Aquatic Biology (Wuhan, China) and was maintained in ASW medium [21] at 25 °C and a light intensity of 100 µmol m^{-2} s^{-1}.

4.2. Experimental Design

4.2.1. Experiments to Determine the Optimal Concentrations of Exogenous Substances

To investigate the effects of exogenous substances added to the initial culture media on biomass and the accumulation of bioactive substances, we conducted experiments with CG, MG, and BT. CG and MG were purchased from Shanghai Aladdin Biochemical Technology Company (Shanghai, China). and BT was purchased from Jiangsu Rishengchang Biotechnology Company (Nanjing, China). BT is a mixture of amino acids, polypeptides, and proteins; the detailed composition of BT has been reported previously [30].

For single-factor experiments, the six CG concentrations (0, 2, 3, 4, 5, and 6 g L^{-1}), six MG concentrations (0, 10, 11, 12, 13, and 14 g L^{-1}), and six BT concentrations (0, 0.5, 1, 1.5, 2, and 2.5 g L^{-1}) were used. For the experiments, *P. purpureum* in the logarithmic growth phase were inoculated into 500 mL Erlenmeyer flasks containing 200 mL ASW medium. All culture media were pre-sterilized by autoclaving at 121 °C, 1 bar, for 20 min. The inoculation amount was 10% of the liquid volume load. After inoculation, the flasks were placed on a shaker at a speed of 170 rpm. *P. purpureum* was cultivated at 25 °C with continuous light (100 μmol m^{-2} s^{-1}) for 24 days. The continuous light was provided by white LED lights (Philips Lighting, Shanghai, China), and the light intensity was measured by illuminometer. The biomass and the phycoerythrin and polysaccharide concentrations were measured every four days. Three parallel experimental replicates were established for all experimental treatment groups.

4.2.2. Response Surface Experiments for Three Exogenous Substances

Based on the results of the single-factor experiments, a three-factor, three-level experiment was designed (Table 4). The single-factor experiments demonstrated that the addition of exogenous substances enhanced the biomass, phycoerythrin, and polysaccharide concentrations in *P. purpureum*. Therefore, the Box-Behnken Design (BBD) [50], a well-known statistical design for experiments, was chosen as the response surface method for optimization. The optimal concentrations of each factor, as determined by the single-factor experiments, were chosen as the central point of the BBD. The relationships between the dependent (biomass, phycoerythrin, and polysaccharide concentrations) and independent variables (A: CG, B: MG, and C: BT) were quantitatively determined. The experimental design and mathematical model were created using Design-Expert software (version 13.0.1.0), and the statistical analysis was conducted using the same platform.

Table 4. Independent variables (CG, MG and BT concentration) and the levels of each treatment used in the Box-Behnken design.

Parameters (g L^{-1})	Lable	Coded Levels and Concentrations		
		−1	0	+1
CG	A	2	4	6
MG	B	8	12	16
BT	C	0.5	2	3.5

CG, calcium gluconate; MG, magnesium gluconate; BT, polypeptides.

According to the BBD, 17 sets of experiments were conducted (Table 2). Different concentrations of CG, MG, and BT were added to the ASW medium in each experimental group. The experimental conditions were the same as those detailed previously. It is worth noting that there were two parallel experimental replicates for each treatment setup in the response-surface experiments. validation experiments were conducted using the optimal conditions predicted by the RSM model with the aim of maximizing biomass, phycocyanin, and polysaccharide concentrations.

4.3. Dry Cell Weight

Dry cell weight (DCW) was determined using the dry-weight method [51]. In brief, a dry weighing disc was first weighed, with its weight represented as M_1. Then, 5 mL of each microalgae solution (V) was harvested by centrifuging the cells at 8000 r min^{-1} for 5 min, washing them with deionized water once, centrifuging them again, then drying them in a 100 °C oven until they reached a constant weight. Then, the weighing disk was cooled and weighed, with its weight recorded as M_2. Finally, the DCW was calculated as follows:

$$DCW = \frac{M_2 - M_1}{V} \tag{1}$$

where DCW is the biomass concentration (g L^{-1}); M_2 represents the weight of dried weighing disc with the sample (g); M_1 represents the weight of pre-dried empty weighing disc (g); and V is the sampling volume (L).

4.4. Phycoerythrin Concentration

The concentration of phycoerythrin (PE) was determined spectrophotometrically. Firstly, a 5 mL sample of the *P. purpureum* culture was centrifuged at 8000 rpm for 5 min. After centrifugation, the supernatant was carefully drained off. Then, 5 mL of 0.1 mol L^{-1} phosphate buffer (pH 6.8) was added to resuspend the precipitated biomass. To break the microalgae cells and release the PE, the resuspended cells were subjected to three freeze-thaw cycles. Then, the mixture was centrifuged again at 5000 rpm for 5 min to collect the supernatant. The concentration of PE was determined by measuring the absorbance of the supernatant at 564 nm, 592 nm, and 455 nm. The concentration of phycoerythrin was determined using the following formula [52]:

$$PE = [(OD_{564} - OD_{592}) - (OD_{455} - OD_{592}) \times 0.2] \times 0.12 \tag{2}$$

where PE is the concentration of phycoerythrin in the microalgal solution (g L^{-1}) and OD_{564}, OD_{592}, and OD_{455} are the absorbances at 564, 592, and 455 nm, respectively.

4.5. Polysaccharide Concentration

A 5 mL suspension of *P. purpureum* culture was subjected to three freeze-thaw cycles. The resulting extract solution was centrifuged at 5000 rpm for 5 min. The supernatant was collected so the polysaccharide concentration could be measured. The polysaccharide concentration of *P. purpureum* was determined using the sulfuric acid-phenol method [53]. The concentration of polysaccharides (Y, g L^{-1}) was determined using a standard absorbance curve (R^2 = 0.995).

$$Y = \frac{A_{490} + 0.0173}{7.1137} \tag{3}$$

4.6. Statistical Analysis

After all experiments had been conducted, the relationship between the dependent and independent variables was explained by the second-degree polynomial as shown by the following equation:

$$y = \beta_0 + \sum_{i=1}^{n} b_i x_i + \sum_{i=2}^{n} b_{ii} x_i^2 + \sum_{j=i+1}^{n} b_{ij} x_i x_j \tag{4}$$

where y is the response; β_0 is the intercept; β_i, β_{ii}, and β_{ij} are the regression coefficients of different variables in linear and quadratic equations; n is the number of studied variables; and x_i and x_j represent independent variables.

Statistical differences were observed among the experiments, as determined by analysis of variance (ANOVA) and tests conducted in Design-Export software (version 13.0.1.0). Origin software was used for data analysis, and the results were expressed as mean $\pm$ standard deviation (mean $\pm$ SD).

5. Conclusions

In this study, the optimal amounts of CG, MG, and BT to add to the medium in order to maximize biomass and the production of phycoerythrin and polysaccharides in *P. purpureum* were determined using RSM forecasting models. According to the forecasting models, the biomass was primarily influenced by MG, while phycoerythrin concentration was mainly influenced by BT. Meanwhile, the concentration of polysaccharides was influenced by the interactive effects between CG and BT and between MG and BT. The maximum concentrations of biomass, phycoerythrin, and polysaccharides (5.97 g L^{-1}, 102.95 mg L^{-1}, and 1.42 g L^{-1}) surpassed their predicted values and were reached when the microalgae were cultured under the optimal conditions indicated by the models. Hence, CG, MG, and BT can be considered as exogenous additives to greatly promote *P. purpureum* growth and the synthesis of phycocyanin and polysaccharides. As a next step, it is important to further increase the production of algal biomass, phycoerythrin, and polysaccharides by utilizing a larger-scale photobioreactor and optimizing light intensity during culture in the optimal medium identified in the present study.

Supplementary Materials: The following supporting information can be downloaded at https://www.mdpi.com/article/10.3390/md22030138/s1, Table S1: Analysis of variance for a quadratic model of biomass concentration; Table S2: Analysis of variance for a quadratic model of phycoerythrin concentration; Table S3: Analysis of variance for a quadratic model of polysaccharide concentration; Table S4: Effects of different exogenous substances on phycoerythrin and polysaccharide contents.

Author Contributions: Conceptualization: S.Y., A.-H.Z. and J.H.; methodology: S.Y. and T.Y.; software: S.Y. and T.Y.; formal analysis: S.Y. and J.H.; investigation: S.Y. and A.-H.Z.; writing—original draft preparation: S.Y.; writing—review and editing: S.Y., J.H., X.Z., Y.H., M.P. and B.F.; supervision: J.H.; Funding acquisition: A.-H.Z. and J.H.; Visualization: A.-H.Z. and S.Y. All authors have read and agreed to the published version of the manuscript.

Funding: This work was supported by the Marine Science and Technology Innovation Project of Jiangsu Province (JSZRHYKJ202309), Jiangsu Innovation Center of Marine Bioresources (China) (No. 822153216), National Natural Science Foundation of China (No. 32273157), and Jiangsu Coastal Development Group Co., Ltd. Marine biological high value development and utilization project (No. 2023YHTZZZ02).

Data Availability Statement: All data generated or analyzed in the present study are available on reasonable request.

Conflicts of Interest: The authors declare no competing financial interests.

References

1. Kathiresan, S.; Sarada, R.; Bhattacharya, S.; Ravishankar, G.A. Culture media optimization for growth and phycoerythrin production from *Porphyridium purpureum*. *Biotechnol. Bioeng.* **2007**, *96*, 456–463. [CrossRef]
2. Juin, C.; Chérouvrier, J.-R.; Thiéry, V.; Gagez, A.-L.; Bérard, J.-B.; Joguet, N.; Kaas, R.; Cadoret, J.-P.; Picot, L. Microwave-Assisted Extraction of Phycobiliproteins from *Porphyridium purpureum*. *Appl. Biochem. Biotechnol.* **2015**, *175*, 1–15. [CrossRef]
3. Durmaz, Y.; Monteiro, M.; Bandarra, N.; Gökpinar, Ş.; Işik, O. The effect of low temperature on fatty acid composition and tocopherols of the red microalga, *Porphyridium cruentum*. *J. Appl. Phycol.* **2007**, *19*, 223–227. [CrossRef]
4. Lu, Q.; Li, H.; Xiao, Y.; Liu, H. A state-of-the-art review on the synthetic mechanisms, production technologies, and practical application of polyunsaturated fatty acids from microalgae. *Algal Res.* **2021**, *55*, 102281. [CrossRef]
5. Arad, S.; Levy-Ontman, O. Red microalgal cell-wall polysaccharides: Biotechnological aspects. *Curr. Opin. Biotechnol.* **2010**, *21*, 358–364. [CrossRef] [PubMed]
6. Lawrenz, E.; Fedewa, E.J.; Richardson, T.L. Extraction protocols for the quantification of phycobilins in aqueous phytoplankton extracts. *J. Appl. Phycol.* **2011**, *23*, 865–871. [CrossRef]
7. Borovkov, A.B.; Gudvilovich, I.N.; Lelekov, A.S.; Avsiyan, A.L. Effect of specific irradiance on productivity and pigment and protein production of *Porphyridium purpureum* (Rhodophyta) semi-continuous culture. *Bioresour. Technol.* **2023**, *374*, 128771. [CrossRef]
8. Yabuta, Y.; Fujimura, H.; Kwak, C.S.; Enomoto, T.; Watanabe, F. Antioxidant Activity of the Phycoerythrobilin Compound Formed from a Dried Korean Purple Laver (*Porphyra* sp.) during in vitro Digestion. *Food Sci. Technol. Res.* **2010**, *16*, 347–352. [CrossRef]
9. Wang, C.; Shen, Z.; Li, L.; Li, Y.; Zhao, H.; Jiang, X. Immunomodulatory activity of R-phycoerythrin from *Porphyra haitanensis* via TLR4/NF-κB-dependent immunocyte differentiation. *Food Funct.* **2020**, *11*, 2173–2185. [CrossRef]
10. Cai, C.; Wang, Y.; Li, C.; Guo, Z.; Jia, R.; Wu, W.; Hu, Y.; He, P. Purification and photodynamic bioactivity of phycoerythrin and phycocyanin from *Porphyra yezoensis* Ueda. *J. Ocean Univ. China* **2014**, *13*, 479–484. [CrossRef]
11. Simovic, A.; Combet, S.; Cirkovic Velickovic, T.; Nikolic, M.; Minic, S. Probing the stability of the food colourant R-phycoerythrin from dried Nori flakes. *Food Chem.* **2022**, *374*, 131780. [CrossRef]
12. Serive, B.; Kaas, R.; Bérard, J.-B.; Pasquet, V.; Picot, L.; Cadoret, J.-P. Selection and optimisation of a method for efficient metabolites extraction from microalgae. *Bioresour. Technol.* **2012**, *124*, 311–320. [CrossRef]
13. Patel, A.K.; Laroche, C.; Marcati, A.; Ursu, A.V.; Jubeau, S.; Marchal, L.; Petit, E.; Djelveh, G.; Michaud, P. Separation and fractionation of exopolysaccharides from *Porphyridium cruentum*. *Bioresour. Technol.* **2013**, *145*, 345–350. [CrossRef]
14. Morais, M.G.; Santos, T.D.; Moraes, L.; Vaz, B.S.; Morais, E.G.; Costa, J.A.V. Exopolysaccharides from microalgae: Production in a biorefinery framework and potential applications. *Bioresour. Technol. Rep.* **2022**, *18*, 101006. [CrossRef]
15. Medina-Cabrera, E.V.; Rühmann, B.; Schmid, J.; Sieber, V. Characterization and comparison of *Porphyridium sordidum* and *Porphyridium purpureum* concerning growth characteristics and polysaccharide production. *Algal Res.* **2020**, *49*, 101931. [CrossRef]
16. Wijesekara, I.; Pangestuti, R.; Kim, S.-K. Biological activities and potential health benefits of sulfated polysaccharides derived from marine algae. *Carbohydr. Polym.* **2011**, *84*, 14–21. [CrossRef]

17. Sosa-Hernández, J.E.; Rodas-Zuluaga, L.I.; Castillo-Zacarías, C.; Rostro-Alanís, M.; de la Cruz, R.; Carrillo-Nieves, D.; Salinas-Salazar, C.; Fuentes Grunewald, C.; Llewellyn, C.A.; Olguín, E.J.; et al. Light Intensity and Nitrogen Concentration Impact on the Biomass and Phycoerythrin Production by *Porphyridium purpureum*. *Mar. Drugs* **2019**, *17*, 460. [CrossRef] [PubMed]

18. García, A.B.; Longo, E.; Murillo, M.C.; Bermejo, R. Using a B-Phycoerythrin Extract as a Natural Colorant: Application in Milk-Based Products. *Molecules* **2021**, *26*, 297. [CrossRef] [PubMed]

19. Hsieh-Lo, M.; Castillo, G.; Ochoa-Becerra, M.A.; Mojica, L. Phycocyanin and phycoerythrin: Strategies to improve production yield and chemical stability. *Algal Res.* **2019**, *42*, 101600. [CrossRef]

20. Jiao, K.; Chang, J.; Zeng, X.; Ng, I.S.; Xiao, Z.; Sun, Y.; Tang, X.; Lin, L. 5-Aminolevulinic acid promotes arachidonic acid biosynthesis in the red microalga *Porphyridium purpureum*. *Biotechnol. Biofuels* **2017**, *10*, 168. [CrossRef]

21. You, T.; Barnett, S.M. Effect of light quality on production of extracellular polysaccharides and growth rate of *Porphyridium cruentum*. *Biochem. Eng. J.* **2004**, *19*, 251–258. [CrossRef]

22. Jiao, K.; Xiao, W.; Xu, Y.; Zeng, X.; Ho, S.-H.; Laws, E.A.; Lu, Y.; Ling, X.; Shi, T.; Sun, Y.; et al. Using a trait-based approach to optimize mixotrophic growth of the red microalga *Porphyridium purpureum* towards fatty acid production. *Biotechnol. Biofuels* **2018**, *11*, 273. [CrossRef]

23. Li, S.; Ji, L.; Shi, Q.; Wu, H.; Fan, J. Advances in the production of bioactive substances from marine unicellular microalgae *Porphyridium* spp. *Bioresour. Technol.* **2019**, *292*, 122048. [CrossRef]

24. Sánchez-Saavedra, M.d.P.; Castro-Ochoa, F.Y.; Nava-Ruiz, V.M.; Ruiz-Güereca, D.A.; Villagómez-Aranda, A.L.; Siqueiros-Vargas, F.; Molina-Cárdenas, C.A. Effects of nitrogen source and irradiance on *Porphyridium cruentum*. *J. Appl. Phycol.* **2018**, *30*, 783–792. [CrossRef]

25. Seemashree, M.H.; Chauhan, V.S.; Sarada, R. Phytohormone supplementation mediated enhanced biomass production, lipid accumulation, and modulation of fatty acid profile in *Porphyridium purpureum* and *Dunaliella salina* cultures. *Biocatal. Agric. Biotechnol.* **2022**, *39*, 102253. [CrossRef]

26. Gorain, P.C.; Bagchi, S.K.; Mallick, N. Effects of calcium, magnesium and sodium chloride in enhancing lipid accumulation in two green microalgae. *Environ. Technol.* **2013**, *34*, 1887–1894. [CrossRef] [PubMed]

27. Croft, M.T.; Warren, M.J.; Smith, A.G. Algae need their vitamins. *Eukaryot. Cell* **2006**, *5*, 1175–1183. [CrossRef] [PubMed]

28. Kumar, M.; Dahuja, A.; Tiwari, S.; Punia, S.; Tak, Y.; Amarowicz, R.; Bhoite, A.G.; Singh, S.; Joshi, S.; Panesar, P.S.; et al. Recent trends in extraction of plant bioactives using green technologies: A review. *Food Chem.* **2021**, *353*, 129431. [CrossRef] [PubMed]

29. Guihéneuf, F.; Stengel, D.B. Towards the biorefinery concept: Interaction of light, temperature and nitrogen for optimizing the co-production of high-value compounds in *Porphyridium purpureum*. *Algal Res.* **2015**, *10*, 152–163. [CrossRef]

30. Yao, T.; Huang, J.; Su, B.; Wei, L.; Zhang, A.; Zhang, D.-F.; Zhou, Y.; Ma, G.-s. Enhanced phycocyanin production of *Arthrospira maxima* by addition of mineral elements and polypeptides using response surface methodology. *Front. Mar. Sci.* **2022**, *9*, 1057201. [CrossRef]

31. Cañete-Rodríguez, A.M.; Santos-Dueñas, I.M.; Jiménez-Hornero, J.E.; Ehrenreich, A.; Liebl, W.; García-García, I. Gluconic acid: Properties, production methods and applications—An excellent opportunity for agro-industrial by-products and waste bio-valorization. *Process Biochem.* **2016**, *51*, 1891–1903. [CrossRef]

32. Pang, N.; Gu, X.; Fu, X.; Chen, S. Effects of gluconate on biomass improvement and light stress tolerance of *Haematococcus pluvialis* in mixotrophic culture. *Algal Res.* **2019**, *43*, 101647. [CrossRef]

33. Nemanja, T.; Bojanić, N.; Rakić, D.; Takači, A.; Zeković, Z.; Fišteš, A.; Bodroža-Solarov, M.; Pavlić, B. Defatted wheat germ as source of polyphenols—Optimization of microwave-assisted extraction by RSM and ANN approach. *Chem. Eng. Process.* **2019**, *143*, 107634.

34. Mousavi, M.; Mehrzad, J.; Najafi, M.F.; Zhiani, R.; Shamsian, S.A.A. Nitrate and ammonia: Two key nitrogen sources for biomass and phycocyanin production by *Arthrospira (Spirulina) platensis*. *J. Appl. Phycol.* **2022**, *34*, 2271–2281. [CrossRef]

35. Li, X.; Li, W.; Zhai, J.; Wei, H. Effect of nitrogen limitation on biochemical composition and photosynthetic performance for fed-batch mixotrophic cultivation of microalga *Spirulina platensis*. *Bioresour. Technol.* **2018**, *263*, 555–561. [CrossRef]

36. Arumugam, M.; Agarwal, A.; Arya, M.C.; Ahmed, Z. Influence of nitrogen sources on biomass productivity of microalgae *Scenedesmus bijugatus*. *Bioresour. Technol.* **2013**, *131*, 246–249. [CrossRef]

37. Gladfelter, M.F.; Buley, R.P.; Belfiore, A.P.; Fernandez-Figueroa, E.G.; Gerovac, B.L.; Baker, N.D.; Wilson, A.E. Dissolved nitrogen form mediates phycocyanin content in *cyanobacteria*. *Freshw. Biol.* **2022**, *67*, 954–964. [CrossRef]

38. Li, Y.; Horsman, M.; Wang, B.; Wu, N.; Lan, C.Q. Effects of nitrogen sources on cell growth and lipid accumulation of green alga *Neochloris oleoabundans*. *Appl. Microbiol. Biotechnol.* **2008**, *81*, 629–636. [CrossRef]

39. Ji, L.; Li, S.; Chen, C.; Jin, H.; Wu, H.; Fan, J. Physiological and transcriptome analysis elucidates the metabolic mechanism of versatile *Porphyridium purpureum* under nitrogen deprivation for exopolysaccharides accumulation. *Bioresour. Bioprocess.* **2021**, *8*, 73. [CrossRef]

40. Liu, Y.; Zhou, J.; Liu, D.; Zeng, Y.; Tang, S.; Han, Y.; Jiang, Y.; Cai, Z. A growth-boosting synergistic mechanism of *Chromochloris zofingiensis* under mixotrophy. *Algal Res.* **2022**, *66*, 102812. [CrossRef]

41. Grama, B.S.; Agathos, S.N.; Jeffryes, C.S. Balancing Photosynthesis and Respiration Increases Microalgal Biomass Productivity during Photoheterotrophy on Glycerol. *ACS Sustain. Chem. Eng.* **2016**, *4*, 1611–1618. [CrossRef]

42. Cecchin, M.; Benfatto, S.; Griggio, F.; Mori, A.; Cazzaniga, S.; Vitulo, N.; Delledonne, M.; Ballottari, M. Molecular basis of autotrophic vs mixotrophic growth in *Chlorella sorokiniana*. *Sci. Rep.* **2018**, *8*, 6465. [CrossRef] [PubMed]

43. Li, T.; Xu, J.; Wu, H.; Jiang, P.; Chen, Z.; Xiang, W. Growth and Biochemical Composition of *Porphyridium purpureum* SCS-02 under Different Nitrogen Concentrations. *Mar. Drugs* **2019**, *17*, 124. [CrossRef] [PubMed]

44. Soanen, N.; Da Silva, E.; Gardarin, C.; Michaud, P.; Laroche, C. Improvement of exopolysaccharide production by *Porphyridium marinum*. *Bioresour. Technol.* **2016**, *213*, 231–238. [CrossRef] [PubMed]

45. Velea, S.; Ilie, L.I. Optimization of *Porphyridium purpureum* culture growth using two variables experimental design: Light and sodium bicarbonate. *UPB Sci. Bull.* **2011**, *73*, 81–94.

46. Razaghi, A.; Godhe, A.; Albers, E. Effects of nitrogen on growth and carbohydrate formation in *Porphyridium cruentum*. *Cent. Eur. J. Biol.* **2014**, *9*, 156–162. [CrossRef]

47. Sun, L.; Wang, C.; Shi, Q.; Ma, C. Preparation of different molecular weight polysaccharides from *Porphyridium cruentum* and their antioxidant activities. *Int. J. Biol. Macromol.* **2009**, *45*, 42–47. [CrossRef]

48. Liqin, S.; Wang, C.; Lei, S. Effects of Light Regime on Extracellular Polysaccharide Production by *Porphyridium cruentum* Cultured in Flat Plate Photobioreactors. In Proceedings of the 2nd International Conference on Bioinformatics and Biomedical Engineering, Shanghai, China, 16–18 May 2008; IEEE: Piscataway, NJ, USA, 2008; pp. 1488–1491.

49. Singh, S.; Arad, S.; Richmond, A. Extracellular polysaccharide production in outdoor mass cultures of *Porphyridium* sp. in flat plate glass reactors. *J. Appl. Phycol.* **2000**, *12*, 269–275. [CrossRef]

50. Aslan, N.; Cebeci, Y. Application of Box–Behnken design and response surface methodology for modeling of some Turkish coals. *Fuel* **2007**, *86*, 90–97. [CrossRef]

51. Marjakangas, J.M.; Chen, C.-Y.; Lakaniemi, A.-M.; Puhakka, J.A.; Whang, L.-M.; Chang, J.-S. Selecting an indigenous microalgal strain for lipid production in anaerobically treated piggery wastewater. *Bioresour. Technol.* **2015**, *191*, 369–376. [CrossRef]

52. Beer, S.; Eshel, A. Determining Phycoerythrin and Phycocyanin Concentrations in Aqueous Crude Extracts of Red Algae. *Mar. Freshw. Res.* **1985**, *36*, 785–792. [CrossRef]

53. DuBois, M.; Gilles, K.A.; Hamilton, J.K.; Rebers, P.A.; Smith, F. Colorimetric Method for Determination of Sugars and Related Substances. *Anal. Chem.* **1956**, *28*, 350–356. [CrossRef]

Article

Effect of Iron Concentration on the Co-Production of Fucoxanthin and Fatty Acids in *Conticribra weissflogii*

Ke Peng [†], David Kwame Amenorfenyo [†], Xiangyu Rui, Xianghu Huang, Changling Li and Feng Li *

College of Fisheries, Guangdong Ocean University, Zhanjiang 524088, China; 2112101138@stu.gdou.edu.cn (K.P.);
davidamenorfenyo@yahoo.com (D.K.A.); 2112001098@stu.gdou.edu.cn (X.R.); huangxh@gdou.edu.cn (X.H.);
licl@gdou.edu.cn (C.L.)
* Correspondence: lifeng2318@gdou.edu.cn
[†] These authors contributed equally to this work.

Abstract: The production of fucoxanthin and fatty acids in *Conticribra weissflogii* has been examined, but there is still a lack of understanding regarding the impact of trace elements, including iron, on their co-production. To address this knowledge gap, this study investigated the effects of $FeCl_3 \cdot 6H_2O$ on the growth, fucoxanthin, and fatty acids of *C. weissflogii*. The findings revealed that the highest cell density (1.9×10^6 cells mL^{-1}), cell dry weight (0.89 ± 0.15 g L^{-1}), and total fatty acid concentration (83,318.13 μg g^{-1}) were achieved at an iron concentration of 15.75 mg L^{-1}, while the maximum carotenoid and fucoxanthin contents were obtained at an iron concentration of 3.15 mg L^{-1}. The study demonstrated that the content of the active substance in *C. weissflogii* could be increased by adjusting the iron concentration, providing new information as to the more efficient co-production of fucoxanthin and fatty acids and offering experimental support for large-scale production.

Keywords: *Conticribra weissflogii*; iron; fucoxanthin; fatty acids

Citation: Peng, K.; Amenorfenyo, D.K.; Rui, X.; Huang, X.; Li, C.; Li, F. Effect of Iron Concentration on the Co-Production of Fucoxanthin and Fatty Acids in *Conticribra weissflogii*. *Mar. Drugs* **2024**, *22*, 106. https://doi.org/10.3390/md22030106

Academic Editors: Cecilia Faraloni and Eleftherios Touloupakis

Received: 3 February 2024
Revised: 22 February 2024
Accepted: 22 February 2024
Published: 24 February 2024

1. Introduction

Microalgae, as the main primary producers in marine ecosystems [1], are diverse, widely distributed, and rich in a variety of bioactive substances [2]. These microorganisms play a vital role in the material cycle and energy flow, and have become a valuable resource for commercialization in recent years [3,4]. Their abundant unsaturated fatty acids, proteins, lipids, and carotenoids (including docosahexaenoic acid, eicosapentaenoic acid, β-carotene, astaxanthin, lutein, and fucoxanthin) [2] make them a promising source of biological resources for the food, bait, cosmetics, health care products, and pharmaceutical industries [5–7].

Fucoxanthin is a type of carotenoid, found in brown algae, diatoms, and golden algae, and its unique molecular structure makes it effective in terms of anti-obesity, tumor inhibition, anti-inflammatory, anti-diabetic, cancer prevention, and especially antioxidant effects [8–11]. It has been found that the content of fucoxanthin in microalgae is more than four times that in macroalgae [12]. Although the production of fucoxanthin and fatty acids in *C. weissflogii* has been studied, there is still a gap in our understanding of the influence of trace elements, such as iron, on their co-production.

Some diatoms are rich in eicosapentaenoic acid (EPA), docosahexaenoic acid (DHA), and other highly unsaturated fatty acids in addition to fucoxanthin. Studies have shown that EPA and DHA can improve the survival rate and growth performance of aquatic animals and provide a broad market for human biomedicine [13]. Currently, marine fish are the main source of the essential fatty acids, EPA, and DHA. However, due to rising fish prices and overfishing, diatoms are being considered a promising alternative for the large-scale production of EPA and DHA products, as they can be easily controlled in their growth environment [14].

Conticribra weissflogii is a typical diatom that is easy to culture, has a fast growth rate, and is considered a potential cell factory for the production of fucoxanthin and unsaturated fatty acids [15,16]. Iron is an essential trace element for microalga growth, enzymatic reactions, nitrogen metabolism, and chlorophyll synthesis [17], and iron deficiency or excess inhibits the productivity of microalga cells [18,19]. In addition, the concentration of iron can fluctuate significantly [20], potentially affecting the growth rate of microalgae and the synthesis of secondary metabolites, including fucoxanthin and fatty acids. In controlled cultivation, the optimization of iron levels can be a crucial factor for enhancing productivity and cost-efficiency. However, the specific effects of varying iron concentrations on fucoxanthin and fatty acid co-production in *C. weissflogii* remain poorly understood. This lack of knowledge constrains the ability to effectively harness the full potential of this microalga for the commercial production of these valuable compounds.

The present study aimed to address this gap by systematically investigating the effect of different iron concentrations on the growth dynamics of *C. weissflogii* and the concurrent production of fucoxanthin and fatty acids. Understanding this relationship can enhance our understanding of the metabolic flexibility of this microalga and inform optimization strategies for industrial cultivation processes, ultimately contributing to the sustainable production of high-demand bioactive compounds.

2. Results

2.1. Growth of C. weissflogii under Different Iron Concentrations

As shown in Figure 1a, varied iron concentrations demonstrated different effects on *C. weissflogii*. Although each group exhibited a similar growth pattern between day 2 and day 10, those with 15.75 mg L^{-1} and 31.5 mg L^{-1} iron concentrations showed remarkable growth on day 8 compared with the other three groups, which continued to grow until they reached their maximum growth on day 10. A 15.75 mg L^{-1} iron concentration showed a maximum growth of 1.9×10^6 cells mL^{-1} on day 10 compared with other treatments. However, there was no significant difference ($p < 0.05$) between the other iron treatment groups, except for that with a 0 mg L^{-1} iron concentration treatment.

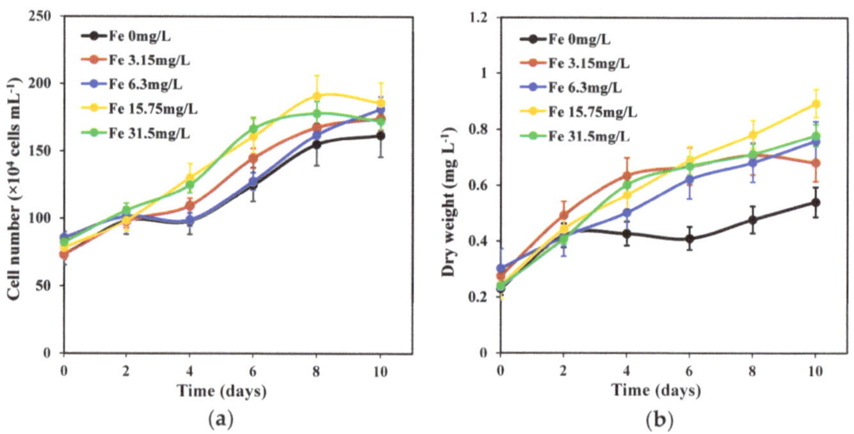

Figure 1. Changes in cell number (**a**) and dry weight (**b**) of *C. weissflogii* cultures (mean ± SD, *n* = 3).

The maximum biomass concentrations (dry weight) of the five iron concentration groups were obtained on the 10th day of continuous growth (Figure 1b). Specifically, treatment groups with a 15.75 mg L^{-1} iron concentration obtained the highest biomass concentration compared with other iron concentration treatment groups, and this was significantly higher than that of the 0 mg L^{-1} iron concentration group. No significant differences were observed between biomass concentrations of 15.75 mg L^{-1} iron, 3.15 mg L^{-1}, 6.3 mg L^{-1}, and 31.5 mg L^{-1} iron concentrations.

2.2. Changes in the Pigment Content of C. weissflogii under Different Iron Concentrations

Iron concentration had a significant effect on carotenoid accumulation in *C. weissflosii* ($p < 0.05$). The results (Figure 2a) showed that of the carotenoid accumulation of the 0.00 mg L^{-1} treatment group was significantly lower than that of the other treatment groups. The 3.15 mg L^{-1} iron concentration groups obtained the maximum carotenoid content on day 8, which was significantly higher than that of the other groups ($p < 0.05$).

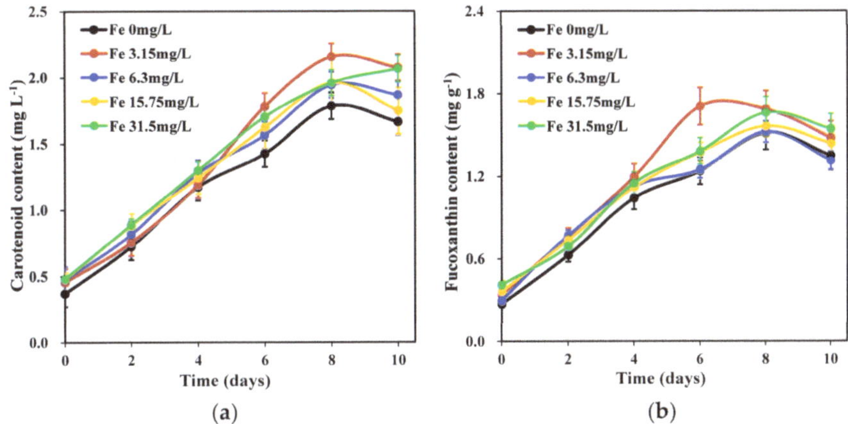

Figure 2. Changes in carotenoid content (**a**) and fucoxanthin content (**b**) of *C. weissflogii* cultures (mean ± SD, *n* = 3).

As shown in Figure 2b, the fucoxanthin concentration of each iron concentration group first increased (from day 2 to day 8) and then decreased (after day 8). The 3.15 mg L^{-1} iron concentration resulted in the maximum fucoxanthin concentration on the 6th day, which was the highest among all treatment groups. However, no significant differences were observed between the experimental groups ($p > 0.05$).

2.3. Effects of Different Iron Concentrations on C. weissflosi Biomass and Fucoxanthin Productivity

Biomass and fucoxanthin productivity on the 4th, 6th, 8th, and 10th days of cultivation were selected for comparative analysis. As depicted in Figure 3, there were significant differences in biomass and fucoxanthin productivity on other days, except for fucoxanthin productivity on day 4. The maximum biomass productivity was observed in the later part of the cultivation period (day 10), while the maximum fucoxanthin productivity was attained on day 6 of the cultivation period. From Figure 3, except for that on day 4, the 15.75 mg L^{-1} iron concentration showed the highest biomass productivity (Figure 3b) compared with that of all other iron concentration treatment groups, and except for the 10th day, the highest fucoxanthin yield in each experimental group was obtained by the group with an iron concentration of 3.15 mg L^{-1}, so it can be determined that the biomass accumulation of *C. weissflosi* is more suitable for an iron concentration of 15.75 mg L^{-1}, and the accumulation of fucoxanthin is more suitable at a 3.15 mg L^{-1} iron concentration.

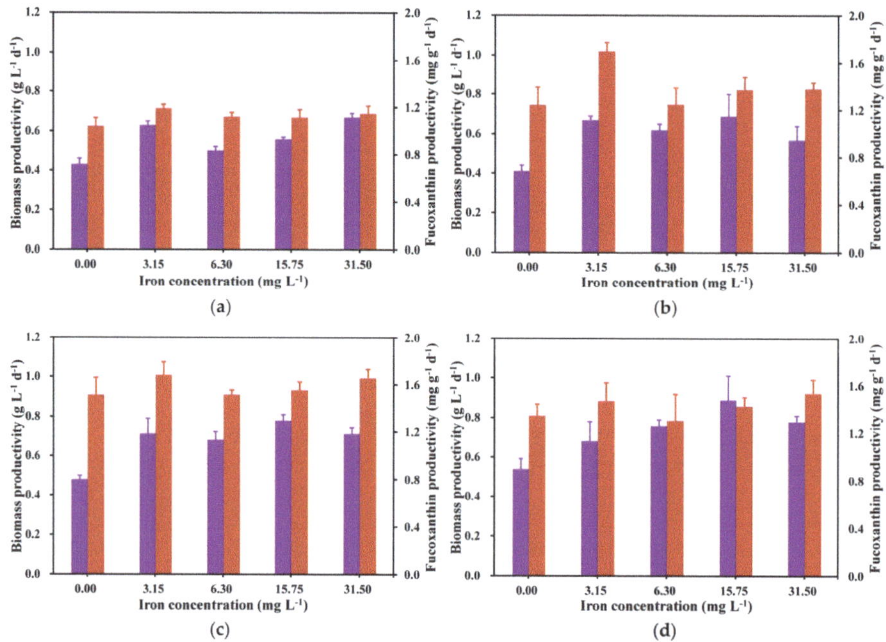

Figure 3. Biomass (purple) and fucoxanthin (red) productivity on day 4 (**a**), day 6 (**b**), day 8 (**c**), and day 10 (**d**) (means ± SD, $n = 3$).

2.4. Fatty Acid Composition and Concentration of C. weissflosi at Different Iron Concentrations

As shown in Table 1, 30 fatty acids were measured under different iron concentrations, including 12 unsaturated fatty acids (UFAs) and 18 saturated fatty acids (SFAs). The main fatty acids were C16:1n7, C20:5n3, C22:6n3, C14:0, and C16:0. Except for that of the group with an iron concentration of 0 mg L^{-1}, the UFA content of each group was higher than that of SFAs, and the UFA content of the group with an iron concentration of 15.75 mg L^{-1} was higher than that of the other groups. The order of fatty acid content from highest to lowest was SFA (40.9–60.4%) > MUFA (monounsaturated fatty acid) (36.9–39.1%) > PUFA (polyunsaturated fatty acid) (18.0–20.2%).

Table 1. Fatty acid composition and concentrations ($\mu g\ g^{-1}$) of *C. weissflogii* under different iron concentrations.

Test Items	0 mg/L	3.15 mg/L	6.3 mg/L	15.75 mg/L	31.5 mg/L
C4:0	18.13	19.57	45.97	15.83	43.11
C6:0	30.33	137.49	30.65	23.48	102.16
C8:0	263.9	112.91	119.47	130.3	269.79
C10:0	384.5	25.29	254.38	310.88	326.63
C11:0	Not detected	25.95	Not detected	35.37	53
C12:0	48.36	45.24	34.13	42.32	40.47
C13:0	Not detected	47	Not detected	Not detected	Not detected
C14:0	8321.98	8272.94	7789.73	8754.77	8622.26
C14:1n5	161.41	154.7	134.28	179.01	159.08
C15:0	950.01	1120.68	1055.68	1224.67	1180.82
C16:0	14,727.31	16,135.16	14,813.51	16,884.4	15,317.65
C16:1n7	22,365.49	25,421.25	24,049.79	26,828.85	24,082.01
C17:0	32.67	20.12	62.36	67.49	55.43
C18:0	501.14	425.01	338.4	470.36	367.42
C18:1n9c	999.17	1013.63	864.56	1102.62	988.37

Table 1. *Cont.*

Test Items	0 mg/L	3.15 mg/L	6.3 mg/L	15.75 mg/L	31.5 mg/L
C18:2n6c	281.33	260.08	198.92	252.58	264.04
C18:3n6	Not detected	170.72	60.06	20.83	27.57
C18:3n3	119.37	78.65	76.48	91.64	78.98
C20:0	57.75	69.59	92.43	126.45	85.36
C20:2	64.61	Not detected	101.77	Not detected	Not detected
C20:3n6	112.49	Not detected	66.83	Not detected	Not detected
C21:0	134.28	121.97	98.69	112.28	138.49
C20:3n3	126.25	Not detected	59.34	164.18	145.83
C20:4n6	123.17	86.67	78.95	97.39	114.95
C20:5n3	9254.11	9465.2	7936.26	10,328.37	10,280.99
C22:0	119.21	107.57	79.92	118.49	108.39
C22:6n3	3511.2	3146.65	2600.33	3647.48	3364.62
C23:0	318.16	Not detected	447.67	Not detected	Not detected
C24:0	1035.7	888.02	634.48	1025.92	967.25
C24:1n9	167.1	37.87	93.67	146.45	28.88
SFA	38,745.22	27,574.51	25,897.47	40,458.73	27,678.23
MUFA	23,693.17	26,683.25	25,142.3	28,256.93	25,258.34
PUFA	13,592.53	13,152.17	11,178.94	14,602.47	14,276.98
UFA	37,285.7	39,835.42	36,321.24	42,859.4	39,535.32
FA	76,030.92	67,409.93	62,218.71	83,318.13	67,213.55

Under the condition of 15.75 mg L-1 of iron, the concentrations of C14:0, C16:1n7, C16:0, EPA, and DHA were higher than those in the other groups, and the total lipid content reached the cumulative maximum. As shown in Figure 4, the EPA concentration of the 15.75 mg L^{-1} group reached 10.3 mg g^{-1} (approximately 12.4% of the total lipids), which was slightly higher than that of the 31.5 mg L^{-1} group and significantly higher than that of the other groups. It was also found that under the same iron concentration of 15.75 mg L^{-1}, the DHA concentration reached the maximum value of approximately 3.6 mg g^{-1}, accounting for 4.4% of the total lipid content.

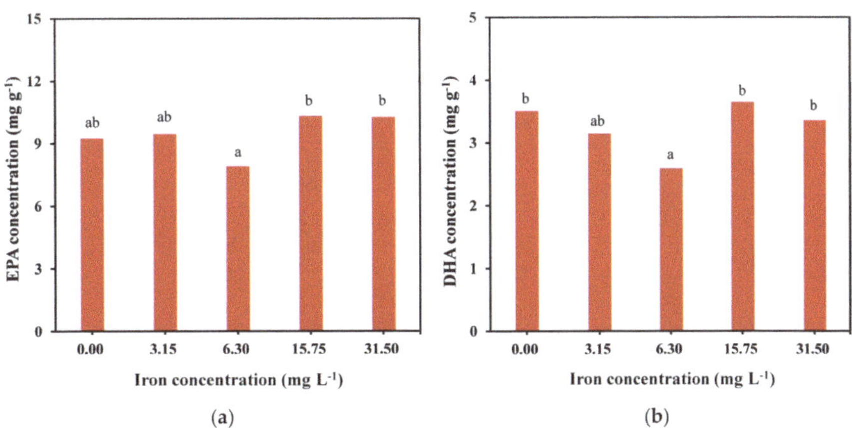

Figure 4. EPA (**a**) and DHA (**b**) concentrations of *C. weissflogii* under different iron concentration culture conditions. Means within the same column of different letters are significantly different at ($p < 0.05$).

3. Discussion

Iron is one of the important trace elements necessary for cell growth metabolism and lipid synthesis in microalgae and is also a limiting factor involved in a variety of enzymatic reactions and transport systems of microalgae, such as redox reactions, oxygen carrier proteins, photosystem II, nitrogen depletion, and chlorophyll synthesis [17,21–23]. Different concentrations of iron ions have different effects on the growth of microalgae, and iron concentrations that are too high or too low will limit the growth and lipid synthesis of microalgae because iron is the carrier of certain oxidoreductase enzymes, while the components of coenzymes in algal cells and iron deficiency can affect a variety of metabolic processes to inhibit cell growth and lipid synthesis [18,19,24]. Liang et al. [25] pointed out that the fastest growth rate and the highest biomass were observed in *Phaeodactylum tricornutum* at a concentration of 0.5 mg L^{-1}, and the highest content of total lipids was observed at a concentration of 0.25 mg L^{-1}. The total lipid content of this alga increased with an increase in the concentration of iron before reaching the optimal concentration of iron, and the synthesis of the total lipids of the microalgae was inhibited when the concentration of iron exceeded the optimal concentration. It has also been noted that iron concentrations as high as 2 mg L^{-1} have positive effects on most microalgal species, but negative effects are usually observed when iron concentrations exceed 40 mg L^{-1} [26]. Zhou et al. [27] reported that the optimal iron concentration for the growth of *Microcystis aeruginosa* ranged from 3 to 12 mg L^{-1}. When the Fe concentration exceeded 12 mg L^{-1}, the concentration of Fe ions in the algal cells increased significantly, resulting in the inhibition of algal growth. Zhou et al. reported that the utilization of iron by *M. aeruginosa* was significantly correlated with its growth status in the optimal concentration range. In this experiment, the cell biomass and fatty acid content of *C. weissflogii* reached the maximum value at an iron concentration of 15.75 mg L^{-1}, indicating that the range of iron concentrations required for growth and lipid synthesis varied among different algal species.

Iron ions can enhance the pigment synthesis and cell membrane stability of microalgae at appropriate concentrations to improve the accumulation ability of active substances [24,28–30]. Wu et al. [31] found that adding an appropriate concentration of ferric chloride during the culture of *Navicula tenera* can significantly promote the accumulation of pigments in *N. tenera*. In another study [32], it was shown that iron-enriched medium stimulated fucoxanthin accumulation in *Nitzschia* sp. and *Nanofrustulum shiloi*. Zhu et al. [33] observed that at an iron concentration of 0.135 mg L^{-1}, the carotenoid content of *P. tricornutum* showed an increase of 5% to 30% compared to that under iron deficiency. In the results of a study [34] on the qualitative and quantitative effects of Fe concentration on the pigment composition of *P. tricornutum*, it was found that both the β-carotenoid and fucoxanthin contents of *P. tricornutum* reached a maximum und a 10 μM Fe concentration; however, a strong decrease in fucoxanthin content was observed in *P. tricornutum* grown under very low Fe concentrations (0.001 and 0.01 μM). Kosakowska et al. [34] suggested that *P. tricornutum* prioritizes the accumulation of diadinoxanthin at the expense of fucoxanthin synthesis to compensate for the sharp decrease in the content of β-carotenoids (another photoprotective pigment) under low-Fe conditions. In the present study, the contents of carotenoids and fucoxanthin showed a trend of increasing and then decreasing in all groups, which may be attributed to the increase in cell density in the later stages of cultivation and the inter-cellular shading effect that reduces the uptake of light energy, which in turn reduces the rate of synthesis of carotenoids and fucoxanthin. In addition, the maximum values of the carotenoid and fucoxanthin contents of *C. weissflogii* were found in the 3.15 mg L^{-1} treatment group, suggesting that this iron concentration may be the optimal concentration for the accumulation of carotenoids and fucoxanthin in *C. weissflogii*. In addition, we observed that the carotenoid content of *C. weissflogii* was the lowest when iron was deficient, indicating that iron deficiency could inhibit the accumulation of carotenoids in *C. weissflogii* to some extent.

Fatty acid double bonds are formed by desaturase, which presents three conserved histidine clusters that are bound to Fe^{2+} to form the active center of the enzyme [35]. Therefore, the concentration of iron may affect the fatty acid composition of microalgae by influencing desaturase activity. Liang et al. [25] found that a high concentration of $FeSO_4$ (1 mg L^{-1}) was conducive to the synthesis of EPA, DHA, and PUFA in *P. tricornutum*. However, another study on the effect of iron on the fatty acid composition of *Tropidoneis maxima* [36] showed a decreasing trend for EPA, DHA, and PUFA in *T. maxima* and the influence of high concentrations of $FeSO_4$ (1 mg L^{-1}). Jiang observed [37] that the PUFA content of *Isochrysis galbana* decreased under Fe^{3+} concentrations above and below 24.5 μM, while the DHA content of *I. galbana* decreased under Fe^{3+} concentrations above and below 60.5 μM. These studies indicate that the effect of iron on the fatty acid composition of microalgae is species-specific. In this study, EPA, DHA, and PUFA contents were highest at an iron concentration of 15.75 mg L^{-1}; an iron concentration above or below this could inhibit polyunsaturated fatty acid synthesis in *C. weissflogii*. In addition, the SFA content was significantly higher than the PUFA content under an iron concentration of 0 mg L^{-1}, which may be due to the lack of Fe^{3+} in the medium, resulting in the inability of the conserved histidine clusters of the fatty acid desaturase to bind with sufficient Fe^{3+} to form the active center of the enzyme, thus affecting the process of saturated fatty acid desaturation.

4. Materials and Methods

4.1. C. weissflogii Strain and Culture Conditions

The *C. weissflogii* strain [15] used in this study was isolated from a shrimp pond in Southern China and preserved at the Laboratory of Algae Resource Development and Aquaculture Environmental Ecological Restoration of Guangdong Ocean University. This strain was grown autotrophically at a temperature of 25 °C in a modified F/2 medium, which contained $NaNO_3$ (75 mg), KH_2PO_4 (5 mg), Na_2SiO_3-$9H_2O$ (20 mg), F/2 trace element solution (1 mL), and F/2 vitamin solution (1 mL) per liter of double-distilled water. The F/2 trace element solution comprised $C_{10}H_{14}N_2Na_2O_8$ (4160 mg), $FeCl_3 \cdot 6H_2O$ (3150 mg), $MnCl_2 \cdot 4H_2O$ (180 mg), $ZnSO_4 \cdot 4H_2O$ (22 mg), $CuSO_4 \cdot 5H_2O$ (10 mg), $H_4MoNa_2O_6$ (6 mg), and $CoCl_2 \cdot 6H_2O$ (4160 mg) per liter of double-distilled water. The F/2 vitamin solution was formulated with biotin (0.5 mg), vitamin B_{12} (0.5 mg), and vitamin B_1 (100 mg) per liter of double-distilled water. The *C. weissflogii* culture was maintained under a continuous light intensity of 30 μmol m^{-2} s^{-1} and mixed with continuous aeration in 5 L flasks with filtered seawater added to the F/2 culture medium.

4.2. Experimental Setup

In this experiment, the effect of Fe ($FeCl_3 \cdot 6H_2O$) was evaluated using varying concentrations from 0 mg L^{-1} to 31.5 mg L^{-1}. $FeCl_3 \cdot 6H_2O$ was firstly configured into a mother liquor with a concentration of 3.15 g L^{-1}. For the experiment, 0 mL, 1 mL, 2 mL, 5 mL, and 10 mL of $FeCl_3 \cdot 6H_2O$ mother liquor were added to the iron-free f/2 medium to represent the 0-fold iron, 1-fold iron, 2-fold iron, 5-fold iron, and 10-fold iron treatment groups, respectively. The volume of the mother cultures was 700 mL with the initial inoculation density of approximately 6×10^5 cells mL^{-1}. The experiment was performed under LED light (30 ± 2 μmol m^{-2} s^{-1}), aeration (0.4 L min^{-1}), temperature (25 ± 2 °C), pH (8.0 ± 0.2), and salinity (25), and cultured in a 1 L glass cylindrical photobioreactor (inner diameter, 5 cm) for 10 days. All cultures were mixed via continuous bubbling with 100% filtered air. Illumination was provided form the side by a T8 LED Tube light (white) with a light–dark cycle of 24 h: 0 h for 10 days. All treatments were carried out in triplicates.

4.3. Analytical Methods

After each sampling, the total cells of *C. weissflogii* were counted using a Neubauer improved cell counting chamber (25 mm × 16 mm) under an Olympus BX53 light microscope.

Dry weight (DW) was determined by filtering a 10 mL of the algal suspension through a pre-weighted (M_1) acetate membrane (47 mm, nominal pore size 1 um). The algal biomass was rinsed twice with 0.5 M ammonium bicarbonate. The filter membrane was placed in an oven and dried at 80 °C to a constant mass, and the total mass M_2 was measured and recorded. DW is calculated using Equation (1):

$$DW = (M_2 - M_1)/10 \tag{1}$$

The carotenoid content was determined via the ethanol extraction method [38,39]. Algal cells were collected via centrifugation (5000 rpm for 10 min). Then, after addition of a 95% volume fraction of ethanol to the algal cells (10 mL), the mixture was treated under dark conditions for 24 h. The supernatant was collected via centrifugation (5000 rpm, 10 min), and the optical density values (D) of the supernatant at 480 nm, 510 nm, and 750 nm were measured using a spectrophotometer. The carotenoid content was calculated using Equation (2):

$$\rho(Carotenoids) = 7.6 \times [(D_{480} - D_{750}) - 1.49 \times (D_{510} - D_{750})] \tag{2}$$

Fucoxanthin content was measured using an organic solvent extraction method [40]. *C. weissflogii* cells (80 mL) were collected via centrifugation (5000 r min^{-1}, 10 min) at 4 °C and placed in a freeze dryer for 2 days. The freeze-dried algal cells were ground into a powder, then anhydrous ethanol was added so that the material-to-liquid ratio was 1 g:40 mL and extracted twice, each time for 1 h, in the dark at 60 °C. The supernatant was collected via centrifugation at 5000 r min^{-1} for 10 min, and then the absorbance was measured at 445 nm using a UV spectrophotometer (D$_{445}$). The fucoxanthin content was calculated using Equation (3):

$$C(Fucoxanthin) = (1000 \times D_{445} \times N \times V) / (A' \times M \times 100) \tag{3}$$

N is the dilution ratio; V is the volume of the crude extract; A′ is the theoretical absorption value of a solute, which is 1600; M is the sample mass.

The fatty acid composition was determined via gas chromatography [41]. The appropriate amount of the sample was weighed in a glass tube, 0.5 mol L^{-1} of NaOH methanol solution, shocked uniformly, was added and placed in a 60 °C heated water bath for 20 min of saponification; after sufficient saponification, it was removed and cooled. Boron trifluoride methanol complexing solution, shocked uniformly, was added to a 60 °C water bath for 6 min for methylation; after cooling, isooctane was added for extraction and filtered with a 0.45 nm organic filtration membrane, and the supernatant was put into the injection vials for determination.

The prepared samples were analyzed with an Agilent 7890A gas chromatograph with the parameters and measurement procedures described in the authors' previous studies [15]. A single fatty acid methyl ester standard solution and a fatty acid methyl ester mixed standard solution were injected into the gas chromatograph, and the peaks were characterized. The parameters were as follows: a capillary column (DB-23MS, column length 60 m; internal diameter 0.25 mm; film thickness 0.15 μm), the split injection mode, a split ratio of 35:1; nitrogen as the carrier gas, an inlet temperature of 270 °C, an initial temperature of 100 °C, a duration of 13 min, temperatures of 100 °C~180 °C at a heating rate of 10 °C min^{-1} for 6 min, temperatures of 180 °C~200 °C at a heating rate of 1 °C min^{-1} for 20 min, temperatures of 200 °C~230 °C at a heating rate of 4 °C min^{-1} for 10.5 min, and FID as the detector. Under the above chromatographic conditions, the fatty acid standard solution and the sample solution were injected into the gas chromatograph and quantified according to the peak area of the chromatogram. Sensitivity analysis was used to check the robustness of the fatty acid concentration.

Statistical analysis was performed using the SPSS for Windows statistical software package (IBM SPSS v26.0; Chicago, IL, USA). One-way analysis of variance (ANOVA) was used to test for significant differences ($p < 0.05$) between treatments, with the results presented as the mean ± SD (standard deviation).

5. Conclusions

This study demonstrated that the co-production of fucoxanthin and fatty acids in *C. weissflogii* could be considerably enhanced by adjusting the iron concentration in the growth media. These findings indicate that an optimal iron concentration of 15.75 mg L^{-1} is beneficial for achieving the highest cell density and total fatty acid content, while a different optimal iron concentration of 3.15 mg L^{-1} resulted in the highest carotenoid and fucoxanthin production. These results provide a foundation for the advancement of large-scale microalgae cultivation, particularly for the efficient and targeted production of valuable bioactive substances. This study adds to the broader understanding of microalgal biotechnology and highlights the potential of microalgae as a sustainable source of health-promoting compounds for aquatic organisms and humans. Furthermore, this study offers a valuable experimental framework and support for large-scale industrial production, which could have significant implications for nutrition, aquaculture, and the development of functional foods and nutraceuticals.

Author Contributions: Data curation, K.P. and X.R.; formal analysis, K.P., X.R., and F.L.; funding acquisition, F.L.; investigation, K.P. and X.R.; methodology, X.R. and F.L.; project administration, F.L.; re-sources, X.H., F.L., and C.L.; writing—original draft, K.P., D.K.A., and F.L.; writing—review and editing, D.K.A. and F.L. All authors have read and agreed to the published version of the manuscript.

Funding: This research was mainly supported by the Program for Scientific Research Start-up Funds of Guangdong Ocean University (grant number 060302022103).

Data Availability Statement: Data are contained within the article.

Conflicts of Interest: The authors declare no conflicts of interest.

References

1. Jónasdóttir, S.H. Fatty acid profiles and production in marine phytoplankton. *Mar. Drugs* **2019**, *17*, 151. [CrossRef] [PubMed]
2. Zhou, L.; Li, K.; Duan, X.; Hill, D.; Barrow, C.; Dunshea, F.; Marin, G.; Suleria, H. Bioactive compounds in microalgae and their potential health benefits. *Food Biosci.* **2022**, *49*, 101932. [CrossRef]
3. Khan, M.I.; Shin, J.H.; Kim, J.D. The promising future of microalgae: Current status, challenges, and optimization of a sustainable and renewable industry for biofuels, feed, and other products. *Microb. Cell Fact.* **2018**, *17*, 36. [CrossRef] [PubMed]
4. Dolganyuk, V.; Belova, D.; Babich, O.; Prosekov, A.; Ivanova, S.; Katserov, D.; Patyukov, N.; Sukhikh, S. Microalgae: A promising source of valuable bioproducts. *Biomolecules* **2020**, *10*, 1153. [CrossRef] [PubMed]
5. Ampofo, J.; Abbey, L. Microalgae: Bioactive composition, health benefits, safety, and prospects as potential high-value ingredients for the functional food industry. *Foods* **2022**, *11*, 1744. [CrossRef]
6. Fernández, F.G.A.; Reis, A.; Wijffels, R.H.; Barbosa, M.; Verdelho, V.; Llamas, B. The role of microalgae in the bioeconomy. *New Biotechnol.* **2021**, *61*, 99–107. [CrossRef] [PubMed]
7. Yarnold, J.; Karan, H.; Oey, M.; Hankamer, B. Microalgal aquafeeds as part of a circular bioeconomy. *Trends Plant Sci.* **2019**, *24*, 959–970. [CrossRef]
8. Maeda, H.; Hosokawa, M.; Sashima, T.; Miyashita, K. Dietary combination of fucoxanthin and fish oil attenuates the weight gain of white adipose tissue and decreases blood glucose in obese/diabetic KK-Ay mice. *J. Agric. Food Chem.* **2007**, *55*, 7701–7706. [CrossRef]
9. Zhang, H.; Tang, Y.; Zhang, Y.; Zhang, S.; Qu, J.; Wang, X.; Kong, R.; Han, C.; Liu, Z. Fucoxanthin: A promising medicinal and nutritional ingredient. *Evid. Based Complement. Altern. Med.* **2015**, *2015*, 723515. [CrossRef]
10. Yoong, K.L.; Chun-Yen, C.; Sunita, V.; Jo-Shu, C. Producing fucoxanthin from algae—Recent advances in cultivation strategies and downstream processing. *Bioresour. Technol.* **2022**, *344*, 126170.
11. Pajot, A.; Hao Huynh, G.; Picot, L.; Marchal, L.; Nicolau, E. Fucoxanthin from algae to human, an extraordinary bioresource: Insights and advances in up and downstream processes. *Mar. Drugs* **2022**, *20*, 222. [CrossRef] [PubMed]
12. Khaw, Y.S.; Yusoff, F.M.; Tan, H.T.; Noor Mazli, N.A.I.; Nazarudin, M.F.; Shaharuddin, N.A.; Omar, A.R.; Takahashi, K. Fucoxanthin production of microalgae under different culture factors: A systematic review. *Mar. Drugs* **2022**, *20*, 592. [CrossRef]
13. Lu, Q.; Li, H.; Xiao, Y.; Liu, H. A state-of-the-art review on the synthetic mechanisms, production technologies, and practical application of polyunsaturated fatty acids from microalgae. *Algal Res.* **2021**, *55*, 102281. [CrossRef]
14. Tyagi, R.; Rastogi, R.P.; Babich, O.; Awasthi, M.K.; Tiwari, A. New perspectives of omega-3 fatty acids from diatoms. *Syst. Microbiol. Biomanuf* **2023**. [CrossRef]
15. Rui, X.; Amenorfenyo, D.K.; Peng, K.; Li, H.; Wang, L.; Huang, X.; Li, C.; Li, F. Effects of different nitrogen concentrations on co-production of fucoxanthin and fatty acids in *Conticribra weissflogii*. *Mar. Drugs* **2023**, *21*, 106. [CrossRef]
16. Marella, T.K.; Tiwari, A. Marine diatom *Thalassiosira weissflogii* based biorefinery for co-production of eicosapentaenoic acid and fucoxanthin. *Bioresour. Technol.* **2020**, *307*, 123245. [CrossRef] [PubMed]
17. Ece, P.; Ebubekir, Y.; Mahmut, A. Effect of different iron sources on sustainable microalgae-based biodiesel production using *Auxenochlorella protothecoides*. *Renew. Energy* **2020**, *162*, 1970–1978.
18. Rizwan, M.; Mujtaba, G.; Lee, K. Effects of iron sources on the growth and lipid/carbohydrate production of marine microalga *Dunaliella tertiolecta*. *Biotechnol. Bioproc. Eng.* **2017**, *22*, 68–75. [CrossRef]

19. Terauchi, A.M.; Peers, G.; Kobayashi, M.C.; Niyogi, K.K.; Merchant, S.S. Trophic status of *Chlamydomonas reinhardtii* influences the impact of iron deficiency on photosynthesis. *Photosynth. Res.* **2010**, *105*, 39–49. [CrossRef]
20. Yuan, X.; Liang, L.; Liu, K.; Xie, L.; Huang, L.; He, W.; Chen, Y.; Xue, T. Spent yeast as an efficient medium supplement for fucoxanthin and eicosapentaenoic acid (EPA) production by *Phaeodactylum tricornutum*. *J. Appl. Phycol.* **2020**, *32*, 59–69. [CrossRef]
21. Sunda, W.; Huntsman, S. Interrelated influence of iron, light, and cell size on marine phytoplankton growth. *Nature* **1997**, *390*, 389–392. [CrossRef]
22. Chen, M.; Dei, R.C.H.; Wang, W.X.; Guo, L. Marine diatom uptake of iron bound with natural colloids of different origins. *Mar. Chem.* **2003**, *81*, 177–189. [CrossRef]
23. Naito, K.; Matsui, M.; Imai, I. Ability of marine eukaryotic red tide microalgae to utilize insoluble iron. *Harmful Algae* **2005**, *4*, 1021–1032. [CrossRef]
24. Liu, Z.Y.; Wang, G.C.; Zhou, B.C. Effect of iron on growth and lipid accumulation in *Chlorella vulgaris*. *Bioresour. Technol.* **2008**, *99*, 4717–4722. [CrossRef] [PubMed]
25. Liang, J.; Jiang, X.; Li, Y.; Han, Q. Effects of nitrogen, phosphorus and iron on the growth, total lipid content and fatty acid composition of *Phaeodactylum tricornutum* mutant strain. *Chin. J. Ecol.* **2016**, *35*, 189–198. (In Chinese)
26. Sajjadi, B.; Chen, W.Y.; Aziz, A.; Raman, A.; Ibrahim, S. Microalgae lipid and biomass for biofuel production: A comprehensive review on lipid enhancement strategies and their effects on fatty acid composition. *Renew. Sustain. Energy Rev.* **2018**, *97*, 200–232. [CrossRef]
27. Zhou, H.; Hu, T.; Xuan, Y.; Zhou, J. Study on influence of typical microscale metal elements on growth of *Microcystis aeruginosa*. *J. Water Resour. Water Eng.* **2016**, *27*, 9–13.
28. Mohsen, G.; Behzad, R.; Berat, Z.H. Effects of macro and micronutrients on neutral lipid accumulation in oleaginous microalgae. *Biofuels* **2018**, *9*, 147–156.
29. Mohit, S.R.; Shashi, B.; Sudhakar, D.R.; Sanjeev, K.P. Effect of iron oxide nanoparticles on growth and biofuel potential of *Chlorella* spp. *Algal Res.* **2020**, *49*, 101942.
30. Mohit, S.R.; Sanjeev, K.P. Resolving the dilemma of iron bioavailability to microalgae for commercial sustenance. *Algal Res.* **2021**, *59*, 102458.
31. Wu, S.; Zhang, X.; Zheng, W.; Li, J. Optimization of culture media for spaceflight and non-spaceflight *Navicula tenera*. *J. Plant Resour. Environ.* **2006**, *15*, 33–37. (In Chinese)
32. Sahin, M.S.; Khazi, M.I.; Demirel, Z.; Dalay, D.M. Variation in growth, fucoxanthin, fatty acids profile and lipid content of marine diatoms *Nitzschia* sp. and *Nanofrustulum shiloi* in response to nitrogen and iron. *Biocatal. Agric. Biotechnol.* **2019**, *17*, 390–398. [CrossRef]
33. Zhu, M.; Mu, X.; Li, R.; Lü, R. The effects of iron on growth, photosynthesis, and biochemical composition of a marine algae *Phaeodactylum tricornutum*. *Acta Oceanol. Sin.* **2000**, *22*, 110–116. (In Chinese)
34. Kosakowska, A.; Lewandowska, J.; Stoń, J.; Burkiewicz, K. Qualitative and quantitative composition of pigments in *Phaeodactylum tricornutum* (Bacillariophyceae) stressed by iron. *Biometals* **2004**, *17*, 45–52. [CrossRef] [PubMed]
35. Wei, D.; Zhang, X.C. New advances on fatty acid desaturase and regulation of these genes' expression by ecological factors in microalgae cells. *Mar. Sci.* **2000**, *24*, 42–46. (In Chinese)
36. Liang, J.; Jiang, X.; Zhang, Z.; Han, Q. Effects of phosphorus iron and silicon on the growth, total lipid content and fatty acid composition of *Tropidoneis maxima*. *China Oils Fats* **2016**, *41*, 69–74. (In Chinese)
37. Jiang, H. Studies on Optimizing Production of Polyunsaturated Fatty Acids by Marine Microalgae. Master Dissertation, Shantou University, Guangdong, China, 2003. (In Chinese).
38. Li, F.; Rui, X.; Amenorfenyo, D.K.; Pan, Y.; Huang, X.; Li, C. Effects of temperature, light, and salt on the production of fucoxanthin from *Conticribra weissflogii*. *Mar. Drugs* **2023**, *21*, 495. [CrossRef]
39. Parsons, T.R.; Maita, Y.; Lalli, C.M. Determination of chlorophylls and total carotenoids: Spectrophotometric method—Science Direct. In *A Manual of Chemical & Biological Methods for Seawater Analysis*; Pergamon Press: Oxford, UK, 1984; pp. 101–104.
40. Xu, R.R.; Gong, Y.F.; Chen, W.T.; Li, S.R.; Chen, R.Y.; Zheng, X.Y.; Chen, X.M.Z.; Wang, H.Y. Effects of LED monochromatic light quality of different colors on fucoxanthin content and expression levels of related genes in *Phaeodactylum Tricornutum*. *Acta Opt. Sin.* **2019**, *9*, 299–307.
41. Zhang, W.Y.; Gao, B.Y.; Li, A.F.; Zhang, C.W. Effects of different culture conditions on growth and accumulation of bioactive components by *Phaeodactylum tricornutum*. *Mar. Sci.* **2016**, *40*, 57–65.

Article

Description and Characterization of the *Odontella aurita* OAOSH22, a Marine Diatom Rich in Eicosapentaenoic Acid and Fucoxanthin, Isolated from Osan Harbor, Korea

Sung Min An, Kichul Cho, Eun Song Kim, Hyunji Ki, Grace Choi and Nam Seon Kang *

Department of Microbial Resources, National Marine Biodiversity Institute of Korea, Seocheon 33662, Republic of Korea; sman@mabik.re.kr (S.M.A.); kichul.cho@mabik.re.kr (K.C.); kes2523@mabik.re.kr (E.S.K.); hki@mabik.re.kr (H.K.); gchoi@mabik.re.kr (G.C.)
* Correspondence: kang3610@mabik.re.kr

Abstract: Third-generation biomass production utilizing microalgae exhibits sustainable and environmentally friendly attributes, along with significant potential as a source of physiologically active compounds. However, the process of screening and localizing strains that are capable of producing high-value-added substances necessitates a significant amount of effort. In the present study, we have successfully isolated the indigenous marine diatom *Odontella aurita* OAOSH22 from the east coast of Korea. Afterwards, comprehensive analysis was conducted on its morphological, molecular, and biochemical characteristics. In addition, a series of experiments was conducted to analyze the effects of various environmental factors that should be considered during cultivation, such as water temperature, salinity, irradiance, and nutrients (particularly nitrate, silicate, phosphate, and iron). The morphological characteristics of the isolate were observed using optical and electron microscopes, and it exhibited features typical of *O. aurita*. Additionally, the molecular phylogenetic inference derived from the sequence of the small-subunit 18S rDNA confirmed the classification of the microalgal strain as *O. aurita*. This isolate has been confirmed to contain 7.1 mg g^{-1} dry cell weight (DCW) of fucoxanthin, a powerful antioxidant substance. In addition, this isolate contains 11.1 mg g^{-1} DCW of eicosapentaenoic acid (EPA), which is one of the nutritionally essential polyunsaturated fatty acids. Therefore, this indigenous isolate exhibits significant potential as a valuable source of bioactive substances for various bio-industrial applications.

Keywords: carotenoid pigments; fatty acids; *Odontella aurita*; culture condition; ultrastructure; 18S rDNA

Citation: An, S.M.; Cho, K.; Kim, E.S.; Ki, H.; Choi, G.; Kang, N.S. Description and Characterization of the *Odontella aurita* OAOSH22, a Marine Diatom Rich in Eicosapentaenoic Acid and Fucoxanthin, Isolated from Osan Harbor, Korea. *Mar. Drugs* **2023**, *21*, 563. https://doi.org/10.3390/md21110563

Academic Editors: Cecilia Faraloni and Eleftherios Touloupakis

Received: 26 September 2023
Revised: 24 October 2023
Accepted: 24 October 2023
Published: 27 October 2023

1. Introduction

Microalgae contain various functional substances [1]. Unlike resources such as terrestrial plants and seaweeds, they can be cultured in large quantities, which has the advantage of securing raw materials in a stable manner [2]. Microalgae contain a significant amount of useful high-value substances, including unsaturated fatty acids (such as omega-3 fatty acids), natural pigments (such as astaxanthin, lutein, and fucoxanthin), and polysaccharides and oligosaccharides (such as fucoidan, alginic acid, and carrageenan) [3]. They are used in various bio-industries such as the food, health and functional food, cosmetics, and pharmaceutical industries [4]. However, despite the high diversity of microalgae, only a few, such as *Chlorella* (Chlorophyta) and *Spirulina* (Cyanobacteria), are used as raw food materials and functional ingredients [5]. Since the development of useful materials derived from microalgae has been limited to certain species, there is strong potential for discovering new sources for biomaterial development in the future.

Diatoms (Bacillariophyta) are the dominant group of microalgae in marine environments [6]. They are the primary producers in coastal ecosystems [7]. They play a significant role in the biogeochemical cycles of carbon and silicate [8]. Since diatoms have high nutritional value and industrial potential, numerous studies have been conducted on various diatom species, including *Cylindrotheca closterium*, *Nanofrustulum shiloi*, *Nitzschia laevis*, *Odontella aurita*, *Phaeodactylum tricornutum*, *Skeletonema costatum*, and *Thalassiosira weissflogii* [9–15]. Among the various species, *Odontella aurita* has garnered significant attention. This species is the type species of the genus *Odontella* and is classified under the order Eupodiscales. The size of this species exhibits significant variability, ranging from 10 to 100 μm [16]. Additionally, it has the ability to form colonies characterized by their ribbon-like shape [16]. This species exhibits a global distribution, being found in both benthic and planktonic forms [17]. It has been observed to occasionally form blooms during the winter and early spring [17]. This species is of significant interest due to its remarkable capacity to accumulate eicosapentaenoic acid (EPA) and fucoxanthin [18]. Polyunsaturated fatty acids (PUFAs), such as EPA, have been recognized for their ability to provide a variety of health benefits to individuals, and traditionally, these fatty acids have been primarily obtained from fish oil [19]. As awareness of animal welfare and the promotion of vegan culture continue to grow, microalgae are being recognized as a viable alternative to fish oil [20]. Among these microalgae, *O. aurita* stands out due to its high content of EPA, comprising approximately 25–26% of its total fatty acid composition [18]. Consequently, *O. aurita* is being explored as a promising source of EPA [21]. Fucoxanthin, a marine xanthophyll present in brown algae and diatoms, exhibits a diverse array of bioactivities, including anti-oxidant, anti-cancer, and anti-obesity properties [22]. Additionally, it has been reported that the substance contains a variety of beneficial components, including fiber, phytosterols, protein, and minerals [23,24].

In 2002, the Agence Française de Sécurité Sanitaire des Aliments (AFSSA) granted approval for the consumption of *Odontella aurita*, citing its substantial equivalence to other edible seaweeds that had already been approved under EC Regulation 258/97 [25]. Since then, *O. aurita* has been officially designated as a Novel Food in the European Union (EU) [26]. According to the regulations set by the EU, the entire biomass of *O. aurita* can be utilized in certain food products, subject to maximum content limitations [27]. This particular species is one of the few commercially available options, even though it has not been officially recognized as safe for consumption by the United States Food and Drug Administration (US FDA) under the "generally recognized as safe" (GRAS) category [28]. Thus, it possesses the potential to be developed for feed, food, and functional material with high value-added properties [29,30]. The French company Innovalg has successfully cultivated this species on a large scale in raceway ponds and subsequently commercialized it as a dietary supplement [31]. The supplement is available in the form of capsules that contain dried cells. The species has not yet undergone human nutrition tests; however, it has demonstrated the potential to mitigate the risk of metabolic syndrome in mice that were fed a high-fat diet [21]. In addition, the lipophilic extract derived from this particular species has been recognized for its ability to mitigate the effects of skin aging, making it a popular choice as a cosmetic ingredient [29].

Acquiring new strains is of importance because the biological attributes of specific microalgae can vary depending on their habitat and strain, even within the same species [32]. Additionally, the regulations on Access and Benefit Sharing (ABS) under the Nagoya Protocol have recognized a variety of biomaterials, including microalgae, as valuable resources [33]. Per these regulations, any profits derived from these resources must be distributed among the resource providers. This has the potential to result in increased production expenses, thereby hindering the process of industrialization for the species [34]. Therefore, it can be argued that the exploration and cultivation of indigenous strains hold significant importance.

The objective of this research is to identify and analyze the indigenous *Odontella aurita* strain that was isolated from the coastal waters of Sonyang-myeon, Yangyang-gun, Gangwon-do, Republic of Korea. In addition, this study aims to determine the optimal culture conditions for each factor, including temperature, salinity, irradiance, and nutrients concentration, which have an impact on the growth of this species and analyze the composition of fatty acids and carotenoid pigments to examine its potential applications in various industries.

2. Results and Discussion

2.1. Morphological Identification of Strain OAOSH22

Our morphological observations revealed that our isolate exhibited several features that are consistent with the characteristics of the genus *Odontella*. These features include bipolar valves, two elevations at the apices with rimmed ocelli at the summits, two types of pore occlusion, a distinct expanded hyaline valve margin with an upturned rim, rimoportulae located in the subcentral position, valvocopula extending beneath the flange, and chain formation [35]. The morphological characteristics of the isolate are outlined in detail below and illustrated in Figure 1. The cells were strongly silicified, with an apical axis measuring 25–51 μm ($n = 16$). Numerous small circular or elliptical chloroplasts were observed adhering to the cell wall (Figure 1A). The cells typically exhibited colony formation characterized by a zigzag pattern, with a single horn connecting them, or a linear colony formation, with both horns serving as points of connection (Figure 1A). Valves were more or less elliptical (bipolar), with two obtuse horns (elevations) and an ocellus at each pole. There was also a distinct convex area between the horns (Figure 1A,B,F; arrowhead). The valve mantle became increasingly constricted towards the edge and greatly curved outward from the edge again (Figure 1F; arrow). Valves were found to be embedded within the girdle band (Figure 1E,F). Two or more labiate processes (up to 14 observed) with spine-like external tubes were located in the central convex area of each valve (Figure 1D–F). The areolae were arranged radially from the center of the valve (11 in 10 μm, $n = 7$) and were occluded by two types of vela (Figure 1C; arrow and arrowhead). The surface of the valve exhibited a multitude of small spines (Figure 1C,D,F).

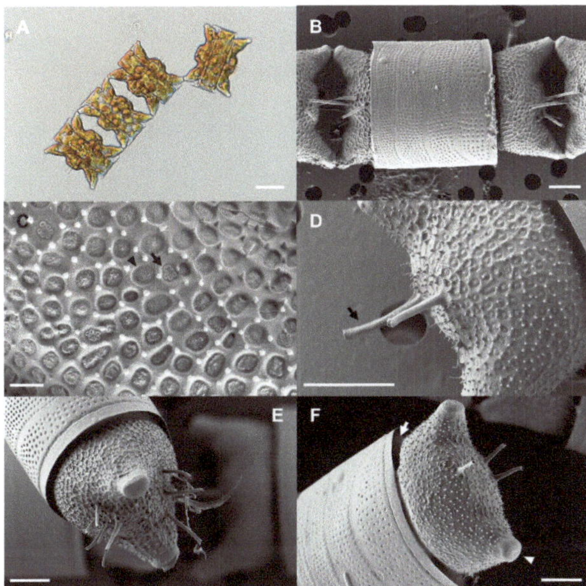

Figure 1. Light and scanning electron microscopy micrographs of *Odontella aurita* OAOSH22. (**A**) Cells form colonies with one or two horns connected. (**B**) External whole frustule view. (**C**) Detail of external areolae occluded by two types of velum (arrow and arrowhead). (**D**) Detail of external valve central area showing two labiate processes with spine-like external tubes (arrow). (**E**) Valve with 14 labiate processes. (**F**) Embedded valve in the girdle band (arrow) and two obtuse horns with ocelli at apices (arrowhead). Scale bars: (**A**) = 20 μm, (**B,D–F**) = 5 μm, and (**C**) = 1 μm.

Microscopic observations revealed that strain OAOSH22 exhibited the characteristic morphological traits of *O. aurita* (Lyngbye) C.A. Agardh 1832 [35]. The size of *O. aurita* cells exhibits significant variability [16]. Therefore, it can be confusing to differentiate between *O. aurita* and other species that share similar morphological features, such as *O. obtusa* and *Hobaniella longicuris*. *O. obtusa* exhibits shorter and more obtuse horns, displaying greater inflection at the base and a lower elevation at the center of the valve compared to *O. aurita* [17]. In contrast to *O. aurita*, *H. longicruris* exhibits elongated and slender horns with minimal curvature at the base. Additionally, it possesses dome-shaped areolae [35].

2.2. Molecular Identification of Strain OAOSH22

The length of the trimmed and assembled 18S rDNA sequences for strain OAOSH22 was determined to be 1684 base pairs (bp). The sequences obtained as a result of this study have been submitted to GenBank under the accession number OP502635. A BLASTn search was conducted to determine the similarity of the 18S rDNA sequence of strain OAOSH22. The results revealed a high level of identity, with 99.6% similarity (query cover of 100% and E-value of 0), when compared to the 18S rDNA sequence of *Odontella aurita* (MW750334). Based on the BLASTn search results, we conducted phylogenetic analyses using maximum likelihood (ML), and Bayesian inference (BI) methods were used to confirm the taxonomic classification of strain OAOSH22 within the order Eupodiscales, which includes the genus *Odontella*. Strain OAOSH22 exhibited a close relationship with *O. aurita*, as evidenced by strong bootstrap values (ML bootstrap = 98% and BI posterior probabilities = 100%) (Figure 2). Finally, the strain OAOSH22 was identified as *Odontella aurita* via the analysis of morphological characteristics and sequencing data. The strain was deposited in the Korean Collection for Type Cultures (KCTC 15114BP).

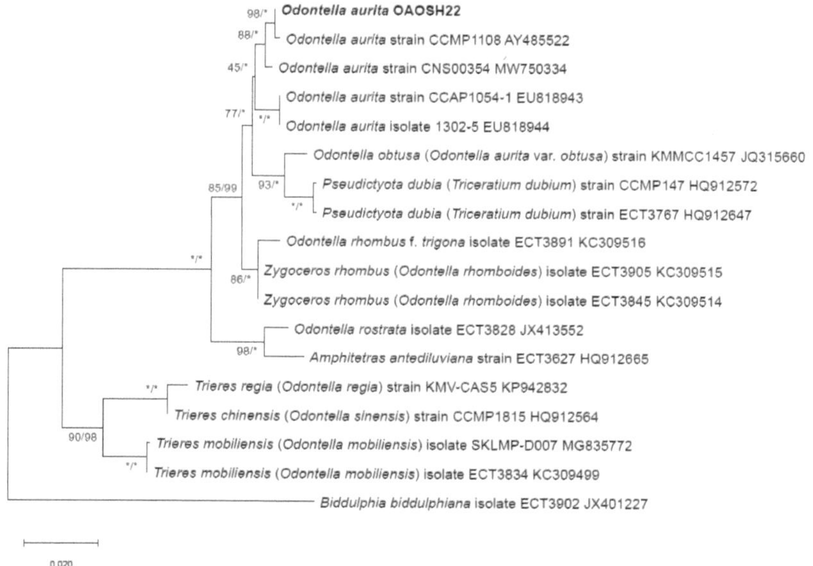

Figure 2. ML and BI phylogenetic tree of 18S rRNA gene from Eupodiscales species. The values on each node indicate ML bootstrap and Bayesian posterior probabilities (%), respectively. The asterisk (*) indicates 100.

Accurate identification of the microalgae species used for food is of utmost importance. Some microalgae have the ability to synthesize toxins, which can potentially lead to severe health complications [36]. Therefore, accurately identifying the microalgae and demonstrating that it belongs to a species that has been previously recognized as safe for consumption ensures that it is suitable for human consumption. In addition, accurate species identification holds significant importance from a quality control standpoint. Because each species of microalgae possesses distinct nutritional profiles and properties, precise identification of the target microalgae is crucial in ensuring that the final product achieves the intended nutritional content and properties [36]. In fact, the standardization of species identification is one of the research recommendations outlined in the Phycomorph European Guidelines for the Sustainable Aquaculture of Seaweeds [37]. In conclusion, precise species identification plays a crucial and indispensable role in ensuring the safety, quality, and nutritional value of microalgae during the industrialization process.

2.3. Optimization of Culture Conditions for Strain OAOSH22

To assess the influence of various factors on the growth of *Odontella aurita* OAOSH22 and identify the optimal cultivation conditions for each factor, an analysis was conducted to examine the growth response under different conditions of irradiance, temperature, salinity, and nutrient concentration at the laboratory scale (Figures 3 and 4). The growth rate and statistical analysis results for each treatment are depicted in Figure S1.

The optimal irradiance (E_k) required to saturate photosynthesis in *O. aurita* OAOSH22 was determined via the rapid light curve method to be 76.5 µmol photons $m^{-2}\,s^{-1}$. Furthermore, the ETR_{max} was determined to be 5.29 (Figure 3). The growth of microalgae and biomass production are more significantly influenced by suitable irradiance rather than nutrient availability, as supported by previous studies [38,39]. At higher light intensities, where saturation occurs, additional illumination does not enhance the rate of photosynthesis. When microalgae are exposed to excessively intense light, it can result in photo-oxidative damage to the photosynthetic machinery via the generation of singlet oxygen. This subsequently reduces the efficiency and speed of photosynthesis, a phenomenon known as photoinhibition [40,41]. Furthermore, low levels of irradiance can hinder growth rates. Various studies have shown that certain species of microalgae are capable of achieving their highest growth rates when exposed to irradiances below 100 µmol photons $m^{-2}\,s^{-1}$ [42]. Additionally, it has been observed that photoinhibition can occur even at irradiance levels ranging from 100 to 200 µmol photons $m^{-2}\,s^{-1}$, which is significantly lower than the typical intensity of sunlight [43,44]. The results of this investigation showed similarities to the findings of previous studies. However, irradiances below 100 µmol photons $m^{-2}\,s^{-1}$ may be considered suitable for laboratory-scale cultivation, as supported by the findings of this study. Conversely, when it comes to large-scale cultivations beyond the pilot scale, it may be necessary to increase light intensities in order to counteract the self-shading effects [45].

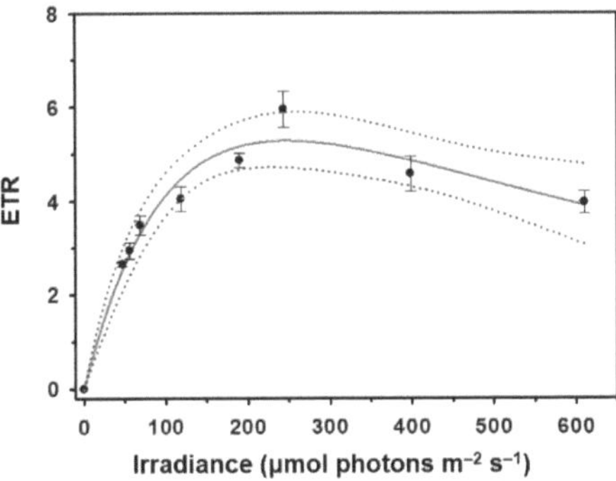

Figure 3. The rapid light-response curve of *Odontella aurita* OAOSH22. Solid lines indicate best fit according to model of Platt et al. [46], and blue dotted lines represent 95% confidence intervals (r^2 = 0.96). Symbols and error bars represent the mean $\pm$ SE (n = 3).

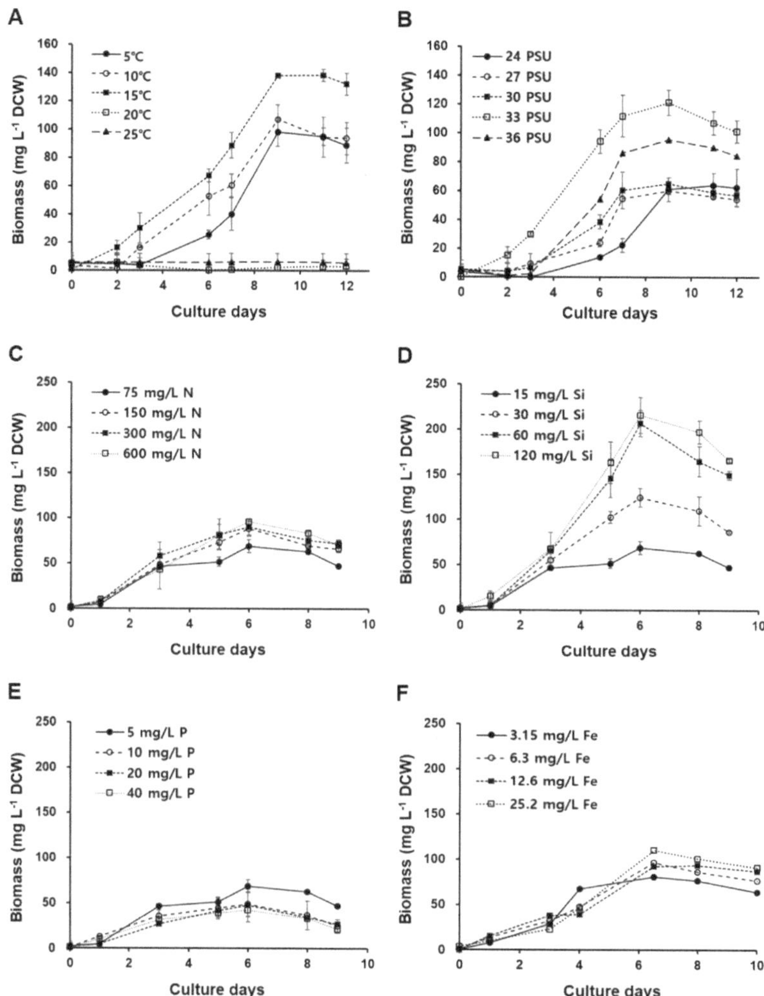

Figure 4. Growth curves of *Odontella aurita* OAOSH22 under different conditions of temperature (**A**), salinity (**B**), nitrate (**C**), silicate (**D**), phosphate (**E**), and iron (**F**). Symbols and error bars represent the mean ± SE (n = 3).

The growth curves of *O. aurita* OAOSH22 at temperatures of 5, 10, 15, 20, and 25 °C, respectively, are illustrated in Figure 4A. Biomass production reached its maximum value of 138.7 mg L^{-1} after a 9-day incubation period at a temperature of 15 °C ($p < 0.05$). However, no growth was observed at either 5 °C or 25 °C. Temperature is a critical determinant in the growth and development of microalgae [47]. Various aspects are influenced by it, including the growth rate, cell size, biochemical composition, and nutrient requirements [22]. *O. aurita* is a common species frequently found in temperate regions and present throughout the year. The species under consideration is classified as tychopelagic and is primarily distributed in coastal regions [48]. It primarily inhabits the seafloor during the summer and autumn, and can exert dominance in water columns from late winter to early spring [16,48]. This species was observed to thrive in a temperature range of −1.8 to approximately 26.0 °C, with the most favorable temperature for growth reported to be between −1.5 and 6.0 °C [49]. Martens [50] conducted a study at the Sylt–Rømø tidal basin, where it was found that a low temperature of −2 °C was the main factor responsible for the bloom of *O. aurita*. However,

Baars [49] proposed that the species' normal growth is best achieved at temperatures below 20 °C. In contrast, Pasquet et al. [51] conducted a study to investigate the impact of temperature on chlorophyll–fluorescent photosynthesis parameters and found that this particular species is capable of tolerating temperatures as high as 28 °C. Temperature critically affects photosynthesis efficiency via its impact on enzyme kinetics [52]. Lower temperatures impair enzymatic processes associated with photosynthesis, while moderate temperature elevations enhance respiratory rates [47]. In contrast, extreme temperatures suppress both metabolic and respiratory functions in microalgae [53]. Optimal microalgal growth is attained when energy production in the thylakoid membranes aligns with energy consumption in the Calvin cycle [47]. Environmental variations, especially temperature shifts, can disrupt this balance, leading to adjustments in the photosynthetic components, including altered structural dimensions and Rubisco activity [47]. At lower temperatures, carboxylase activity significantly diminishes. In contrast, at higher temperatures, some photosynthetic enzymes may cease to function. However, the tolerance of microalgae to these temperatures varies by species [47].

In order to achieve optimal growth of our strain, it is crucial to consider the salinity of the medium. The growth curves of *O. aurita* OAOSH22 at various salinities (24, 27, 30, 33, and 36 psu) are depicted in Figure 4B. Biomass production reached its peak at 138.2 mg L^{-1} on day 9 of the experiment at a salinity level of 33 psu ($p < 0.05$). Additionally, biomass production exhibited comparable levels within the range of 24 to 30 psu. There is limited existing research on the correlation between growth and salinity levels in *O. aurita*. However, McQuoid [54] found that low salinity levels below 15 psu could have a detrimental impact on the germination process of *O. aurita*. Salinity stress can significantly impact the growth and biochemical composition of microalgae. Indeed, salt stress has been identified as a primary factor affecting both the growth and biochemical composition of these organisms [55]. Different microalgae species exhibit preferences for specific salinity ranges, which are often associated with their natural habitats. This salinity affects osmotic and ionic balances, subsequently impacting growth, photosynthesis, and metabolite production [56]. For example, when *Chlorella vulgaris* is grown under varying salinity levels, it exhibits distinct metabolite profiles, characterized by variations in lipids, proteins, and carbohydrates [56]. Furthermore, extreme salinity can induce osmotic stress, potentially restricting the activity of ATP synthase and thus influencing crucial metabolic pathways [57].

In the present study, the biomass production of *O. aurita* OAOSH22 increased to approximately 100 mg L^{-1} (Figure 4C), which represents a 1.2-fold increase compared to the control, when the concentration of nitrate in the medium was doubled ($p < 0.05$). However, when the nitrate concentrations were doubled or higher, there was little to no increase in biomass production. Therefore, a concentration of 150 mg N L^{-1}, which is twice the amount of nitrate found in the standard F/2 medium, appears to be adequate for the growth of *O. aurita* OAOSH22. Previous studies have consistently reported a strong correlation between the concentration of nitrate and the biomass of microalgae [58,59]. Similar findings have been observed in studies focusing on *O. aurita* [18,45,60]. However, Xia et al. [61] found that the biomass of *O. aurita* was produced at similar levels (approximately 4 g L^{-1}) under both high (18 mM) and low (6 mM) nitrate concentrations when cultured at 100 μmol photons $m^{-2} s^{-1}$. Additionally, they observed that biomass production increased approximately 1.5-fold under high nitrate concentrations compared to low concentrations when cultured at 300 μmol photons $m^{-2} s^{-1}$. This observation demonstrates that providing sufficient nutrients alone may not guarantee optimal growth outcomes, as the fulfillment of basic physical environmental conditions is also crucial in determining growth effects.

In the present study, we observed that the using a silicate-enriched medium significantly enhanced the biomass production of *O. aurita* OAOSH22 (Figure 4D). The growth of *O. aurita* exhibited a significant increase with the rise in silicate concentration in the medium ($p < 0.05$). The maximum biomass of 216 mg L^{-1} was observed on the 6th day of culture when the silicate concentration was 8 times higher than that of the standard F/2 medium. Xia et al. [60] demonstrated that an increase in silicate concentration positively correlated with the biomass production of *O. aurita*. However, contrary to the findings of the present study, no significant difference in biomass production was observed across varying silicate concentrations (27.3–104.2 mg L^{-1}). Silicates play a crucial role as vital nutrients in promoting diatom growth and are indispensable for the development of their cell walls composed of silica [62]. Therefore, the presence of silicates can have a substantial impact on the growth of diatoms [63]. When the availability of silicate is limited, a majority of diatoms experience disruptions in their cell cycles during the G1/S or G2/M phases, resulting in thinner frustules [62,64]. Additionally, the imposition of silicate restriction resulted in a reduction in the fucoxanthin content within *Phaeodactylum tricornutum* [65]. Conversely, it is imperative to appropriately adjust the concentration of silicate based on the target indicator material to be utilized, as research has shown that the restriction of silicate strongly promotes lipid accumulation in diatom cells [66].

In the case of phosphate, the biomass production of *O. aurita* exhibited no significant variation across different treatment concentrations, as depicted in Figure 4E. Phosphorus constitutes only 1% of the dry weight of microalgal cells; however, it plays a crucial role in limiting microalgal growth in natural ecosystems [22]. However, the impact of phosphorus on the growth of microalgae is relatively less significant compared to nitrogen. Additionally, it has been observed that beyond a certain concentration, phosphorus does not contribute to the growth and biomass production of microalgae [67,68]. Lu et al. [69] reported a negative correlation between phosphate concentration and biomass production in their study on the culture of *Nitzschia laevis*. As a result, it seems that there is no need to provide additional phosphate supply beyond the phosphate concentration present in the F/2 medium is not required for the growth of *O. aurita* OAOSH22.

Iron plays a crucial role in various metabolic processes that regulate photosynthesis via enzymatic reactions. It is a key component of cytochromes b and c, which function as electron transporters in both the photosynthetic and respiratory chain. This involvement of iron positively impacts the growth rate of diatoms [39,70]. The study conducted by Sahin et al. [12] demonstrated that *Nanofrustulum shiloi* exhibited 1.3- and 1.1-fold increases in response to an iron-rich environment. Contrarily, the limitation of iron frequently leads to an elevation in the silica composition of diatoms [71]. This, in turn, can cause a reduction in the concentration of silicate in the medium, ultimately resulting in the inhibition of diatom growth. In the present study, *O. aurita* OAOSH22 showed no significant variation in biomass production compared to the control group when given additional iron supplementation, (Figure 4F).

2.4. Carotenoid Content of Strain OAOSH22

The concentration of major carotenoid pigments in *Odontella aurita* OAOSH22 is depicted in Figure 5. Additionally, the LC chromatogram can be found in Figure S2. The main carotenoid pigment found in the isolate obtained in this study was fucoxanthin, with a content of 7.10 ± 0.47 mg g^{-1} DCW. It was also found to contain a small amount of diadinoxanthin (0.98 ± 0.06 mg g^{-1} DCW) and diatoxanthin (1.37 ± 0.04 mg g^{-1} DCW).

Fucoxanthin, a xanthophyll pigment derived from carotenoids, is a naturally occurring pigment. According to Matsuno [72], it is estimated that this particular carotenoid contributes to over 10% of the overall carotenoid production in nature and holds the highest prevalence among carotenoids in marine ecosystems. In the context of fucoxanthin production, it is highly probable that commercially viable microalgae species would include diatoms (Bacillariophyta), Prymnesiales (Haptophyta), and Chrysophyceae (Ochrophyta) [73]. One notable chemotaxonomic characteristic of diatoms is their high concentration of fucoxanthin, which is also found in brown algae [74]. The fucoxanthin content in diatoms is approximately 1–6%, which is over 100 times greater than that found in brown algae [75]. Moreau et al. [29] conducted a study on the anticancer activity of fucoxanthin against bronchopulmonary cancer and epithelial cancer and reported that *O. aurita* is a significant source of fucoxanthin. Fucoxanthin exhibits health-promoting effects attributed to its potent antioxidant properties [76]. Additionally, it demonstrates anti-obesity, anti-diabetic, anti-cancer, anti-angiogenic, anti-inflammatory, anti-metastatic, and anti-Alzheimer's disease activity [18,77]. Due to its diverse physiological activities, fucoxanthin has found extensive applications in the food, pharmaceutical, and cosmetic industries. It has gained significant attention as a functional material with anti-obesity properties [78]. Fucoxanthin has been scientifically proven to possess superior anti-cancer, anti-microbial, and free radical-scavenging properties compared to widely used compounds such as β-carotene and astaxanthin [79,80]. Based on the observed physiological activity, commercially available fucoxanthin-based products derived from microalgae, such as Fucovital and BrainPhyt[TM], have been developed and marketed [81]. Fucovital is a health supplement developed by Algatechnologies. It is the first food additive to obtain US FDA approval in 2017 in recognition of its liver function improvement effect and is the first microalgae-derived fucoxanthin-containing product released on the market [22]. BrainPhyt[TM] is a health supplement developed by Microphyt, France, and was listed on the US FDA's New Dietary Ingredients (NDI) in 2019 [22]. It can help improve cognitive ability and short- and long-term memory by alleviating oxidative and inflammatory stress in the brain. Consequently, it is anticipated that there will be a rise in demand in the future [22].

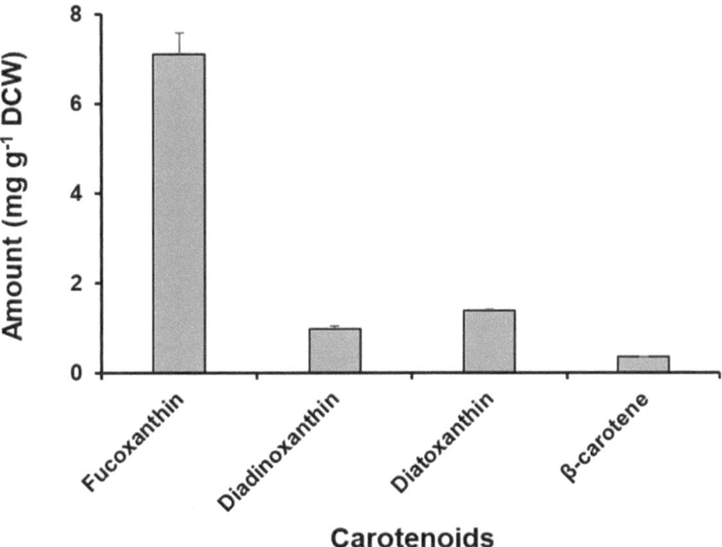

Figure 5. Concentrations of the major carotenoid pigments in *Odontella aurita* OAOSH22.

Xia et al. [60] conducted a study on the fucoxanthin content of *O. aurita* and found that it reached up to 21.7 mg g^{-1} (dry weight). The actual content varied depending on the optimal culture conditions, specifically a light intensity of 100 μmol photons m^{-2} s^{-1} and a nitrate supply of 6 mM. This finding was significant, as it represented the highest reported fucoxanthin content in diatoms [82]. Although the fucoxanthin content observed in our study was lower than that reported in previous studies focusing on *O. aurita*, it was found to be comparable to or higher than the levels found in other species such as *Chaetoceros gracilis*, *Cylindrotheca closterium*, *Nitzschia laevis*, and *Phaeodactylum tricornutum* [68,83,84]. The composition of high-value-added substances can vary among different strains, even within the same species [32]. This variation is influenced by various culture conditions including light intensity [60], temperature [85], salinity [58], nutrient concentration [60], and culture media [86]. For instance, it has been observed that with an increase in light intensity, there is a corresponding increase in microalgal biomass [73]. However, it has also been noted that this increase in light intensity leads to a decrease in fucoxanthin production [87]. Light intensity exceeding 150 µmol photons $m^{-2} s^{-1}$ has been found to induce the synthesis of photoprotective pigments, namely diadinoxanthin and diatoxanthin [88]. Therefore, further investigation into the optimal culture conditions is required in order to enhance the content of fucoxanthin.

2.5. Fatty Acids Composition of Strain OAOSH22

The composition of fatty acids in *Odontella aurita* OAOSH22 consisted of saturated fatty acids (SFAs) (42.5%), monounsaturated fatty acids (MUFAs) (37.8%), and polyunsaturated fatty acids (PUFAs) (19.7%). The main fatty acids synthesized by this strain were palmitoleic acid (C16:1, 36.4 ± 1.4%), palmitic acid (hexadecanoic acid, C16:0, 25.8 ± 1.0%), eicosapentaenoic acid (EPA, timnodonic acid, C20:5ω3, 17.7 ± 3.3%), and myristic acid (tetradecanoic acid, C14:0, 15.6 ± 1.2%) (Table 1).

Table 1. Fatty acid composition (% total fatty acids) of *Odontella aurita* OAOSH22.

	Fatty Acids		Amount (mg g^{-1} DCW)	Composition (%)
SFA	Myristic acid	C14:0	9.65 ± 0.23	15.61 ± 1.20
	Palmitic acid	C16:0	15.96 ± 0.27	25.76 ± 0.95
	Stearic acid	C18:0	0.68 ± 0.02	1.10 ± 0.09
MUFA	Palmitoleic acid	C16:1n7	22.52 ± 0.36	36.34 ± 1.36
	Oleic acid	C18:1n9	0.93 ± 0.02	1.50 ± 0.05
PUFA	Linoleic acid	C18:2n6 cis	0.75 ± 0.03	1.21 ± 0.01
	Gamma-linolenic acid (GLA)	C18:3n6	0.30 ± 0.03	0.48 ± 0.03
	Arachidonic acid (AA)	C20:4n6	0.22 ± 0.06	0.33 ± 0.09
	Eicosapentaenoic acid (EPA)	C20:5n3	11.07 ± 2.63	17.66 ± 3.28

The distribution of fatty acids exhibits significant variation among different microalgae taxa as well as within species. Diatoms are commonly known to possess a significant concentration of various fatty acids, including myristic acid (C14:0), palmitic acid (C16:0), palmitoleic acid (C16:1), stearic acid (C18:0), oleic acid (C18:1), and EPA (C20:5ω3) [13,89]. These fatty acids play a significant role in various industries, including food, pharmaceuticals, cosmeceuticals, aquaculture, and biofuel [77]. In particular, PUFAs, represented by EPA and docosahexaenoic acid (DHA), have garnered significant interest. PUFAs refers to unsaturated fatty acids containing 18 or more carbon and two or more double bonds [90]. PUFAs, such as omega-3 or omega-6 unsaturated fatty acids, play crucial roles in various physiological processes within the human body. However, these fatty acids are either not naturally synthesized (e.g., linoleic acid and α-linolenic acid) or are synthesized in limited quantities (e.g., EPA, DHA, and arachidonic acid). Consequently, it is necessary to obtain these PUFAs via dietary intake [91]. EPA offers a range of nutritional and health advantages, including its anti-inflammatory, anti-microbial, anti-cancer, vision and cardiovascular-protective, anti-Parkinsonian syndrome, and anti-Alzheimer's disease effects [18,77].

Odontella aurita is a representative EPA-rich species among microalgae and is known to have an EPA content of more than 20% total fatty acids [18,45,51,92]. The fatty acid composition of *O. aurita* OAOSH22 was similar to that of previous studies, but the content of EPA was slightly lower, measuring at 17.7%. Several previous studies have documented that a deficiency of silicate in the growth medium stimulates lipid synthesis and leads to an increase in EPA levels [65,93]. Hence, it is hypothesized that the high concentration of silicate in the medium used in this study had a negative effect on the EPA content.

Currently, the primary source of EPA is derived from oily fish species such as salmon, mackerel, pilchard, herring, and trout [19]. However, diatoms present a promising alternative source of EPA, offering the advantage of meeting vegan dietary requirements [85].

In particular, the species *O. aurita* has already been commercially utilized for food in Europe, suggesting that it holds significant potential in the food and functional food industries [21,31]. In 2018, SAS Odontella, a French startup, introduced Solmon, a vegan food product that incorporates the extract of *Odontella aurita* to maintain the flavor profile of smoked salmon while also being a significant source of omega-3 fatty acids [94].

Myristic acid, which is the main fatty acid present in this strain, acts as a stabilizing agent for various proteins, including those involved in immune system function and anti-cancer properties [95]. Additionally, it has extensive applications in the beauty industry as a fragrance, surfactant, detergent, and emulsifier [96]. Palmitoleic acid has been documented to exhibit antibacterial properties [77], and has recently been suggested as a potential food ingredient for managing obesity [97].

3. Materials and Methods

3.1. Sample Collection and Isolation

A sample was collected from the coastal water at the port of Osan (38°5′25.51″ N, 128°39′53.36″ E) in Yangyang-gun, Gangwon-do, Republic of Korea, on 16 February 2022 by the Survey on Marine Bio-Resources. Cell isolation was conducted using the capillary method with a Pasteur pipette while being observed under an Eclipse Ti-U inverted microscope (Nikon, Tokyo, Japan). The isolated cells were subsequently transferred to a cell culture flask (SPL Life Sciences, Pocheon, Republic of Korea) containing F/2 medium supplemented with silicate (Sigma Aldrich Co., St. Louis, MO, USA) and a 0.2% antibiotic mixture (Penicillin-streptomycin-neomycin) (Sigma Aldrich Co.). To assess the ability of the isolated strain to grow on a solid medium, the monoculture strain was inoculated onto a 1% agar plate (Bacto Agar, BD Difco Ltd., Detroit, MI, USA) supplemented with F/2 medium. Culture strains cultivated in both liquid and solid media were periodically transferred to fresh medium at intervals of three weeks and two months, respectively. The culture strains were then incubated at a temperature of 17 °C, a 14:10 h light/dark cycle, and an irradiance of 40 μmol photons $m^{-2} s^{-1}$.

3.2. Morphological Identification

The culture strain was harvested via centrifugation at $2100 \times g$ for 5 min and mixed with glycerol gelatin (Sigma Aldrich Co.) to be mounted on a slide. The mixed sample was placed dropwise on a glass slide and fixed in position with a coverslip. Finally, the margin of the coverslip was sealed with CoverGrip coverslip sealant (Biotium, Hayward, CA, USA). The slide was examined using a Nikon Eclipse Ni light microscope. For scanning electron microscopy (SEM), the cultured strain was fixed in 5% Lugol's solution, filtered through a polycarbonate membrane with a pore size of 3 μm and a diameter of 25 mm (Advantec, Tokyo, Japan), and washed three times with sterile distilled water. The membrane was dehydrated in a graded series of ethanol (10%, 30%, 50%, 70%, 90%, and 100%) and finally dried using tetramethylsilane (Sigma Aldrich Co.). The membrane was mounted on a stub and sputter-coated with gold using an MC1000 ion sputter (Hitachi, Tokyo, Japan). The cells and surface morphology were observed using a high-resolution Zeiss Sigma 500 VP field-emission scanning electron microscope (FE-SEM, Carl Zeiss, Oberkochen, Germany).

3.3. Molecular Identification

The culture medium, which contained the strain, was transferred into a 50 mL conical tube and subjected to centrifugation at a speed of $5370 \times g$ for a duration of 5 min. The supernatant was subsequently removed. Genomic DNA extraction was performed using the DNeasy PowerSoil Pro Kit (Qiagen Inc., Hilden, Germany) according to the manufacturer's instructions. Polymerase chain reaction (PCR) amplification was conducted using the Diatom9F [98]/EukBR [99] primer pairs in order to amplify the 18S rRNA sequence. PCR analysis was conducted following the protocol outlined by Raja et al. [100]. The PCR product underwent purification using ExoSAP-IT Express PCR Product Cleanup Reagent (Thermo Fisher Scientific, MA, USA) and was subsequently sequenced by Cosmogenetech

Co., Ltd. (Seoul, Republic of Korea). The sequence underwent trimming, assembly, and alignment using Geneious Prime v.2022.2.2 (Biomatters Ltd., Auckland, New Zealand). The data set of 18S rRNA sequences was compiled, comprising the genetic sequences of 16 species belonging to the Eupodiscales order, as retrieved from GenBank. *Biddulphia biddulphiana* (JX401227) was utilized as an outgroup. Phylogenetic analyses were performed using maximum likelihood (ML) and Bayesian inference (BI) methods. Randomized Axelerated Maximum Likelihood (RAxML) v.8.2.10 [101] and MrBayes version 3.2.7 [102] were used for the ML and BI analyses, respectively. ML and BI analyses were conducted using the methodologies outlined by An et al. [103].

3.4. Determination of Optimal Culture Conditions

The isolate was cultured under various temperature, salinity, and nutrient conditions to determine the optimal culture conditions. The experimental conditions for water temperature, salinity, and irradiance were set based on previous research findings [38,49,51,54]. Additionally, the concentrations of each nutrient were set at levels that were 2, 4, and 8 times higher, respectively, than those found in the commonly used F/2 medium for culturing marine microalgae. To determine the optimal growth temperature, the temperature experiment compared growth at temperatures ranging from 5 to 25 °C. The salinity experiment, on the other hand, compared growth at salinities ranging from 24 to 36 psu using F/2 media that contained silicate. These experiments were conducted using a multi-thermo incubator (MTI-202B, Eyela, Tokyo, Japan). In addition, to confirm the growth characteristics based on nutrient concentrations, the isolate was cultured at 17 °C using a medium enriched with each nutrient (nitrate, silicate, phosphate, and iron) (Sigma Aldrich Co.). The concentration of each nutrient in the standard F/2 medium was set to the control level. Optimal conditions for each factor were determined via daily growth tests. Detailed experimental conditions for each factor are shown in Table 2. Samples for all tests, with the exception of the test aimed at determining optimal irradiance, were carried out by introducing 30 mL of medium containing the strain into a 25 cm^2 cell culture flask. A dimmable LED panel was used as the light source for the cultivation process. Irradiance was measured using the HD2102.2 Portable Luxmeter Data Logger, which was equipped with LP471PAR Quantum Radiometric Probe (Delta OHM, Caselle, Padova, Italy).

Table 2. Experimental conditions for determining optimal cultivation conditions for each factor. The minimum condition for each nutrient utilized in the experiments is its presence in a standard F/2 medium.

	Experimental Conditions
Temperature (°C)	5, 10, 15, 20, 25
Salinity (psu)	24, 27, 30, 33, 36
Nutrients (mg L^{-1})	
Nitrate (NaNO$_3$)	75, 150, 300, 600
Silicate (Na$_2$SiO$_3 \bullet$9H$_2$O)	15, 30, 60, 120
Phosphate (NaH$_2$PO$_4 \bullet$H$_2$O)	5, 10, 20, 40
Iron (FeCl$_3 \bullet$6H$_2$O)	3.15, 6.3, 12.6, 25.2

To determine the optimal growth irradiance, the pulse amplitude modulation (PAM) fluorometry technique was used in this study. The PAM technique is commonly employed to assess parameters associated with the photosynthetic efficiency of microalgae using chlorophyll fluorescence quenching analysis [104]. RLCs obtained using the PAM technique offer comprehensive insights into the saturation characteristics of electron transport and the overall photosynthetic capacity of microalgal strains [105]. This information can be used to determine the optimal level of irradiance required for cultivating specific types of microalgae [106]. It can also be used to estimate the maximum productivity of the culture when provided with the optimal irradiance [107]. The determination of the light saturation coefficient (E_k), which indicates the point at which photosynthesis reaches saturation, involves considering two factors: the maximum electron transport rate (ETR_{max}) and the initial slope (α) of the RLC. The initial slope of a graph represents the quantum efficiency of photosynthetic electron transport [108]. E_k can be regarded as the ideal irradiance level for the cultivation of microalgal strains [109]. A 3.0 mL quantity of the culture strain was placed in a DUAL-K25 quartz glass cuvette, which was supplied with the Dual-PAM and dark-adapted for 30 min before PAM measurement. Rapid light curve (RLC, ETR versus irradiance curve) was conducted at eight incremental irradiances (47, 56, 69, 119, 190, 244, 398, and 610 µmol photons m^{-2} s^{-1}) of actinic light using the Dual-PAM-100 (Heinz Walz Gmbh, Effeltrich, Germany) equipped with an Optical Unit ED-101US/MD. The light curve was fitted according to the model of Platt et al. [46] to determine the maximum electron transport rate (ETR_{max}), the initial slope of the curve (α), and the irradiance at which ETR saturation occurs (E_k). Data processing was performed following the method described in Ralph and Gademann [105] using SigmaPlot v.12.3 (Systat Software Inc., San Jose, CA, USA).

3.5. Determination of Biomass

Biomass production was calculated based on the equation derived from the calibration curve between chlorophyll fluorescence (with an excitation wavelength of 440 nm and emission wavelength of 680 nm) and dry cell weight (DCW) (Figure S3). To obtain a calibration curve of chlorophyll fluorescence versus biomass weight, we measured the chlorophyll fluorescence of five pre-cultures of algae with different cell densities using a microplate reader (Synergy H1, BioTek, Winooski, VT, USA). Each sample was then placed in a pre-weighed tube and centrifuged at 8400× g. After the supernatant was removed, the sample was washed once with distilled water to remove salt and then centrifuged again. The cell pellet was lyophilized using a benchtop freeze dryer for 24 h (Freezezone 4.5, Labconco, Kansas City, MO, USA). Finally, the DCW was determined by subtracting the weight of the pre-weighed tube from the total weight of the tube containing the pellet.

3.6. Carotenoids Analysis

Carotenoid was analyzed using a slight modification of the method from Kang et al. [110]. To quantify the carotenoid content, algal biomass concentration was calculated using a pre-determined conversion equation mentioned earlier. Subsequently, algal cells were collected by centrifugation at 33,600× g for 2 min. The supernatant was then removed, and the pigments were extracted from the cells using 100% methanol and algal cells disrupted via ultrasonic water bath (DAIHAN Scientific, Republic of Korea) for 3 min at 60 °C. The resulting supernatant was then filtered through a 0.2 µm PTFE membrane filter (Millipore, Billerica, MA, USA). Carotenoid pigments were analyzed using an Agilent 1260 Infinity HPLC system (Agilent, Waldbronn, Germany) equipped with a Spherisorb 5.0 µm ODS1 4.6 × 250 mm cartridge column (Waters, St. Louis, MO, USA) at 40 °C. Chromatograms were identified by comparing them to carotenoid standards including fucoxanthin (Sigma Aldrich, St. Louis, MO, USA), diadinoxanthin, diatoxanthin, and β-carotene (DHI, Hørsholm, Denmark), and the concentration of each pigment was calculated using the peak area of the standard pigments.

3.7. Fatty Acid Analysis

Cells were harvested via centrifugation at 16,500× g for fatty acid analysis. After + as much supernatant as possible was removed, the pellets were lyophilized via freeze-drying at −110 °C under a vacuum for 24 h. Fatty acid extraction was performed following the methods described by Garces and Mancha [111]. The fatty acid composition was analyzed using a 7890A gas chromatograph (Agilent, Wilmington, DE, USA) equipped with a flame ionization detector (280 °C, H2 35 mL min^{-1}, air 350 mL min^{-1}, He 30 mL min^{-1}) and a DB-23 column (60 mm × 0.25 mm × 0.25 µm film thickness; Agilent). The initial GC oven temperature was set at 80 °C and maintained for 3 min. The temperature was ramped up at a rate of 15 °C min^{-1} to 200 °C and held for 8 min. It was then ramped at a rate of

1 °C min^{-1} to 215 °C and held for 8 min. After that, it was ramped at a rate of 2 °C min^{-1} to 250 °C and held for 5 min. Finally, it was ramped at a rate of 50 °C min^{-1} to 80 °C and held for 3 min. The sample (2 μL) was injected with a split ratio of 10:1. The injector and detector temperatures were set at 250 °C and 280 °C, respectively. The identification of fatty acids was performed by comparing their retention time to the retention time of standards (Supelco 37-component FAME mix; Supelco, Bellefonte, PA, USA) and an internal standard (pentadecanoic acid; Sigma Aldrich Co.). Fatty acid analysis was conducted at the National Instrumentation Center for Environmental Management (NICEM) at Seoul National University in the Republic of Korea.

3.8. Statistics Analysis

All experimental procedures were conducted in triplicate. First, the data were subjected to analysis in order to evaluate its normality through the utilization of the Shapiro–Wilk test. Once the normality of the data was confirmed, the mean differences between treatments were examined via one-way analysis of variance (ANOVA) with Tukey's post hoc test ($p < 0.05$) using the SPSS v.14.0 software (SPSS Inc., Chicago, IL, USA). The mean and standard deviations are reported, and distinct letters are used to indicate a statistically significant difference at a significance level of $p < 0.05$.

4. Conclusions

In the present study, we obtained an indigenous strain of *Odontella aurita* OAOSH22 and conducted an analysis of its fundamental characteristics and optimal culture conditions. As observed in previous studies on this species, the current isolate also exhibited elevated levels of fucoxanthin and EPA. This particular species has obtained certification and has been utilized as a cosmetic ingredient not only in Korea but also in numerous other countries. Furthermore, it possesses significant potential as a material for food and health functional products. The composition of high-value-added substances is subject to the influence of diverse cultural conditions, including light intensity, temperature, salinity, nutrient concentration, and culture media. Therefore, additional investigation is required to enhance the synthesis of valuable compounds. In addition, in order to establish optimal conditions for large-scale cultivation in an industrial environment, follow-up research to reduce cultivation costs, such as exploring alternatives to expensive media, must also be conducted.

Supplementary Materials: The following supporting information can be downloaded at: https://www.mdpi.com/article/10.3390/md21110563/s1, Figure S1: Relative growth rates of *Odontella aurita* OAOSH22 under different conditions of temperature (a), salinity (b), nitrate (c), silicate (d), phosphate (e), and iron (f). Symbols and error bars represent the mean $\pm$ SE ($n = 3$). The same letter indicates homologous groups as recognized via the Tukey HSD test at $p = 0.05$. Figure S2: The chromatogram, acquired via high-performance liquid chromatography with photodiode array detection (HPLC-PDA), of the carotenoid extract derived from *Odontella aurita* OAOSH22. Figure S3: Calibration curve for *Odontella aurita* OAOSH22 based on fluorescence excitation of chlorophyll.

Author Contributions: Conceptualization, S.M.A.; Data curation, K.C. and N.S.K.; Formal analysis, E.S.K. and H.K.; Funding acquisition, G.C.; Investigation, S.M.A., K.C., E.S.K. and H.K.; Methodology, S.M.A.; Project administration, G.C.; Writing—original draft, S.M.A.; Writing—review and editing, K.C. and N.S.K. All authors have read and agreed to the published version of the manuscript.

Funding: This research was also supported by the development of useful materials derived from marine microorganisms and microalgae (2023M00600) funded by the National Marine Biodiversity Institute of Korea (MABIK).

Institutional Review Board Statement: Not applicable.

Data Availability Statement: Data are contained within the article.

Conflicts of Interest: The authors declare no conflict of interest.

References

1. Dineshbabu, G.; Goswami, G.; Kumar, R.; Sinha, A.; Das, D. Microalgae–nutritious, sustainable aqua- and animal feed source. *J. Funct. Foods* **2019**, *62*, 103545. [CrossRef]
2. Nwoba, E.G.; Ogbonna, C.N.; Ishika, T.; Vadiveloo, A. Microalgal pigments: A source of natural food colors. In *Microalgae Biotechnology for Food, Health and High Value Products*; Alam, M.A., Xu, J.L., Wang, Z., Eds.; Springer: Singapore, 2020; pp. 81–123.
3. Orejuela-Escobar, L.; Gualle, A.; Ochoa-Herrera, V.; Philippidis, G.P. Prospects of microalgae for biomaterial production and environmental applications at biorefineries. *Sustainability* **2021**, *13*, 3063. [CrossRef]
4. Chandra, R.; Iqbal, H.M.; Vishal, G.; Lee, H.S.; Nagra, S. Algal biorefinery: A sustainable approach to valorize algal-based biomass towards multiple product recovery. *Bioresour. Technol.* **2019**, *278*, 346–359. [CrossRef] [PubMed]
5. Priyadarshani, I.; Rath, B. Commercial and industrial applications of micro algae—A review. *J. Algal Biomass Util.* **2012**, *3*, 89–100.
6. Boyd, P.W.; Strzepek, R.; Fu, F.; Hutchins, D.A. Environmental control of open-ocean phytoplankton groups: Now and in the future. *Limnol. Oceanogr.* **2010**, *55*, 1353–1376. [CrossRef]
7. Malviya, S.; Scalco, E.; Audic, S.; Vincent, F.; Veluchamy, A.; Poulain, J.; Wincker, P.; Iudicone, D.; Vargas, C.; Bittner, L.; et al. Insights into global diatom distribution and diversity in the world's ocean. *Proc. Natl. Acad. Sci. USA* **2016**, *133*, E1516–E1525. [CrossRef]
8. Dugdale, R.C.; Wilkerson, F.P.; Minas, H.J. The role of silicate pump in driving new production. *Deep Sea Res. Part I* **1995**, *42*, 697–719. [CrossRef]
9. Shah, M.R.; Lutzu, G.A.; Alam, A.; Sarker, P.; Chowdhury, K.; Parsaeimehr, A.; Liang, Y.; Daroch, M. Microalgae in aquafeeds for a sustainable aquaculture industry. *J. Appl. Phycol.* **2018**, *30*, 197–213. [CrossRef]
10. Wang, S.; Verma, S.K.; Hakeem Said, I.; Thomsen, L.; Ullrich, M.S.; Kuhnert, N. Changes in the fucoxanthin production and protein profiles in *Cylindrotheca closterium* in response to blue light-emitting diode light. *Microb. Cell Factories* **2018**, *17*, 1–13. [CrossRef]
11. Lu, X.; Liu, B.; He, Y.; Guo, B.; Sun, H.; Chen, F. Novel insights into mixotrophic cultivation of *Nitzschia laevis* for co-production of fucoxanthin and eicosapentaenoic acid. *Bioresour. Technol.* **2019**, *294*, 122145. [CrossRef]
12. Sahin, M.S.; Khazi, M.I.; Demirel, Z.; Dalay, M.C. Variation in growth, fucoxanthin, fatty acids profile and lipid content of marine diatoms *Nitzschia* sp. and *Nanofrustulum shiloi* in response to nitrogen and iron. *Biocatal. Agric. Biotechnol.* **2019**, *17*, 390–398. [CrossRef]
13. Marella, T.K.; Tiwari, A. Marine diatom *Thalassiosira weissflogii* based biorefinery for co-production of eicosapentaenoic acid and fucoxanthin. *Bioresour. Technol.* **2020**, *307*, 123245. [CrossRef] [PubMed]
14. Stiefvatter, L.; Lehnert, K.; Frick, K.; Montoya-Arroyo, A.; Frank, J.; Vetter, W.; Schmid-Staiger, U.; Bischoff, S.C. Oral Bioavailability of Omega-3 Fatty Acids and Carotenoids from the Microalgae *Phaeodactylum tricornutum* in Healthy Young Adults. *Mar. Drugs* **2021**, *19*, 700. [CrossRef] [PubMed]
15. Zhang, H.; Gong, P.; Cai, Q.; Zhang, C.; Gao, B. Maximizing fucoxanthin production in *Odontella aurita* by optimizing the ratio of red and blue light-emitting diodes in an auto-controlled internally illuminated photobioreactor. *Bioresour. Technol.* **2022**, *344*, 126260. [CrossRef] [PubMed]
16. Kraberg, A.; Baumann, M.; Dürselen, C.D. *Coastal Phytoplankton: Photo guide for Northern European Seas*; Verlag Dr. Friedrich Pfeil: Munich, Germany, 2010; pp. 96–97.
17. Lavigne, A.S.; Sunesen, I.; Sar, E.A. Morphological, taxonomic and nomenclatural analysis of species of *Odontella, Trieres* and *Zygoceros* (*Triceratiaceae, Bacillariophyta*) from Anegada Bay (Province of Buenos Aires, Argentina). *Diatom Res.* **2015**, *30*, 307–331. [CrossRef]
18. Xia, S.; Gao, B.; Fu, J.; Xiong, J.; Zhang, C. Production of fucoxanthin, chrysolaminarin, and eicosapentaenoic acid by *Odontella aurita* under different nitrogen supply regimes. *J. Biosci. Bioeng.* **2018**, *126*, 723–729. [CrossRef]
19. Pike, I.H.; Jackson, A. Fish oil: Production and use now and in the future. *Lipid. Technol.* **2010**, *22*, 59–61. [CrossRef]
20. Gohara-Beirigo, A.K.; Matsudo, M.C.; Cezare-Gomes, E.A.; de Carvalho, J.C.M.; Danesi, E.D.G. Microalgae trends toward functional staple food incorporation: Sustainable alternative for human health improvement. *Trends Food Sci. Technol.* **2022**, *125*, 185–199. [CrossRef]
21. Haimeur, A.; Ulmann, L.; Mimouni, V.; Guéno, F.; Pineau-Vincent, F.; Meskini, N.; Tremblin, G. The role of *Odontella aurita*, a marine diatom rich in EPA, as a dietary supplement in dyslipidemia, platelet function and oxidative stress in high-fat fed rats. *Lipids Health Dis.* **2012**, *11*, 1–13. [CrossRef]
22. Khaw, Y.S.; Yusoff, F.M.; Tan, H.T.; Noor Mazli, N.A.I.; Nazarudin, M.F.; Shaharuddin, N.A.; Omar, A.R.; Takahashi, K. Fucoxanthin Production of Microalgae under Different Culture Factors: A Systematic Review. *Mar. Drugs* **2022**, *20*, 592. [CrossRef]
23. Bernaerts, T.M.; Gheysen, L.; Kyomugasho, C.; Kermani, Z.J.; Vandionant, S.; Foubert, I.; Hendrickx, M.E.; Van Loey, A.M. Comparison of microalgal biomasses as functional food ingredients: Focus on the composition of cell wall related polysaccharides. *Algal Res.* **2018**, *32*, 150–161. [CrossRef]
24. Terriente-Palacios, C.; Castellari, M. Levels of taurine, hypotaurine and homotaurine, and amino acids profiles in selected commercial seaweeds, microalgae, and algae-enriched food products. *Food Chem.* **2022**, *368*, 130770. [CrossRef] [PubMed]
25. Avis de l'Agence Française de Sécurité Sanitaire des Aliments Relatif à la Demande D'évaluation de la Démonstration de L'équivalence en Substance D'une Microalgue *Odontella aurita* avec des Algues Autorisées (AFSSA Saisine no. 2001-SA-0082). Available online: https://www.anses.fr/fr/system/files/AAAT2001sa0082.pdf (accessed on 12 September 2023).

26. Summary of Notifications Received by the Commission until 31 December 2004 Pursuant to Article 5 of Regulation (EC) No 258/97 of the European Parliament and of the Council (2005/C 208/2). Available online: https://eur-lex.europa.eu/legal-content/EN/TXT/?uri=CELEX%3A52005XC0825%2801%29&qid=1674696011584 (accessed on 12 September 2023).

27. Commission Implementing Regulation (EU) 2017/2470 of 20 December 2017 Establishing the Union List of Novel Foods in Accordance with Regulation (EU) 2015/2283 of the European Parliament and of the Council on Novel Foods. Available online: https://eur-lex.europa.eu/eli/reg_impl/2017/2470/oj (accessed on 12 September 2023).

28. US Food and Drug Administration. GRAS Notices. Available online: https://www.cfsanappsexternal.fda.gov/scripts/fdcc/cfc/XMLService.cfc?method=downloadxls&set=GRASNotices (accessed on 12 September 2023).

29. Moreau, D.; Tomasoni, C.; Jacquot, C.; Kaas, R.; Le Guedes, R.; Cadoret, J.P.; Muller-Feuga, A.; Kontiza, I.; Viagas, C.; Roussis, V.; et al. Cultivated microalgae and carotenoid fucoxanthin from *Odontella aurita* as potent anti-proliferative agents in bronchopulmonary and epithelial cell lines. *Environ. Toxicol. Pharmacol.* **2006**, *22*, 97–103. [CrossRef] [PubMed]

30. Zhao, W.; Yao, R.; He, X.S.; Liao, Z.H.; Liu, Y.T.; Gao, B.Y.; Zhang, C.W.; Niu, J. Beneficial contribution of the microalga *Odontella aurita* to the growth, immune response, antioxidant capacity, and hepatic health of juvenile golden pompano (*Trachinotus ovatus*). *Aquaculture* **2022**, *555*, 738206. [CrossRef]

31. Mimouni, V.; Ulmann, L.; Pasquet, V.; Mathieu, M.; Picot, L.; Bougaran, G.; Cadoret, J.P.; Morant-Manceau, A.; Schoefs, B. The potential of microalgae for the production of bioactive molecules of pharmaceutical interest. *Curr. Pharm. Biotechnol.* **2012**, *13*, 2733–2750. [CrossRef]

32. Benedetti, M.; Vecchi, V.; Barera, S.; Dall'Osto, L. Biomass from microalgae: The potential of domestication towards sustainable biofactories. *Microb. Cell Factories* **2018**, *17*, 1–18. [CrossRef]

33. Novoveská, L.; Ross, M.E.; Stanley, M.S.; Pradelles, R.; Wasiolek, V.; Sassi, J.F. Microalgal carotenoids: A review of production, current markets, regulations, and future direction. *Mar. Drugs* **2019**, *17*, 640. [CrossRef]

34. Rumin, J.; Gonçalves de Oliveira Junior, R.; Bérard, J.B.; Picot, L. Improving microalgae research and marketing in the European Atlantic area: Analysis of major gaps and barriers limiting sector development. *Mar. Drugs* **2021**, *19*, 319. [CrossRef]

35. Sims, P.A.; Williams, D.M.; Ashworth, M. Examination of type specimens for the genera *Odontella* and *Zygoceros* (*Bacillariophyceae*) with evidence for the new family *Odontellaceae* and a description of three new genera. *Phytotaxa* **2018**, *382*, 1–56. [CrossRef]

36. Prüser, T.F.; Braun, P.G.; Wiacek, C. Microalgae as a novel food. Potential and legal framework. *Ernahr. Umsch* **2021**, *68*, 78–85.

37. Mendes, M.C.; Navalho, S.; Ferreira, A.; Paulino, C.; Figueiredo, D.; Silva, D.; Gao, F.; Gama, F.; Bombo, G.; Jacinto, R.; et al. Algae as food in Europe: An overview of species diversity and their application. *Foods* **2022**, *11*, 1871. [CrossRef] [PubMed]

38. He, Q.; Yang, H.; Wu, L.; Hu, C. Effect of light intensity on physiological changes, carbon allocation and neutral lipid accumulation in oleaginous microalgae. *Bioresour. Technol.* **2015**, *191*, 219–228. [CrossRef]

39. Moreno, C.M.; Lin, Y.; Davies, S.; Monbureau, E.; Cassar, N.; Marchetti, A. Examination of gene repertoires and physiological responses to iron and light limitation in Southern Ocean diatoms. *Polar Biol.* **2018**, *41*, 679–696. [CrossRef]

40. Maltsev, Y.; Maltseva, K.; Kulikovskiy, M.; Maltseva, S. Influence of light conditions on microalgae growth and content of lipids, carotenoids, and fatty acid composition. *Biology* **2021**, *10*, 1060. [CrossRef]

41. Bashir, F.; Rehman, A.U.; Szabó, M.; Vass, I. Singlet oxygen damages the function of Photosystem II in isolated thylakoids and in the green alga *Chlorella sorokiniana*. *Photosynth. Res.* **2021**, *149*, 93–105. [CrossRef] [PubMed]

42. Richardson, K.; Beardall, J.; Raven, J.A. Adaptation of unicellular algae to irradiance. an analysis of strategies. *N. Phytol.* **1983**, *93*, 157–191. [CrossRef]

43. Kirk, J.T.O. *Light and Photosynthesis in Aquatic Ecosystems*, 2nd ed.; Cambridge University Press: Cambridge, UK, 1994; p. 509.

44. Masojídek, J.; Koblížek, M.; Torzillo, G. Photosynthesis in Microalgae. In *Handbook of Microalgal Culture: Biotechnology and Applied Phycology*; Richmond, A., Ed.; Blackwell Science Ltd.: Oxford, UK, 2004; pp. 20–39.

45. Roleda, M.Y.; Slocombe, S.P.; Leakey, R.J.; Day, J.G.; Bell, E.M.; Stanley, M.S. Effects of temperature and nutrient regimes on biomass and lipid production by six oleaginous microalgae in batch culture employing a two-phase cultivation strategy. *Bioresour. Technol.* **2013**, *129*, 439–449. [CrossRef]

46. Platt, T.G.C.L.; Gallegos, C.L.; Harrison, W.G. Photoinhibition of photosynthesis in natural assemblages of marine phytoplankton. *J. Mar. Res.* **1980**, *38*, 687–701.

47. Ras, M.; Steyer, J.P.; Bernard, O. Temperature effect on microalgae: A crucial factor for outdoor production. *Rev. Environ. Sci. Biotechnol.* **2013**, *12*, 153–164. [CrossRef]

48. Hoppenrath, M.; Elbrächter, M.; Drebes, G. *Marine Phytoplankton Selected Microphytoplankton Species from the North Sea around Helgoland and Sylt*; Schweizerbart Sche Vlgsb.: Stuttgart, Germany, 2009; p. 264.

49. Baars, J.W.M. Autecological investigations on marine diatoms. 4: *Biddulphia aurita* (Lyngb.) Brebisson et Godey—A succession of spring diatoms. *Hydrobiol. Bull.* **1985**, *19*, 109–116. [CrossRef]

50. Martens, P. Effects of the severe winter 1995/96 on the biological oceanography of the Sylt-Rømø tidal basin. *Helgol. Mar. Res.* **2001**, *55*, 166–169. [CrossRef]

51. Pasquet, V.; Ulmann, L.; Mimouni, V.; Guihéneuf, F.; Jacquette, B.; Morant-Manceau, A.; Tremblin, G. Fatty acids profile and temperature in the cultured marine diatom *Odontella aurita*. *J. Appl. Phycol.* **2014**, *26*, 2265–2271. [CrossRef]

52. Peng, X.; Meng, F.; Wang, Y.; Yi, X.; Cui, H. Effect of pH, temperature, and CO_2 concentration on growth and lipid accumulation of *Nannochloropsis* sp. MASCC 11. *J. Ocean Univ. China* **2020**, *19*, 1183–1192. [CrossRef]

53. Breuer, G.; Lamers, P.P.; Martens, D.E.; Draaisma, R.B.; Wijffels, R.H. Effect of light intensity, pH, and temperature on triacylglycerol (TAG) accumulation induced by nitrogen starvation in *Scenedesmus obliquus*. *Bioresour. Technol.* **2013**, *143*, 1–9. [CrossRef] [PubMed]

54. McQuoid, M.R. Influence of salinity on seasonal germination of resting stages and composition of microplankton on the Swedish west coast. *Mar. Ecol. Prog. Ser.* **2005**, *289*, 151–163. [CrossRef]

55. Pandit, P.R.; Fulekar, M.H.; Karuna, M.S.L. Effect of salinity stress on growth, lipid productivity, fatty acid composition, and biodiesel properties in *Acutodesmus obliquus* and *Chlorella vulgaris*. *Environ. Sci. Pollut. Res.* **2017**, *24*, 13437–13451. [CrossRef]

56. Haris, N.; Manan, H.; Jusoh, M.; Khatoon, H.; Katayama, T.; Kasan, N.A. Effect of different salinity on the growth performance and proximate composition of isolated indigenous microalgae species. *Aquac. Rep.* **2022**, *22*, 100925. [CrossRef]

57. Allakhverdiev, S.I.; Nishiyama, Y.; Takahashi, S.; Miyairi, S.; Suzuki, I.; Murata, N. Systematic analysis of the relation of electron transport and ATP synthesis to the photodamage and repair of photosystem II in *Synechocystis*. *Plant Physiol.* **2005**, *137*, 263–273. [CrossRef]

58. Wang, H.; Zhang, Y.; Chen, L.; Cheng, W.; Liu, T. Combined Production of Fucoxanthin and EPA from Two Diatom Strains *Phaeodactylum tricornutum* and *Cylindrotheca fusiformis* Cultures. *Bioprocess Biosyst. Eng.* **2018**, *41*, 1061–1071. [CrossRef]

59. Zarrinmehr, M.J.; Farhadian, O.; Heyrati, F.P.; Keramat, J.; Koutra, E.; Kornaros, M.; Daneshvar, E. Effect of nitrogen concentration on the growth rate and biochemical composition of the microalga, *Isochrysis galbana*. *Egypt. J. Aquat. Res.* **2019**, *46*, 153–158. [CrossRef]

60. Xia, S.; Wan, L.; Li, A.; Sang, M.; Zhang, C. Effects of nutrients and light intensity on the growth and biochemical composition of a marine microalga *Odontella aurita*. *Chin. J. Oceanol. Limnol.* **2013**, *31*, 1163–1173. [CrossRef]

61. Xia, S.; Wang, K.; Wan, L.; Li, A.; Hu, Q.; Zhang, C. Production, characterization, and antioxidant activity of fucoxanthin from the marine diatom *Odontella aurita*. *Mar. Drugs* **2013**, *11*, 2667–2681. [CrossRef] [PubMed]

62. Martin-Jézéquel, V.; Hildebrand, M.; Brzezinski, M.A. Silicon metabolism in diatoms: Implications for growth. *J. Phycol.* **2000**, *36*, 821–840. [CrossRef]

63. Mao, X.; Chen, S.H.Y.; Lu, X.; Yu, J.; Liu, B. High silicate concentration facilitates fucoxanthin and eicosapentaenoic acid (EPA) production under heterotrophic condition in the marine diatom *Nitzschia laevis*. *Algal Res.* **2020**, *52*, 102086. [CrossRef]

64. Huysman, M.J.; Vyverman, W.; De Veylder, L. Molecular regulation of the diatom cell cycle. *J. Exp. Bot.* **2014**, *65*, 2573–2584. [CrossRef] [PubMed]

65. Patel, A.; Matsakas, L.; Hrůzová, K.; Rova, U.; Christakopoulos, P. Biosynthesis of nutraceutical fatty acids by the oleaginous marine microalgae *Phaeodactylum tricornutum* utilizing hydrolysates from organosolv-pretreated birch and spruce biomass. *Mar. Drugs* **2019**, *17*, 119. [CrossRef]

66. Hildebrand, M.; Davis, A.K.; Smith, S.R.; Traller, J.C.; Abbriano, R. The place of diatoms in the biofuels industry. *Biofuels* **2012**, *3*, 221–240. [CrossRef]

67. Yang, M.; Zhao, W.; Xie, X. Effects of nitrogen, phosphorus, iron and silicon on growth of five species of marine benthic diatoms. *Acta Ecol. Sin.* **2014**, *34*, 311–319. [CrossRef]

68. Sun, P.; Wong, C.C.; Li, Y.; He, Y.; Mao, X.; Wu, T.; Ren, Y.; Chen, F. A Novel Strategy for Isolation and Purification of Fucoxanthinol and Fucoxanthin from the Diatom *Nitzschia laevis*. *Food Chem.* **2019**, *277*, 566–572. [CrossRef]

69. Lu, X.; Sun, H.; Zhao, W.; Cheng, K.W.; Chen, F.; Liu, B. A hetero-photoautotrophic two-stage cultivation process for production of fucoxanthin by the marine diatom *Nitzschia laevis*. *Mar. Drugs* **2018**, *16*, 219. [CrossRef]

70. Kosakowska, A.; Lewandowska, J.; Stoń, J.; Burkiewicz, K. Qualitative and quantitative composition of pigments in *Phaeodactylum tricornutum* (Bacillariophyceae) stressed by iron. *BioMetals* **2004**, *17*, 45–52. [CrossRef] [PubMed]

71. Wilken, S.; Hoffmann, B.; Hersch, N.; Kirchgessner, N.; Dieluweit, S.; Rubner, W.; Hoffmann, L.J.; Merkel, R.; Peeken, I. Diatom frustules show increased mechanical strength and altered valve morphology under iron limitation. *Limnol. Oceanogr.* **2011**, *56*, 1399–1410. [CrossRef]

72. Matsuno, T. Aquatic animal carotenoids. *Fish. Sci.* **2001**, *67*, 771–783. [CrossRef]

73. Leong, Y.K.; Chen, C.Y.; Varjani, S.; Chang, J.S. Producing fucoxanthin from algae–Recent advances in cultivation strategies and downstream processing. *Bioresour. Technol.* **2022**, *344*, 126170. [CrossRef]

74. Peng, J.; Yuan, J.P.; Wu, C.F.; Wang, J.H. Fucoxanthin, a marine carotenoid present in brown seaweeds and diatoms: Metabolism and bioactivities relevant to human health. *Mar. Drugs* **2011**, *9*, 1806–1828. [CrossRef]

75. Yang, R.; Wei, D.; Xie, J. Diatoms as cell factories for high-value products: Chrysolaminarin, eicosapentaenoic acid, and fucoxanthin. *Crit. Rev. Biotechnol.* **2020**, *40*, 993–1009. [CrossRef]

76. Maeda, H.; Fukuda, S.; Izumi, H.; Saga, N. Anti-oxidant and fucoxanthin contents of brown alga Ishimozuku (*Sphaerotrichia divaricata*) from the West Coast of Aomori, Japan. *Mar. Drugs* **2018**, *16*, 255. [CrossRef]

77. Mal, N.; Srivastava, K.; Sharma, Y.; Singh, M.; Rao, K.M.; Enamala, M.K.; Chandrasekhar, K.; Chavali, M. Facets of diatom biology and their potential applications. *Biomass Convers. Biorefin.* **2022**, *12*, 1959–1975. [CrossRef]

78. Gammone, M.A.; D'Orazio, N. Anti-obesity activity of the marine carotenoid fucoxanthin. *Mar. Drugs* **2015**, *13*, 2196–2214. [CrossRef]

79. Neumann, U.; Derwenskus, F.; Flaiz Flister, V.; Schmid-Staiger, U.; Hirth, T.; Bischoff, S.C. Fucoxanthin, a carotenoid derived from *Phaeodactylum tricornutum* exerts antiproliferative and antioxidant activities in vitro. *Antioxidants* **2019**, *8*, 183. [CrossRef]

80. Karpiński, T.M.; Ożarowski, M.; Alam, R.; Łochyńska, M.; Stasiewicz, M. What Do We Know about Antimicrobial Activity of Astaxanthin and Fucoxanthin? *Mar. Drugs* **2021**, *20*, 36. [CrossRef] [PubMed]

81. Pajot, A.; Hao Huynh, G.; Picot, L.; Marchal, L.; Nicolau, E. Fucoxanthin from algae to human, an extraordinary bioresource: Insights and advances in up and downstream processes. *Mar. Drugs* **2022**, *20*, 222. [CrossRef] [PubMed]

82. Marella, T.K.; López-Pacheco, I.Y.; Parra-Saldívar, R.; Dixit, S.; Tiwari, A. Wealth from waste: Diatoms as tools for phycoremediation of wastewater and for obtaining value from the biomass. *Sci. Total Environ.* **2020**, *724*, 137960. [CrossRef] [PubMed]

83. Pasquet, V.; Chérouvrier, J.R.; Farhat, F.; Thiéry, V.; Piot, J.M.; Bérard, J.B.; Kaas, R.; Serive, B.; Patrice, T.; Cadoret, J.P.; et al. Study on the microalgal pigments extraction process: Performance of microwave assisted extraction. *Process Biochem.* **2011**, *46*, 59–67. [CrossRef]

84. Kim, S.M.; Kang, S.W.; Kwon, O.N.; Chung, D.; Pan, C.H. Fucoxanthin as a major carotenoid in *Isochrysis* aff. *galbana*: Characterization of extraction for commercial application. *J. Korean Soc. Appl. Biol. Chem.* **2012**, *55*, 477–483.

85. Gao, F.; Teles, I.; Wijffels, R.H.; Barbosa, M.J. Process optimization of fucoxanthin production with *Tisochrysis lutea*. *Bioresour. Technol.* **2020**, *315*, 123894. [CrossRef] [PubMed]

86. Tokushima, H.; Inoue-Kashino, N.; Nakazato, Y.; Masuda, A.; Ifuku, K.; Kashino, Y. Advantageous characteristics of the diatom *Chaetoceros gracilis* as a sustainable biofuel producer. *Biotechnol. Biofuels* **2016**, *9*, 1–19. [CrossRef]

87. Li, Y.; Sun, H.; Wu, T.; Fu, Y.; He, Y.; Mao, X. Storage Carbon Metabolism of *Isochrysis zhangjiangensis* under Different Light Intensities and Its Application for Co-Production of Fucoxanthin and Stearidonic Acid. *Bioresour. Technol.* **2019**, *282*, 94–102. [CrossRef]

88. Lavaud, J.; Rousseau, B.; van Gorkom, H.J.; Etienne, A.L. Influence of the Diadinoxanthin Pool Size on Photoprotection in the Marine Planktonic Diatom *Phaeodactylum tricornutum*. *Plant Physiol.* **2002**, *129*, 1398–1406. [CrossRef]

89. Artamonova, E.Y.; Svenning, J.B.; Vasskog, T.; Hansen, E.; Eilertsen, H.C. Analysis of phospholipids and neutral lipids in three common northern cold water diatoms: *Coscinodiscus concinnus*, *Porosira glacialis*, and *Chaetoceros socialis* by ultra-high performance liquid chromatography-mass spectrometry. *J. Appl. Phycol.* **2017**, *29*, 1241e1249. [CrossRef]

90. Kapoor, B.; Kapoor, D.; Gautam, S.; Singh, R.; Bhardwaj, S. Dietary polyunsaturated fatty acids (PUFAs): Uses and potential health benefits. *Curr. Nutr. Rep.* **2021**, *10*, 232–242. [CrossRef] [PubMed]

91. Mariamenatu, A.H.; Abdu, E.M. Overconsumption of omega-6 polyunsaturated fatty acids (PUFAs) versus deficiency of omega-3 PUFAs in modern-day diets: The disturbing factor for their "balanced antagonistic metabolic functions" in the human body. *J. Lipids* **2021**, *2021*, 8848161. [CrossRef] [PubMed]

92. Guihéneuf, F.; Fouqueray, M.; Mimouni, V.; Ulmann, L.; Jacquette, B.; Tremblin, G. Effect of UV stress on the fatty acid and lipid class composition in two marine microalgae *Pavlova lutheri* (*Pavlovophyceae*) and *Odontella aurita* (*Bacillariophyceae*). *J. Appl. Phycol.* **2010**, *22*, 629–638. [CrossRef]

93. Roessler, P.G. Effects of silicon deficiency on lipid composition and metabolism in the diatom *Cyclotella cryptica*. *J. Phycol.* **1988**, *24*, 394–400. [CrossRef]

94. Barone, G.D.; Cernava, T.; Ullmann, J.; Liu, J.; Lio, E.; Germann, A.T.; Nakielski, A.; Russo, D.A.; Chavkin, T.; Knufmann, K.; et al. Recent developments in the production and utilization of photosynthetic microorganisms for food applications. *Heliyon* **2023**, *9*, e14708. [CrossRef] [PubMed]

95. Vidhyalakshmi, R.; Nachiyar, C.V.; Kumar, G.N.; Sunkar, S.; Badsha, I. Production, characterization and emulsifying property of exopolysaccharide produced by marine isolate of *Pseudomonas fluorescens*. *Biocatal. Agric. Biotechnol.* **2018**, *16*, 320–325. [CrossRef]

96. Pushpabharathi, N.; Jayalakshmi, M.; Amudha, P.; Vanitha, V. Identification of bioactive compounds in *Cymodocea serrulata*-a seagrass by gas chromatography–mass spectroscopy. *Asian J. Pharm. Clin. Res.* **2018**, *11*, 317–320.

97. Tovar, R.; Gavito, A.L.; Vargas, A.; Soverchia, L.; Hernandez-Folgado, L.; Jagerovic, N.; Baixeras, E.; Ciccocioppo, R.; de Fonseca, F.R.; Decara, J. Palmitoleoylethanolamide is an efficient anti-obesity endogenous compound: Comparison with oleylethanolamide in diet-induced obesity. *Nutrients* **2021**, *13*, 2589. [CrossRef]

98. Lynch, E.D.; Lee, M.K.; Morrow, J.E.; Welcsh, P.L.; LeoÂn, P.E.; King, M.C. Nonsyndromic Deafness DFNA1 associated with mutation of a human homolog of the *Drosophila* gene diaphanous. *Science* **1997**, *278*, 1315–1318. [CrossRef]

99. Medlin, L.; Elwood, H.J.; Stickel, S.; Sogin, M.L. The characterization of enzymatically amplified eukaryotic 16S-like rRNA-coding regions. *Gene* **1988**, *71*, 491–499. [CrossRef]

100. Raja, R.; Iswarya, S.H.; Balasubramanyam, D.; Rengasamy, R. PCR-identification of *Dunaliella salina* (Volvocales, Chlorophyta) and its growth characteristics. *Microbiol. Res.* **2007**, *162*, 168–176. [CrossRef] [PubMed]

101. Stamatakis, A. RAxML version 8: A tool for phylogenetic analysis and post-analysis of large phylogenies. *Bioinformatics* **2014**, *30*, 1312–1313. [CrossRef] [PubMed]

102. Ronquist, F.; Huelsenbeck, J.P. MrBayes 3: Bayesian phylogenetic inference under mixed models. *Bioinformatics* **2003**, *19*, 1572–1574. [CrossRef] [PubMed]

103. An, S.M.; Noh, J.H.; Kim, J.H.; Kang, N.S. Ultrastructural and Molecular Characterization of *Surirella atomus* Hustedt 1955 (Bacillariophyta, Surirellalceae), A Newly Recorded Species in Korea. *Ocean Polar Res.* **2021**, *43*, 245–253.

104. Elisabeth, B.; Rayen, F.; Behnam, T. Microalgae culture quality indicators: A review. *Crit. Rev. Biotechnol.* **2021**, *41*, 457–473. [CrossRef]

105. Ralph, P.J.; Gademann, R. Rapid light curves: A powerful tool to assess photosynthetic activity. *Aquat. Bot.* **2005**, *82*, 222–237. [CrossRef]

106. Bhola, V.; Desikan, R.; Santosh, S.K.; Subburamu, K.; Sanniyasi, E.; Bux, F. Effects of parameters affecting biomass yield and thermal behaviour of *Chlorella vulgaris*. *J. Biosci. Bioeng.* **2011**, *111*, 377–382. [CrossRef]
107. Barceló Villalobos, M. Optimización de la Producción de Microalgas en Reactores Abiertos de Escala Industrial. Ph.D. Thesis, University of Almería, Almería, Spain, 2020.
108. Consalvey, M.; Perkins, R.G.; Paterson, D.M.; Underwood, G.J. PAM fluorescence: A beginners guide for benthic diatomists. *Diatom Res.* **2005**, *20*, 1–22. [CrossRef]
109. Malapascua, J.R.; Jerez, C.G.; Sergejevová, M.; Figueroa, F.L.; Masojídek, J. Photosynthesis monitoring to optimize growth of microalgal mass cultures: Application of chlorophyll fluorescence techniques. *Aquat. Biol.* **2014**, *22*, 123–140. [CrossRef]
110. Kang, N.S.; Cho, K.; An, S.M.; Kim, E.S.; Ki, H.; Lee, C.H.; Choi, G.; Hong, J.W. Taxonomic and Biochemical Characterization of Microalga *Graesiella emersonii* GEGS21 for Its Potential to Become Feedstock for Biofuels and Bioproducts. *Energies* **2022**, *15*, 8725. [CrossRef]
111. Garces, R.; Mancha, M. One-step lipid extraction and fatty acid methyl esters preparation from fresh plant tissues. *Anal. Biochem.* **1993**, *211*, 139–143. [CrossRef] [PubMed]

 *marine drugs*

Article

Scenedesmus rubescens Heterotrophic Production Strategies for Added Value Biomass

Gonçalo Espírito Santo [1], Ana Barros [1], Margarida Costa [2], Hugo Pereira [3], Mafalda Trovão [1], Helena Cardoso [1], Bernardo Carvalho [3], Maria Soares [1], Nádia Correia [1], Joana T. Silva [1], Marília Mateus [4,*] and Joana L. Silva [1]

[1] Allmicroalgae Natural Products S.A., R&D Department, Rua 25 de Abril s/n, 2445-413 Pataias, Portugal; moisesgoncalo.16@gmail.com (G.E.S.); ana.barros@allmicroalgae.com (A.B.); mafs.8@hotmail.com (M.T.); helena.cardoso@allmicroalgae.com (H.C.); marjsoares@gmail.com (M.S.); nadiasgcorreia@gmail.com (N.C.); joanatlfsilva@gmail.com (J.T.S.); joana.g.silva@allmicroalgae.com (J.L.S.)

[2] Microalgae Section, Norwegian Institute for Water Research (NIVA), Økernveien 94, 0579 Oslo, Norway; costa.anamarg@gmail.com

[3] GreenCoLab—Associação Oceano Verde, University of Algarve, Campus de Gambelas, 8005-139 Faro, Portugal; hugopereira@greencolab.com (H.P.); bernardocarvalho@greencolab.com (B.C.)

[4] iBB—Institute for Bioengineering and Biosciences, Instituto Superior Técnico, Universidade de Lisboa, Av. Rovisco Pais, 1049-001 Lisbon, Portugal

* Correspondence: marilia.mateus@tecnico.ulisboa.pt

Abstract: Microalgae attract interest worldwide due to their potential for several applications. *Scenedesmus* is one of the first in vitro cultured algae due to their rapid growth and handling easiness. Within this genus, cells exhibit a highly resistant wall and propagate both auto- and heterotrophically. The main goal of the present work is to find scalable ways to produce a highly concentrated biomass of *Scenedesmus rubescens* in heterotrophic conditions. *Scenedesmus rubescens* growth was improved at the lab-scale by 3.2-fold (from 4.1 to 13 g/L of dry weight) through medium optimization by response surface methodology. Afterwards, scale-up was evaluated in 7 L stirred-tank reactor under fed-batch operation. Then, the optimized medium resulted in an overall productivity of 8.63 g/L/day and a maximum biomass concentration of 69.5 g/L. *S. rubescens* protein content achieved approximately 31% of dry weight, similar to the protein content of *Chlorella vulgaris* in heterotrophy.

Keywords: *Scenedesmus rubescens*; heterotrophy; media optimization; stirred-tank reactor; DoE—design of experiment; RSM—response surface methodology

Citation: Santo, G.E.; Barros, A.; Costa, M.; Pereira, H.; Trovão, M.; Cardoso, H.; Carvalho, B.; Soares, M.; Correia, N.; Silva, J.T.; et al. *Scenedesmus rubescens* Heterotrophic Production Strategies for Added Value Biomass. *Mar. Drugs* **2023**, *21*, 411. https://doi.org/10.3390/md21070411

Academic Editors: Cecilia Faraloni and Eleftherios Touloupakis

Received: 26 June 2023
Revised: 14 July 2023
Accepted: 17 July 2023
Published: 19 July 2023

1. Introduction

Microalgae or microphytes are microscopic ancestral living organisms defined as oxygenic photosynthesizers. These organisms comprise over 300,000 species of which approximately 30,000 are documented [1]. Their potential to be used in wastewater treatment and effluent bioremediation has been widely discussed [1], as well as other uses, namely for food and feed applications and added value compound extraction [2].

To overcome prohibitive production costs and to achieve the high purity required for more refined niche markets (such as cosmetics and pharmaceutical industries), it is possible to use biorefinery approaches to extract a wide variety of bioproducts, such as proteins, carbohydrates, carotenoids, and lipids such as DHA (docosahexaenoic acid) and EPA (eicosapentaenoic acid) [1]. Besides being more readily incorporated into commonly used products than whole biomass, microalgae extracts are functional ingredients, conveying bioactive properties to those products [3]. Therefore, it is possible to take full advantage of microalgae's inherent ability to produce valuable compounds, channel the different fractions into specific applications directed at highly refined markets, and make the whole production process economically viable [4].

Microalgae can be produced under autotrophic, mixotrophic, and heterotrophic conditions. However, only a few microalgae, such as *Scenedesmus* sp., *Chlorococcum* sp., *Chlorella* sp., and *Chlamydomonas* sp., grow heterotrophically [5–8]. Under these conditions, microalgae use organic substrates both as energy and as carbon sources [9], and production occurs in closed-stirred reactors, such as industrial fermenters [9], and in axenic conditions [10]. Heterotrophic growth is light-independent and allows faster growth and higher yields. For instance, *Scenedesmus acuminatus* produced heterotrophically yielded 274 g/L of dry biomass [5]. Thus, it decreases the need to occupy large areas for inoculum production [7,9]. Overall, it allows efficient, controlled, reproducible, and reliable year-round production, overcoming major limiting factors of autotrophic cultivation, namely the dependency on weather conditions [11].

The green microalgae *Scenedesmus* sp. (Chlorococcales; Scenedesmaceae) are commonly found in fresh and wastewater streams [12]. These algae are typically characterized by a two-dimensional arrangement of two or more cells in regular aggregates called coenobia [13], and algae from this genus were some of the first cultured in vitro due to their rapid growth and handling easiness [12]. *Scenedesmus* sp., similarly to other coccoid green algae, present highly resistant cell walls exhibiting a characteristic trilaminar structure [14].

Scenedesmus sp. can grow both auto- and heterotrophically and has untapped biotechnological potential. They are considered a valuable source of protein, containing up to 60% [15], and, when stress-induced, *Scenedesmus* sp. direct their metabolism to accumulate lipids by repurposing other energetic components, such as proteins and polysaccharides, a key feature for biofuel development [5,16,17]. Lastly, these microalgae can produce carotenoids such as lutein and astaxanthin [18]. This group of pigments is targeted by food, feed, and cosmetic industries due to their appealing color, aroma, remarkable nutritional composition [19,20], and bioactivity as powerful antioxidants [20,21].

To cultivate microalgae and produce a given metabolite, a combination of parameters must be considered [22], namely nutritional or chemical factors and environmental or physical factors. The first includes chemical elements in the culture medium essential for the cell's metabolism, such as carbon, nitrogen, phosphorus, calcium, sodium, silica, metals such as iron and copper, etc. [23,24]. On the other hand, environmental factors include pH, temperature, agitation, and aeration intensity in the system [23].

Traditionally, culture medium optimization is achieved through an OVAT approach, i.e., "one variable at a time" [25]. Although simple, OVAT becomes time-consuming and inefficient since it does not consider possible interactions between different factors [26]. In addition, this time- and labor-intensive approach comes at increased costs [27] compared to alternative methods such as the design of experiments (DoE). DoE is a statistical performance analysis method that allows the development of a model which can predict some system responses given the change of the variables under study [25]. In addition, DoE determines the importance of the factors (screening) and their interactions (optimization) [28]. It determines the effect of each factor (variable in study) individually or by changing the level of other factors (interactions), which means the level of one factor varies the effect that other factors will have on a specific response [29].

In a complex microbial process, evaluating the interactions between the studied variables is critical for obtaining the optimal operation point. The system responses could be biomass production or biocompound(s) productivity [28].

The present work aimed at obtaining an optimal culture medium to cultivate *Scenedesmus rubescens* under heterotrophic conditions. Medium validation with high cell density and biomass characterization for further potential commercial application were also accessed.

2. Results

2.1. Growth Performance

2.1.1. Preliminary Assays (Carbon and Nitrogen Sources and Working pH)

Preliminary assays, aiming to find a baseline medium for the optimization study, were performed. First, different carbon sources (glucose, acetate, and glycerol) were tested. Glucose was the only one that promoted cell growth, and since it was already used to grow

Scenedesmus acuminatus [5], *Scenedesmus obliquus* [30], and *Chlorella vulgaris* [11], this was the chosen carbon source from this point on.

Two different media, TAP [31] and 5× concentrated Bold's medium [32], were screened using OVAT methodology (Figure 1). Both were supplemented with 20 g/L of glucose. The nitrogen sources were ammonia and nitrates for TAP and Bold's media, respectively.

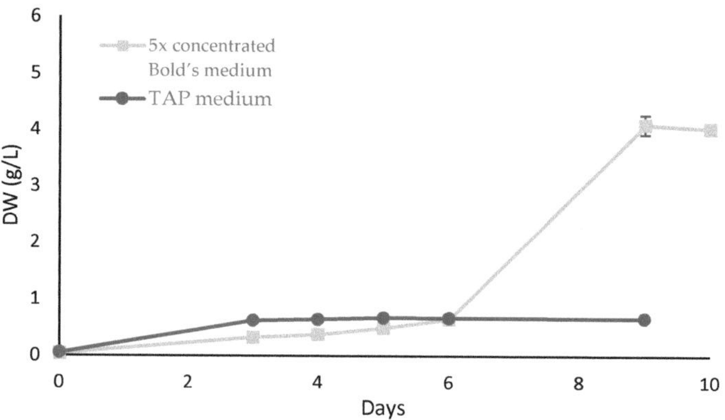

Figure 1. *Scenedesmus rubescens* growth curves under heterotrophic cultivation in 250 mL Erlenmeyer flasks using TAP or 5× concentrated Bold's media supplemented with 20 g/L glucose. The values represent the average and respective standard deviation of 3 individual experiments. SD values were lower than 0.04 g/L.

The highest biomass concentration (4.1 g/L) was reached using Bold's medium while TAP medium only reached 0.81 g/L, as depicted in Figure 1. Comparing the composition of both media, 5× concentrated Bold's medium had a higher concentration of most nutrients, particularly nitrogen and phosphate, which could influence *Scenedesmus* growth as it also affected the growth of *Chlorococcum* sp. and *S. acuminatus* in other published studies [5,6].

Since these media have different nitrogen sources, which could also compromise cell growth [33], the next step was to evaluate *S. rubescens* growth using nitrates (120 mM), ammonia (60 mM), and urea (60 mM) (Figure 2).

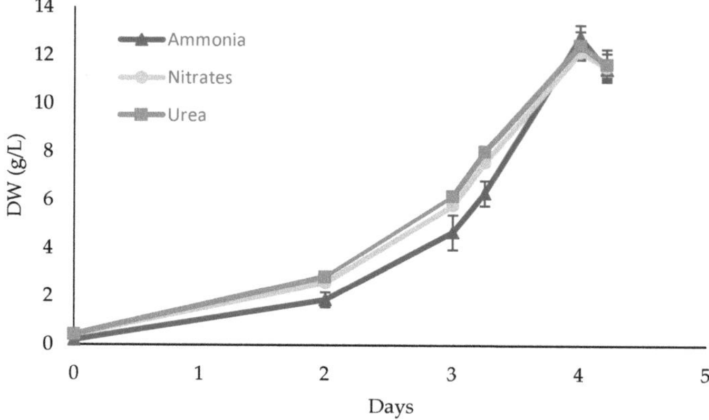

Figure 2. *Scenedesmus rubescens* growth curves using 0037SA medium supplemented with different nitrogen sources and 20 g/L glucose. Cultures were grown heterotrophically in 250 mL Erlenmeyer flasks. The values represent the average and respective standard deviation (SD) of 3 individual experiments. SD values were lower than 0.26 g/L.

No significant differences were found ($p > 0.05$) among treatments, and 13 g/L of dry biomass and 0.91 g/L/day of global productivity were obtained. This result suggests the possibility of using urea and nitrate, which is in agreement with previous studies where *Scenedesmus acuminatus* was supplemented with these two nitrogen sources [5]. However, in this work, ammonia could also be used to control pH in later stages of the scale-up process, suggesting it could also become a promising nitrogen source.

The pH determines the solubility of nutrients and drives many cellular responses, which can significantly influence overall microalgal metabolism [34]. The optimal pH was, therefore, searched. Four pH values were used during the experiments (6.0, 6.5, 7.0, and 8.0, Figure 3). The pH was maintained using 80 mM of PIPES buffer.

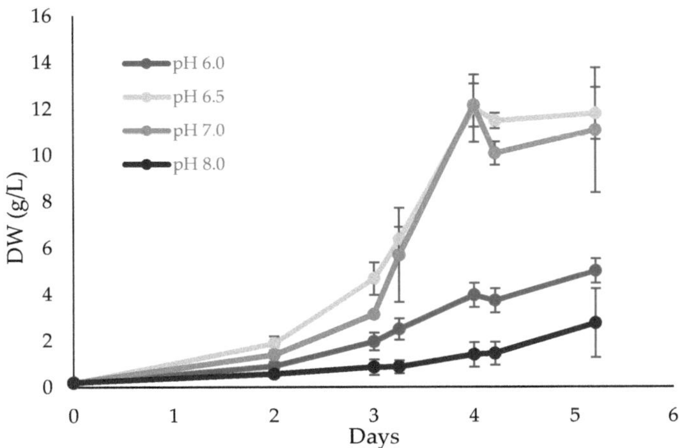

Figure 3. *Scenedesmus rubescens* growth curves using 0037SA medium supplemented with 20 g/L glucose at different pH values. Cultures were grown heterotrophically in 250 mL Erlenmeyer flasks. The values represent the average and respective standard (SD) deviation of 3 individual experiments. SD values are lower than 0.91 g/L.

Under pH 6.5 and 7.0, the culture reached higher cell productivity and growth rate (Table 1). Other studies [5] showed *Scenedesmus acuminatus* achieving a higher concentration at pH 6.0. However, *S. rubescens* growth continues to be favored under a weak acidic/neutral pH environment, unlike *S. acuminatus*. Other resemblant heterotrophic species, such as *Chlorella vulgaris*, have also been cultivated at pH 6.5 in 7 L bench-top fermenters [7].

Table 1. Global biomass productivity and specific growth rate of *Scenedesmus rubescens* grown heterotrophically under different pH values. Different letters indicate significant differences, $p < 0.05$. Values are given as average ± standard deviation ($n = 3$).

Conditions (pH)	Global Productivity (g/L/day)	Growth Rate (day^{-1})
6.0	0.94 ± 0.11 [a]	0.78 ± 0.04 [a]
6.5	2.95 ± 0.26 [b]	1.05 ± 0.01 [b]
7.0	2.98 ± 1.74 [b]	1.04 ± 0.07 [b]
8.0	0.30 ± 0.10 [c]	0.49 ± 0.05 [c]

2.1.2. Culture Medium Screening Using Plackett–Burman Design

As previously mentioned, nutrients are essential for the growth and development of microalgae. In this way, 12 nutrients were studied under different concentrations: N, Mg, Ca, P, Fe, Cu, Zn, Mn, Mo, Co, Ni, and B. Factors and their concentrations were chosen based on previously tested media (TAP and Bold's). Screening was carried out to predict which nutrients influence biomass productivity (Figure 4). Nitrogen sources (nitrates and ammonia) were included to understand their influence on/under different concentrations of other nutrients.

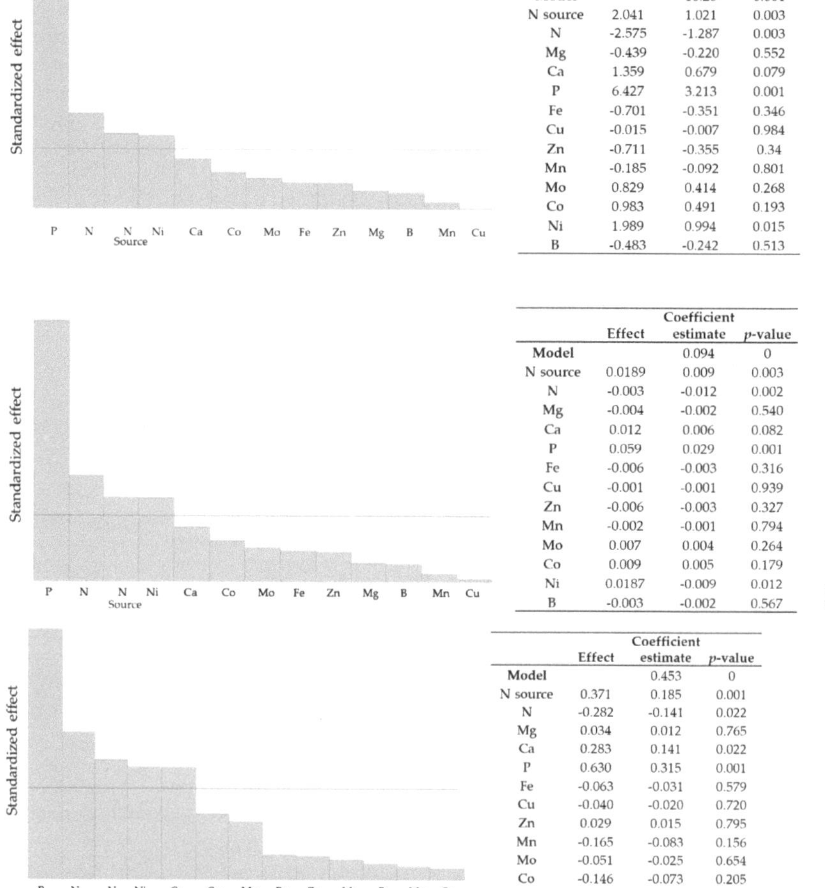

Table A (biomass concentration):

	Effect	Coefficient estimate	p-value
Model		10.28	0.001
N source	2.041	1.021	0.003
N	-2.575	-1.287	0.003
Mg	-0.439	-0.220	0.552
Ca	1.359	0.679	0.079
P	6.427	3.213	0.001
Fe	-0.701	-0.351	0.346
Cu	-0.015	-0.007	0.984
Zn	-0.711	-0.355	0.34
Mn	-0.185	-0.092	0.801
Mo	0.829	0.414	0.268
Co	0.983	0.491	0.193
Ni	1.989	0.994	0.015
B	-0.483	-0.242	0.513

Table B (global productivity):

	Effect	Coefficient estimate	p-value
Model		0.094	0
N source	0.0189	0.009	0.003
N	-0.003	-0.012	0.002
Mg	-0.004	-0.002	0.540
Ca	0.012	0.006	0.082
P	0.059	0.029	0.001
Fe	-0.006	-0.003	0.316
Cu	-0.001	-0.001	0.939
Zn	-0.006	-0.003	0.327
Mn	-0.002	-0.001	0.794
Mo	0.007	0.004	0.264
Co	0.009	0.005	0.179
Ni	0.0187	-0.009	0.012
B	-0.003	-0.002	0.567

Table C (maximum productivity):

	Effect	Coefficient estimate	p-value
Model		0.453	0
N source	0.371	0.185	0.001
N	-0.282	-0.141	0.022
Mg	0.034	0.012	0.765
Ca	0.283	0.141	0.022
P	0.630	0.315	0.001
Fe	-0.063	-0.031	0.579
Cu	-0.040	-0.020	0.720
Zn	0.029	0.015	0.795
Mn	-0.165	-0.083	0.156
Mo	-0.051	-0.025	0.654
Co	-0.146	-0.073	0.205
Ni	0.301	0.151	0.016
B	-0.060	-0.030	0.595

Figure 4. Order bar charts (Pareto charts) and analysis of variance (ANOVA) obtained with the software Minitab® version 19, testing 13 factors for 3 responses: (**A**) biomass concentration, (**B**) global productivity, and (**C**) maximum productivity. Factors above the red line are the most significant factors for all three responses. The model was significant ($p < 0.05$). The cultures were grown heterotrophically in 250 mL Erlenmeyer flasks.

The Plackett–Burman design was used with two coded levels, and 30 runs were employed (Table S1) with the chosen responses: (1) biomass concentration, (2) global productivity, and (3) maximum productivity. Low- and high-level concentrations were defined based on the previously studied culture media (Section 2.1.1).

The nitrogen source was one of the most significant factors affecting cell growth ($p < 0.05$). However, in the previous experiment (Figure 2), there was no significant difference between ammonia and nitrates. Therefore, ammonia was chosen given the convenience regarding pH control in later stages of scale-up. The concentrations of N, P, Ni, and Ca also significantly influenced cell growth (Figure 4). However, calcium concentration only affected maximum productivity (Figure 4C).

2.1.3. Culture Medium Optimization Using Box–Behnken Design

Design-Expert software was used to further optimize the medium composition through Box–Behnken design via the response surface method (RSM). The N, P, Ni, and Ca element concentrations were further optimized (Table 2). In this experimental design, 26 experimental sets were generated with three central points (Table S2). The same responses as before were addressed, including biomass concentration, global productivity, and maximum productivity (Figure 5).

Table 2. Levels of 4 factors used in DoE (with Design-Expert software, version 12): ammonia, phosphate, nickel, and calcium.

Factors (mM)	Coded Levels		
	Low	**Central Point**	**High**
Ammonia (A)	20	40	60
Phosphate (B)	1	5.5	10
Calcium (C)	0.3	1	1.7
Nickel (D)	0	0.01	0.02

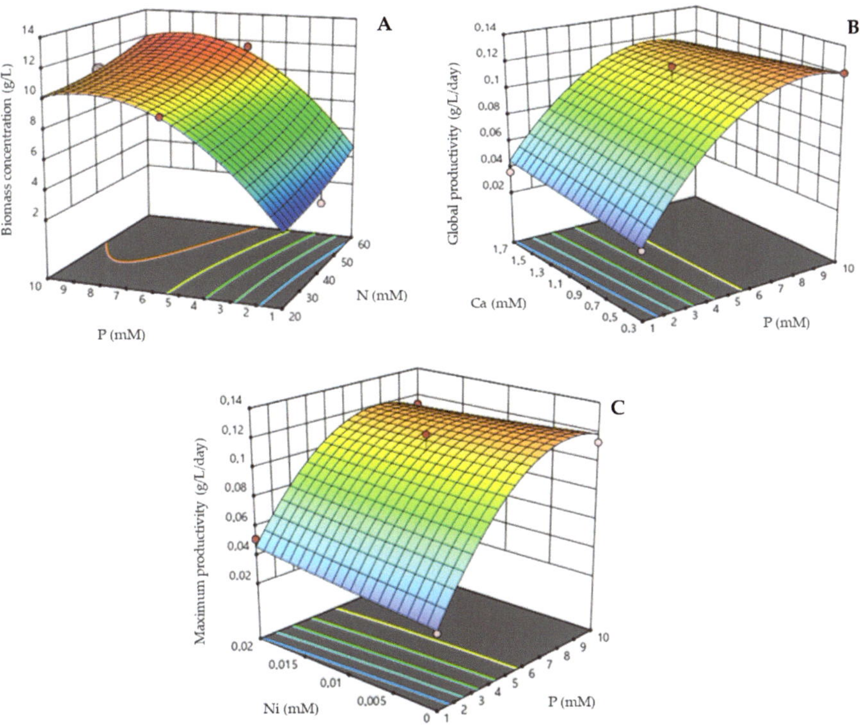

Figure 5. Response surfaces showing the mutual effects of factors present in the culture medium. (**A**) Effects of the interaction between P and N factors for biomass concentration response; Ni was kept at maximum level and Ca was kept at lowest level. (**B**) Effects of the interaction between P and Ca factors for global productivity response; N and Ni were kept at the intermediate levels. (**C**) Effects of the interaction between P and Ni factors for maximum productivity response; N and Ca were kept at the intermediate levels. The cultures were grown heterotrophically in 250 mL Erlenmeyer flasks.

Figure 5 represents the prediction of the interaction among different factors in *S. rubescens* culture medium. In general, the model predicts that P will achieve maximum values to increase all these responses (10 mM). Regarding biomass concentration (Figure 5A), the model shows that P and N concentrations should be near the highest concentrations used (10 and 60 mM, respectively) to achieve higher biomass concentration. The predicted model is represented by Equation (1), $p < 0.05$. Figure 5B characterizes the interaction between Ni and P for global productivity response. N and Ni concentrations at the central values demand P at the highest (10 mM) and Ca at the lowest value (0.3 mM) to obtain the highest global productivity. As a result, the predicted equation was Equation (2) $p < 0.05$.

Finally, P and Ni at the highest level (10 and 0.02 mM, respectively) and N and Ca at the central point induced higher maximum productivity values (Figure 5C). The model predictions are described by Equation (3) ($p < 0.05$).

From the models designed, it was possible to conclude that for an optimized *S. rubescens* culture medium, the highest level for factors N (60 mM), P (10 mM), and Ni (0.02 mM), and lowest value of Ca (0.3 mM) were necessary.

Biomass concentration (g/L):

$$\begin{aligned}
9.74 - 0.0185A + 3.58B + 0.2010C + 0.1098D - 0.0268A \times B - 0.9468A \times C \\
+ 0.2881A \times D - 0.0473B \times C - 0.4322B \times D - 0.5969C \times D \\
+ 0.3750A \times A - 2.85B \times B + 0.3790C \times C + 0.0179D \times D
\end{aligned} \tag{1}$$

Global productivity (g/L/day):

$$\begin{aligned}
0.1101 - 0.0014A + 0.0382B + 0.0004C + 0.0016D - 0.0043A \times B - 0.0035A \times C \\
+ 0.0091A \times D - 0.0062B \times C - 0.0053B \times D - 0.0040C \times D \\
+ 0.0001A \times A - 0.0335B \times B - 0.003C \times C + 0.0005D \times D
\end{aligned} \tag{2}$$

Maximum productivity (g/L/day):

$$0.269 + 0.0266A + 0.131B + 0.001C + 0.001D \tag{3}$$

Finally, to assess the possibility of phosphate being a growth limiting factor, different concentrations were tested, including 10 (control), 50, and 100 mM (Figure 6).

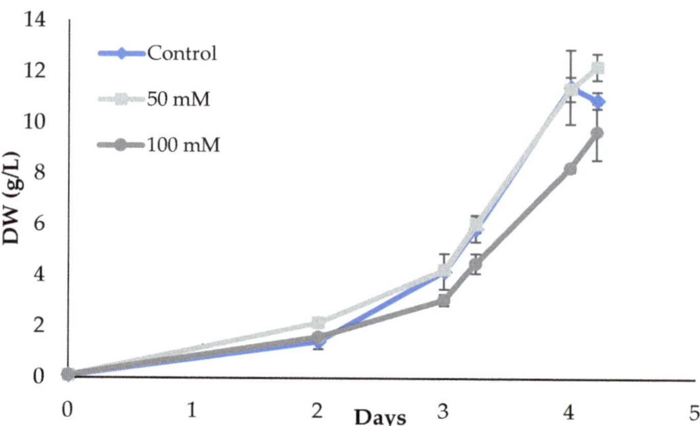

Figure 6. *Scenedesmus rubescens* growth curves using 0037SA medium supplemented with different phosphate concentrations and 20 g/L glucose. Cultures were grown under heterotrophic conditions in 250 mL Erlenmeyer flasks. The values represent the average and respective standard (SD) deviation of 3 individual experiments. SD values are lower than 0.94 g/L.

Alga growth led to similar biomass concentration, comparing the use of 10 and 50 mM of phosphate (11.5 to 12.2 g/L; Table 3). Data suggest there are only growth differences with 100 mM of phosphates, possibly caused by the initial inhibition of cell growth ($p < 0.05$). Comparing global productivity and specific growth rate (Table 3), there were no significant differences between the use of 50 mM and 10 mM phosphate, neither between 50 and

100 mM ($p > 0.05$), but there was a significant difference between 10 and 100 mM. Overall, 50 mM of phosphate was used in the following assays.

Table 3. Biomass concentration, global productivity, and specific growth rate of *Scenedesmus rubescens* under different phosphate concentrations. Different letters indicate significant differences between media, $p < 0.05$. Values are given as average ± standard deviation ($n = 3$).

Concentrations (mM)	Biomass Concentration (g/L)	Global Productivity (g/L/ h)	Specific Growth Rate (day^{-1})
10	11.5	0.119 ± 0.012 [a]	1.18 ± 0.027 [a]
50	12.2	0.115 ± 0.005 [ab]	1.17 ± 0.011 [ab]
100	9.2	0.090 ± 0.011 [b]	1.09 ± 0.005 [b]

In this way, when comparing Bold's medium and 0037SA medium, *S. rubescens* growth was improved by 3.2-fold (from 4.1 to 13 g/L of dry weight), indicating that the medium optimization succeeded.

2.2. Validation of Optimized Medium in Bench-Top Fermenters

The optimized medium resulted in an overall productivity of 8.63 g/L/day and a maximum biomass concentration of 69.5 g/L (Figure 7). This concentration is much higher than what was reported for the same species grown autotrophically, which was 4.1 g/L [33]. However, it is significantly lower than that found for *S. acuminatus* (274 g/L) [5], but the fact that this is a different species should be taken into account. The medium pH in the fermenter from the referenced study was set to 6, rather than 6.5. Additionally, to optimize the biomass concentration, the fermenter feeding was determined by controlling glucose concentration in the range of 0–5 g/L. In the present study, glucose concentration was controlled in the range of 0–20 g/L, and a tighter control may be crucial. Compared to other published data, the cell densities obtained herein represent higher biomass titers than those obtained in other studies with *Aurantiochytrium* sp. (batch) [35], *Chlorella vulgaris* (fed-batch) [7], *Chloroccoccum amblystomatis* (batch) [6], *Nitchia laevis* (fed-batch) [36], and *Schyzochytrium* sp. (fed-batch) [37]. Overall, although different species may behave and respond differently, *S. rubescens* was able to reach a high biomass concentration, in line with other fed-batch heterotrophic microalgae species. Still, further studies are required to optimize *S. rubescens* growth and obtain even higher cell densities and to develop its biotechnology potential for commercial applications.

2.3. Biochemical Analysis

The biochemical composition of the biomass obtained during the validation at the beginning and end of the growth curve (initial and final phase) were analyzed, and the content of proteins, lipids, carbohydrates, and ashes was assessed to understand if the different stages influenced biochemical composition (Table 4).

Table 4. Biomass composition at the beginning and end of the cultivation. Proteins, lipids, carbohydrates, and ashes are presented as the percentage of the biomass dry weight. Different letters indicate significant differences between media ($p < 0.05$). Values are given as average ± standard deviation ($n = 3$).

Sample	Proteins (%)	Lipids (%)	Carbohydrates (%)	Ashes (%)
Beginning of cultivation	32.9 ± 0.25 [a]	13.2 ± 1.90 [a]	51.4 ± 1.90 [a]	2.3 ± 0.35 [a]
End of cultivation	31.2 ± 0.30 [b]	12.3 ± 1.70 [a]	53.5 ± 1.40 [a]	3.2 ± 0.41 [b]

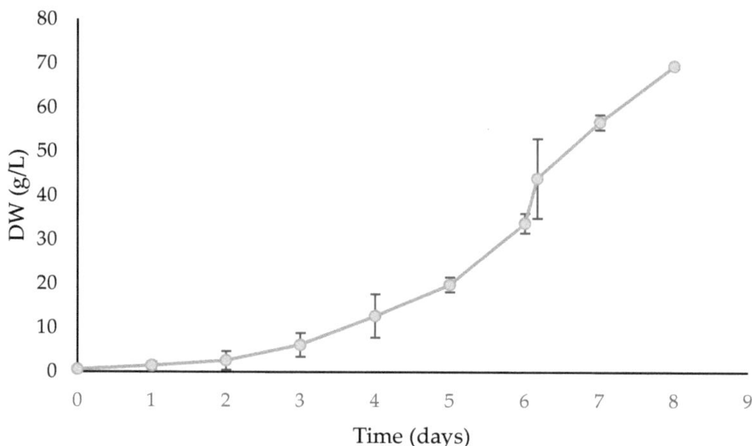

Figure 7. *Scenedesmus rubescens* growth curve in a 7 L bench-top fermenter using the optimized heterotrophic medium. Dissolved oxygen (DO) inside the fermenter was controlled automatically above 40% saturation by increasing aeration up to 5 L/min and stirring speed up to 1200 rpm. The values represent the average and respective standard deviation (SD) of 3 individual experiments. SD values are lower than 2.82 g/L.

S. rubescens biomass displayed 33% and 31% of protein at the initial and final phase of cultivation in the fed-batch fermenter, respectively. These values are comparable to those attained with heterotrophically cultured *Chlorella vulgaris* [7], suggesting that *Scenedesmus* sp. also has great potential to produce biomass for alternative protein markets. Under autotrophic conditions, *S. obliquus*, as other species such as *Chlorella vulgaris* and *Arthrospira platensis*, achieved between 50 and 60% [15], which is significantly higher than *S. rubescens* in heterotrophy. However, heterotrophically produced *S. rubescens* could also subsequently inoculate photobioreactors, where cells would grow autotrophically. This strategy, already used for *Chlorella*, increases production efficiency to obtain a highly concentrated biomass for the inoculation of reactors operating under autotrophic conditions [38] and can be coupled to a second stage of autotrophic cultivation, which would most likely increase the protein and pigment contents and result in higher quality microalgal biomass, as shown in *Chlorella*.

Concerning lipid content, cells grown in the fermenter obtained 13% at the initial and 12% at the final growth phases. This result is in agreement with the lipid content reported in the literature for *S. obliquus* [15]. However, Cheng. et al. 2018 reached 31% [39] by actively inducing lipid production through nitrogen depletion strategies. In another study, also through a nitrogen depletion strategy, *Scenedesmus abudans* achieved high lipid content, between 36% and 67% [40]. Overall, according to the literature, microalgae tend to accumulate lipids when metabolically stressed as a tradeoff of other energetic components, such as proteins and polysaccharides, as reported, for instance, for *Nannocloropsis* sp. [17,41,42].

Lastly, *S. rubescens* obtained a remarkably low ash content (2.3% and 3.2% at initial and final growth phases, respectively) in comparison to that of other microalgae, namely *Arthrospira platensis* (14.5%) [43] and *Nannocloropsis* sp. [44]

Altogether, *S. rubescens* was shown to be a promising source of relevant compounds, such as proteins and lipids, comparable to other commercially available species. Furthermore, depending on the desired commercial application, heterotrophic growth is a promising strategy to obtain high biomass yields or a given metabolite of interest.

3. Discussion

To the best of the authors' knowledge, this represents the first report on *S. rubescens* under heterotrophic conditions. The optimization of culture medium was performed

(Table S3) and compared to the initial Bold's growth medium. When comparing these media, *S. rubescens* growth was successfully improved by 3.2-fold (from 4.1 to 13 g/L of dry weight).

The medium composition resulting from the optimization was also compared to that reported by Jin et al. (2020), designed for *Scenedesmus acuminatus* [5]. While Jin et al. (2020) found the optimum pH at 6.0, the strain used in the present study grew optimally at pH 6.5. In addition, when comparing both media, 0037SA medium is formulated with higher nutrient concentrations, which could have compromised cell growth. *S. acuminatus* is described as reaching a maximum of 274 g/L on a 7.5 L fermenter [5], a biomass concentration value that is significantly higher than the one obtained in the present study. The authors also started to use nitrates (30 mM) as nitrogen sources, and the study was also performed at the laboratory scale; nevertheless, the N-source was replaced by urea at 0.85 g/L in the batch fermenter medium, and the study was concerned with a different species. All these differences most likely influence cell growth significantly.

In a study performed with *Chlorella vulgaris*, biomass reached 175 g/L in a 7 L heterotrophic scale-up phase [7]. Nevertheless, this higher biomass titer was achieved after further medium and abiotic parameter optimization steps.

The optimization strategies in the studies referred to above ([5,7,9]) allow us to hypothesize that there are still opportunities for the further improvement of *Scenedesmus rubescens* biomass productivity in batch and fed-batch bioreactor cultivation such as culture media or growth strategies (C-source concentration control, aeration, or stirring speed, etc.).

Heterotrophically produced *S. rubescens* presented an appealing nutritional profile, and both literature and empirical large-scale production experience suggest there is potential to increase the protein content of this microalgal species under autotrophic conditions. Therefore, one way to develop its commercial application potential could be to combine hetero- and autotrophic cultivation modes, taking advantage of the two metabolic pathways [7,38].

Overall, cultivation conditions were key factors influencing both the growth process and the biochemical profile of the final biomass. Only by learning how to manipulate these variables and understanding the systems' responses does it become possible to grow uncommon microalgae species. The present work demonstrates pilot-scale feasibility of *S. rubescens* production under heterotrophic conditions, shows the derived microalga proximate composition, and highlights strategies for potential commercial applications. Whether aiming at vegetarian/vegan protein substitutes or lipids for biofuel production [39], studies addressing industrial production feasibility open new routes toward commercial application and bring us one step closer to market viability.

4. Materials and Methods

4.1. Microalgae Strain and Culture Media

The axenic *Scenedesmus rubescens* used in this work were obtained from Allmicroalgae's own culture collection (strain code AGF0037SA). This alga was stored in agar slant tubes and subsequently scaled to 250 mL Erlenmeyer flasks. Initially, culture medium was PCB (plate count broth). Throughout experiments, cultures were grown in optimized media. Through the optimization work, the growth medium utilized was continuously updated.

The following media were used for preliminary tests: TAP (Tris-acetate-phosphate) medium [31] and 5× concentrated Bold's Basal Medium [32]. All media were supplemented with 20 g/L glucose. Lastly, 0037SA medium (Table S3) was created and optimized according to the assays described.

Two types of tests were performed to optimize the culture media: OVAT and DoE tests. All the assays were performed using triplicates, except for DoE tests. All culture media were sterilized using filtration through a 0.2 μm pore size PES membrane in a Vacuum Filtration System (VWR, Radnor, PA, USA) and/or autoclaved (Uniclave88 and uniclave77, A.J.Costa, Irmãos, Lda; Cacém, Portugal) at 121 °C for 40 min.

4.2. Growth Assessment

S. rubescens growth was determined using optical density (OD) and dry weight (DW) OD was measured at 600 nm (Figure S1) using a spectrophotometer (Genesis 10S UV–Vis -Thermo Scientific, Waltham, MA, USA). DW was determined using the filtration of culture samples with pre-weighed 0.7 μm GF/C 698 filters (VWR, PA, USA) and dried at 120 °C on a DBS 60–30 electronic moisture analyzer (KERN & SOHN GmbH, Balingen, Germany) These measurements were used to study cell growth, namely specific growth rate, and maximum and overall productivities were calculated.

The specific growth rate (μ) was calculated according to Equation (4):

$$\mu \left(\mathrm{day}^{-1} \right) = \frac{\ln(X2/X1)}{t_2 - t_1} \tag{4}$$

X refers to dry biomass concentration (g/L) at time t_2 and t_1 (days) of cultivation within the exponential growth phase.

Volumetric biomass productivity (Pv) was calculated according to Equation (5):

$$Pv = \frac{Xf - Xi}{tf - ti} \tag{5}$$

where Xf corresponds to final dry biomass concentration, Xi corresponds to initial dry biomass concentration (g L^{-1}), tf corresponds to final time, and ti corresponds to the initial time (h) of cultivation within the exponential growth phase.

4.3. Experimental Trials

All experimental trials for medium optimization were conducted in 250 mL baffled Erlenmeyer flasks with vented caps with a 0.2 μm PTFE membrane (Duran™, Munich, Germany) with a working volume of 50 mL. Cultures were grown in an orbital shaker incubator (SKI 4, ARGOLAB, Carpi, Italy) at 28 °C and 200 rpm (revolutions per minute). All assays ended when cultures reached the stationary phase or carbon source depletion.

Initially, two culture media (5× concentrated Bold's Basal Medium and TAP medium) were tested and the alga's growths compared. Subsequent cultures with the supplementation of different nitrogen sources (ammonia, nitrates, and urea) at different pH values (6.0, 6.5, 7.0, and 8.0) followed and were analyzed. Based on the outcomes, a screening test was carried out to find the impact of different medium composition factors on S. rubescens propagation. Lastly, a Box–Behnken design was conducted to optimize the final culture medium.

Erlenmeyer flask cultures were further scaled-up to inoculate a 7 L bench-top fermenter (New Brunswick BioFlo®/CelliGen®115; Eppendorf AG, Hamburg, Germany) to validate the culture medium. Cultures were grown in a fed-batch regime at 28 °C, and pH was maintained at 6.5 by adding ammonia solution (24% *w/w*), also ensuring a nitrogen source. Glucose and phosphate concentrations were measured twice a day throughout the assays and supplemented to guarantee optimal growth conditions. In the case of glucose, a pure sterile solution of 500 g/L was added in pulses to maintain medium concentration within the range of 0–20 g/L. Additionally, a sterile 2.5 M phosphate buffer solution was added to maintain the medium concentration of 50 mM. Dissolved oxygen (DO) inside the fermenter was controlled automatically above 40% saturation by increasing aeration up to 5 L/min and stirring speed up to 1200 rpm. Samples at the beginning (immediately after inoculation) and end of the fermenter operation time (reported) were collected to analyze the biochemical composition of the respective algal biomasses.

4.4. Nutrient Quantification

The cultures sampled (50 mL) were centrifuged for 10 min at 3500 rpm in VWR Mini Star microcentrifuge (VWR, Radnor, PA, USA). The supernatant was collected to quantify glucose, phosphate, and ammonium concentrations.

When necessary, the supernatant was diluted in a saline solution (10% sodium chloride, 90% distilled water). Freestyle precision Neo kit (Abbott, Witney, Oxon, UK) was used to determine glucose concentration in g/L.

Ammonia and phosphate Sera Tests (Sera, Heinsberg, Germany) were used to determine ammonium and phosphate concentrations, respectively. The supernatant was diluted with distilled water when necessary. The absorbance was measured at the wavelength of 697 nm for ammonium and 716 nm for phosphate. The absorbances were measured using Genesis 10S UV–Vis (Thermo Scientific, Waltham, MA, USA).

4.5. Biomass Characterization

4.5.1. Protein Content

A Vario EL III elemental analyzer (Vario EL, GmbH, Hanau, Germany) was used to quantify the freeze-dried biomass's total carbon, hydrogen, and nitrogen (CNH analysis). The biomass (1 mg) was placed in tiny aluminum capsules and heated at 950 °C. Total protein content was calculated by multiplying the nitrogen amount with a conversion factor of 6.25 [45].

4.5.2. Lipid Content

The lipid content of dry biomass was determined using gravimetry after organic extraction followed by the recovery of clear organic phase and further solvent evaporation [46]. The percentage of lipids was calculated with Equation (6):

$$\% \ lipids = 100 \times \frac{weight \ of \ residue \ from \ evaporated \ clarified \ solvent}{weight \ of \ dry \ biomass \ initially \ put \ into \ the \ evaporated \ extractant \ solvent} \tag{6}$$

4.5.3. Ash Content

A sample of freeze-dried biomass (50 mg) was weighed in a crucible and taken for combustion at 550 °C for 8 h in a JP Selecta Sel horn R9-L furnace (JP Selecta, 22 Barcelona, Spain). The ash content corresponded to the percentual residual weight of the sample after combustion.

4.5.4. Carbohydrate Content

The carbohydrate content of the dry biomass was calculated as the difference to 100% after summing the percentual contents of the other main components analyzed (protein, ash, and lipid contents).

4.6. Statistical Analyses

The statistical tests for OVAT were performed using R software (4.0.2 version) through RStudio 1.3.1073 version (R studio®, Boston, MA, USA). ANOVA analysis was followed by a post hoc Tukey HSD test when comparing three or more conditions. A Student's *t*-test was used to compare groups of independent results. For each test, triplicates, mean, and standard deviation were determined. A statistically significant difference was considered at $p < 0.05$.

The statistical tests for DoE methodology were performed using two software: Minitab (Minitab® version 19, State College, PA, USA), based on a preliminary screening, and Design-Expert (version 12, Stat-Ease®, Minneapolis, MN, USA), based on response surface methodology. Minitab was used for a preliminary screening through the Plackett–Burman method followed by Design-Expert Box–Behnken method. Statistical significance was considered at $p < 0.05$ ANOVA tests. The experimentally observed responses were compared with the predicted values (Y) obtained from the model, given by the polynomial Equation (7), correlating the input variables of the study (A, B, and C):

$$Y = a0 + a1 \ A + a2 \ B + a3 \ C + a4 \ AB + a5 \ AC + a6 \ BC \tag{7}$$

Supplementary Materials: The following supporting information can be downloaded at: https://www.mdpi.com/article/10.3390/md21070411/s1, Table S1: Screening method design in actual level of variables through Minitab® software for *Scenedesmus rubescens*; Table S2: Response functions for optimization of media composition for heterotrophic cultivation of *Scenedesmus rubescens*. Minitab® software, version 19, was used; Table S3: Optimized culture medium developed in this work for 0037SA (macro and micronutrients); Figure S1: Calibration curve. Dry biomass concentration (g L^{-1}) vs. absorbance of *S. rubescens* suspensions (in water) measured at λ = 600 nm for heterotrophic growth.

Author Contributions: Conceptualization, G.E.S., A.B., M.C., H.P., M.M. and J.L.S.; methodology, G.E.S., A.B., M.C., H.P., M.T., B.C., M.S., N.C., J.T.S. and J.L.S.; software, G.E.S., A.B., B.C., N.C., and M.M.; validation, G.E.S., A.B., M.C., H.P. and M.M.; formal analysis, G.E.S., A.B., M.C., H.P., M.M. and J.L.S.; investigation, G.E.S.; resources, J.L.S.; writing—original draft preparation, G.E.S., A.B., M.C., H.P., H.C., B.C., M.S., N.C., J.T.S. and J.L.S.; writing—review and editing, G.E.S., A.B., M.C., H.P., H.C., B.C., M.S., N.C., J.T.S. and J.L.S.; supervision, A.B., M.C., H.P. and J.T.S.; project administration, J.L.S.; funding acquisition, J.L.S. All authors have read and agreed to the published version of the manuscript.

Funding: This research was funded by the AlgaValor project from the European Union's Horizon 2020 research and innovation program (grant agreement nº POCI-01-0247-FEDER-035234; LISBOA-01-0247-FEDER-035234; ALG-01-0247-FEDER-035234), by the Portuguese national budget P2020 in the scope of the project no. 023310–ALGACO2, and from the European Union's Horizon 2020 research and innovation program (grant agreement nº ALG-01-0247-FEDER-069961-Performalgae).

Data Availability Statement: Data is available upon request.

Acknowledgments: The authors would like to acknowledge all members of CCMAR, Green Co-Lab, and Allmicroalgae for the contribution and help given during the work and Performalgae for the funding.

Conflicts of Interest: The authors declare no conflict of interest.

References

1. Mobin, S.; Alam, F. Some Promising Microalgal Species for Commercial Applications: A Review. *Energy Procedia* **2017**, *110*, 510–517. [CrossRef]
2. Khan, M.I.; Shin, J.H.; Kim, J.D. The Promising Future of Microalgae: Current Status, Challenges, and Optimization of a Sustainable and Renewable Industry for Biofuels, Feed, and Other Products. *Microb. Cell Fact.* **2018**, *17*, 36. [CrossRef] [PubMed]
3. Shaima, A.F.; Mohd Yasin, N.H.; Ibrahim, N.; Takriff, M.S.; Gunasekaran, D.; Ismaeel, M.Y.Y. Unveiling Antimicrobial Activity of Microalgae *Chlorella sorokiniana* (UKM2), *Chlorella* sp. (UKM8) and *Scenedesmus* sp. (UKM9). *Saudi J. Biol. Sci.* **2022**, *29*, 1043–1052. [CrossRef] [PubMed]
4. Benemann, J. Microalgae for Biofuels and Animal Feeds. *Energies* **2013**, *6*, 5869–5886. [CrossRef]
5. Jin, H.; Zhang, H.; Zhou, Z.; Li, K.; Hou, G.; Xu, Q.; Chuai, W.; Zhang, C.; Han, D.; Hu, Q. Ultrahigh-Cell-Density Heterotrophic Cultivation of the Unicellular Green Microalga *Scenedesmus acuminatus* and Application of the Cells to Photoautotrophic Culture Enhance Biomass and Lipid Production. *Biotechnol. Bioeng.* **2020**, *117*, 96–108. [CrossRef]
6. Correia, N.; Pereira, H.; Schulze, P.S.C.; Costa, M.M.; Santo, G.E.; Guerra, I.; Trovão, M.; Barros, A.; Cardoso, H.; Silva, J.L.; et al. Heterotrophic and Photoautotrophic Media Optimization Using Response Surface Methodology for the Novel Microalga *Chlorococcum amblystomatis*. *Appl. Sci.* **2023**, *13*, 2089. [CrossRef]
7. Barros, A.; Pereira, H.; Campos, J.; Marques, A.; Varela, J.; Silva, J. Heterotrophy as a Tool to Overcome the Long and Costly Autotrophic Scale-up Process for Large Scale Production of Microalgae. *Sci. Rep.* **2019**, *9*, 13935. [CrossRef]
8. Chen, F.; Johns, M.R. Heterotrophic Growth of *Chlamydomonas reinhardtii* on Acetate in Chemostat Culture. *Process Biochem.* **1996**, *31*, 601–604. [CrossRef]
9. Vuppaladadiyam, A.K.; Prinsen, P.; Raheem, A.; Zhao, M. Microalgae Cultivation and Metabolites Production: A Comprehensive Review. *Biofuels Bioprod. Biorefining* **2018**, *12*, 304–324. [CrossRef]
10. Ende, S.S.W.; Noke, A. Heterotrophic Microalgae Production on Food Waste and By-Products. *J. Appl. Phycol.* **2019**, *31*, 1565–1571. [CrossRef]
11. Liang, Y.; Sarkany, N.; Cui, Y. Biomass and Lipid Productivities of *Chlorella vulgaris* under Autotrophic, Heterotrophic and Mixotrophic Growth Conditions. *Biotechnol. Lett.* **2009**, *31*, 1043–1049. [CrossRef] [PubMed]
12. Pancha, I.; Chokshi, K.; George, B.; Ghosh, T.; Paliwal, C.; Maurya, R.; Mishra, S. Nitrogen Stress Triggered Biochemical and Morphological Changes in the Microalgae *Scenedesmus* sp. CCNM 1077. *Bioresour. Technol.* **2014**, *156*, 146–154. [CrossRef] [PubMed]
13. Çelekli, A.; Balci, M.; Bozkurt, H. Modelling of *Scenedesmus obliquus*; Function of Nutrients with Modified Gompertz Model. *Bioresour. Technol.* **2008**, *99*, 8742–8747. [CrossRef] [PubMed]

14. Dunker, S.; Wilhelm, C. Cell Wall Structure of Coccoid Green Algae as an Important Tradeoff between Biotic Interference Mechanisms and Multidimensional Cell Growth. *Front. Microbiol.* **2018**, *9*, 719. [CrossRef]

15. Becker, E.W. Micro-Algae as a Source of Protein. *Biotechnol. Adv.* **2007**, *25*, 207–210. [CrossRef]

16. Safi, C.; Zebib, B.; Merah, O.; Pontalier, P.Y.; Vaca-Garcia, C. Morphology, Composition, Production, Processing and Applications of *Chlorella vulgaris*: A Review. *Renew. Sustain. Energy Rev.* **2014**, *35*, 265–278. [CrossRef]

17. Choi, W.J.; Chae, A.N.; Song, K.G.; Park, J.; Lee, B.C. Effect of Trophic Conditions on Microalga Growth, Nutrient Removal, Algal Organic Matter, and Energy Storage Products in *Scenedesmus (Acutodesmus) obliquus* KGE-17 Cultivation. *Bioprocess Biosyst. Eng.* **2019**, *42*, 1225–1234. [CrossRef]

18. Aluç, Y.; Kök, O.; Tüzün, I. Profiling the Carotenoids of Microalga (*Scenedesmus obliquus*) Extract by HPLC and Its Antioxidant Capacity. *J. Appl. Biol. Sci.* **2022**, *16*, 206–219. [CrossRef]

19. Ambati, R.R.; Gogisetty, D.; Aswathanarayana, R.G.; Ravi, S.; Bikkina, P.N.; Bo, L.; Yuepeng, S. Industrial Potential of Carotenoid Pigments from Microalgae: Current Trends and Future Prospects. *Crit. Rev. Food Sci. Nutr.* **2019**, *59*, 1880–1902. [CrossRef]

20. Zhang, C. Biosynthesis of Carotenoids and Apocarotenoids by Microorganisms and Their Industrial Potential. *Prog. Carotenoid Res.* **2018**, *5*, 85–105. [CrossRef]

21. Varela, J.C.; Pereira, H.; Vila, M.; León, R. Production of Carotenoids by Microalgae: Achievements and Challenges. *Photosynth. Res.* **2015**, *125*, 423–436. [CrossRef] [PubMed]

22. Mata, T.M.; Almeida, R.; Caetano, N.S. Effect of the Culture Nutrients on the Biomass and Lipid Productivities of Microalgae *Dunaliella tertiolecta*. *Chem. Eng. Trans.* **2013**, *32*, 973–978. [CrossRef]

23. Daliry, S.; Hallajisani, A.; Mohammadi Roshandeh, J.; Nouri, H.; Golzary, A. Investigation of Optimal Condition for *Chlorella vulgaris* Microalgae Growth. *Glob. J. Environ. Sci. Manag.* **2017**, *3*, 217–230. [CrossRef]

24. Lu, L.; Wang, J.; Yang, G.; Zhu, B.; Pan, K. Biomass and Nutrient Productivities of *Tetraselmis chuii* under Mixotrophic Culture Conditions with Various C:N Ratios. *Chin. J. Oceanol. Limnol.* **2017**, *35*, 303–312. [CrossRef]

25. Bowden, G.D.; Pichler, B.J.; Maurer, A. A Design of Experiments (DoE) Approach Accelerates the Optimization of Copper-Mediated 18F-Fluorination Reactions of Arylstannanes. *Sci. Rep.* **2019**, *9*, 11370. [CrossRef]

26. Dejaegher, B.; Vander Heyden, Y. Experimental Designs and Their Recent Advances in Set-up, Data Interpretation, and Analytical Applications. *J. Pharm. Biomed. Anal.* **2011**, *56*, 141–158. [CrossRef]

27. Hron, J. Design Of Experiments For the Analysis And Optimization Of Barcodes Of Food And Agricultural Products. *Agric. Econ.* **2012**, *12*, 549–556. [CrossRef]

28. Hallenbeck, P.C.; Grogger, M.; Mraz, M.; Veverka, D. The Use of Design of Experiments and Response Surface Methodology to Optimize Biomass and Lipid Production by the Oleaginous Marine Green Alga, *Nannochloropsis gaditana* in Response to Light Intensity, Inoculum Size and CO_2. *Bioresour. Technol.* **2015**, *184*, 161–168. [CrossRef]

29. Mahdi, S.; Shahabadi, S.; Reyhani, A. Water Treatment via the Full Factorial Design Methodology. *Sep. Purif. Technol.* **2014**, *132*, 50–61. [CrossRef]

30. Maroneze, M.M.; Zepka, L.Q.; Lopes, E.J.; Pérez-Gálvez, A.; Roca, M. Chlorophyll Oxidative Metabolism during the Phototrophic and Heterotrophic Growth of *Scenedesmus obliquus*. *Antioxidants* **2019**, *8*, 600. [CrossRef]

31. Tap Medium | UTEX Culture Collection of Algae. Available online: https://utex.org/products/tap-medium?variant=3099173689 7626 (accessed on 27 June 2020).

32. BBM Medium | CCALA. Available online: https://ccala.butbn.cas.cz/en/bbm-medium (accessed on 27 June 2020).

33. Lin, Q.; Lin, J. Effects of Nitrogen Source and Concentration on Biomass and Oil Production of a *Scenedesmus rubescens* like Microalga. *Bioresour. Technol.* **2011**, *102*, 1615–1621. [CrossRef] [PubMed]

34. Qiu, R.; Gao, S.; Lopez, P.A.; Ogden, K.L. Effects of PH on Cell Growth, Lipid Production and CO_2 Addition of Microalgae *Chlorella sorokiniana*. *Algal Res.* **2017**, *28*, 192–199. [CrossRef]

35. Trovão, M.; Pereira, H.; Costa, M.; Machado, A.; Barros, A.; Soares, M.; Carvalho, B.; Silva, J.T.; Varela, J.; Silva, J. Lab-Scale Optimization of *Aurantiochytrium* sp. Culture Medium for Improved Growth and DHA Production. *Appl. Sci.* **2020**, *10*, 2500. [CrossRef]

36. Wen, Z.Y.; Jiang, Y.; Chen, F. High Cell Density Culture of the Diatom *Nitzschia laevis* for Eicosapentaenoic Acid Production: Fed-Batch Development. *Process Biochem.* **2002**, *37*, 1447–1453. [CrossRef]

37. Ganuza, E.; Anderson, A.J.; Ratledge, C. High-Cell-Density Cultivation of *Schizochytrium* sp. in an Ammonium/PH-Auxostat Fed-Batch System. *Biotechnol. Lett.* **2008**, *30*, 1559–1564. [CrossRef]

38. Zheng, Y.; Chi, Z.; Lucker, B.; Chen, S. Two-Stage Heterotrophic and Phototrophic Culture Strategy for Algal Biomass and Lipid Production. *Bioresour. Technol.* **2012**, *103*, 484–488. [CrossRef]

39. Cheng, P.; Wang, Y.; Osei-Wusu, D.; Wang, Y.; Liu, T. Development of Nitrogen Supply Strategy for *Scenedesmus rubescens* Attached Cultivation toward Growth and Lipid Accumulation. *Bioprocess Biosyst. Eng.* **2018**, *41*, 435–442. [CrossRef]

40. Mandotra, S.K.; Kumar, P.; Suseela, M.R.; Ramteke, P.W. Fresh Water Green Microalga *Scenedesmus abundans*: A Potential Feedstock for High Quality Biodiesel Production. *Bioresour. Technol.* **2014**, *156*, 42–47. [CrossRef]

41. Beuckels, A.; Smolders, E.; Muylaert, K. Nitrogen Availability Influences Phosphorus Removal in Microalgae-Based Wastewater Treatment. *Water Res.* **2015**, *77*, 98–106. [CrossRef]

42. Suen, Y.J.S.; Hubbard, G.H.; Tornabene, G.T. Total Lipid Production of the Green Alga *Nannochloropsis* sp. QII under Different Nitrogen Regimes. *Production* **1987**, *296*, 289–296. [CrossRef]

43. Seghiri, R.; Kharbach, M.; Essamri, A. Functional Composition, Nutritional Properties, and Biological Activities of Moroccan *Spirulina* Microalga. *J. Food Qual.* **2019**, *2019*, 7219. [CrossRef]
44. Aliyu, A.; Lee, J.G.M.; Harvey, A.P. Microalgae for Biofuel: Isothermal Pyrolysis of a Fresh and a Marine Microalga with Mass and Energy Assessment. *Chem. Eng. J. Adv.* **2023**, *14*, 100474. [CrossRef]
45. Ördög, V.; Stirk, W.A.; Bálint, P.; Lovász, C.; Pulz, O.; van Staden, J. Lipid Productivity and Fatty Acid Composition in *Chlorella* and *Scenepdesmus* Strains Grown in Nitrogen-Stressed Conditions. *J. Appl. Phycol.* **2013**, *25*, 233–243. [CrossRef]
46. Pereira, H.; Barreira, L.; Mozes, A.; Florindo, C.; Polo, C.; Duarte, C.V.; Custádio, L.; Varela, J. Microplate-Based High Throughput Screening Procedure for the Isolation of Lipid-Rich Marine Microalgae. *Biotechnol. Biofuels* **2011**, *4*, 61. [CrossRef]

marine drugs

Review

Genetic Engineering and Innovative Cultivation Strategies for Enhancing the Lutein Production in Microalgae

Bert Coleman [1,†], Elke Vereecke [2,3,4,†], Katrijn Van Laere [2], Lucie Novoveska [5,*] and Johan Robbens [1]

1   Aquatic Environment and Quality, Cell Blue Biotech and Food Integrity, Flanders Research Institute for Agriculture, Fisheries and Food (ILVO), Jacobsenstraat 1, 8400 Ostend, Belgium; bert.coleman@ilvo.vlaanderen.be (B.C.)
2   Plant Sciences Unit, Flanders Research Institute for Agriculture, Fisheries and Food (ILVO), Caritasstraat 39, 9090 Melle, Belgium
3   Department of Plant Biotechnology and Bioinformatics, Ghent University, Technologiepark 71, 9052 Zwijnaarde, Belgium
4   Center for Plant Systems Biology, Flemish Institute for Biotechnology (VIB), Technologiepark 71, 9052 Zwijnaarde, Belgium
5   ScotBio, Livingston EH54 5FD, UK
*   Correspondence: lucie.novoveska@gmail.com
†   These authors contributed equally to this work.

Abstract: Carotenoids, with their diverse biological activities and potential pharmaceutical applications, have garnered significant attention as essential nutraceuticals. Microalgae, as natural producers of these bioactive compounds, offer a promising avenue for sustainable and cost-effective carotenoid production. Despite the ability to cultivate microalgae for its high-value carotenoids with health benefits, only astaxanthin and β-carotene are produced on a commercial scale by *Haematococcus pluvialis* and *Dunaliella salina*, respectively. This review explores recent advancements in genetic engineering and cultivation strategies to enhance the production of lutein by microalgae. Techniques such as random mutagenesis, genetic engineering, including CRISPR technology and multi-omics approaches, are discussed in detail for their impact on improving lutein production. Innovative cultivation strategies are compared, highlighting their advantages and challenges. The paper concludes by identifying future research directions, challenges, and proposing strategies for the continued advancement of cost-effective and genetically engineered microalgal carotenoids for pharmaceutical applications.

Keywords: microalgae; carotenoids; lutein; cultivation strategies; genetic engineering

Citation: Coleman, B.; Vereecke, E.; Van Laere, K.; Novoveska, L.; Robbens, J. Genetic Engineering and Innovative Cultivation Strategies for Enhancing the Lutein Production in Microalgae. *Mar. Drugs* **2024**, *22*, 329. https://doi.org/10.3390/md22080329

Academic Editors: Cecilia Faraloni and Eleftherios Touloupakis

Received: 29 June 2024
Revised: 18 July 2024
Accepted: 19 July 2024
Published: 23 July 2024

Correction Statement: This article has been republished with a minor change. The change does not affect the scientific content of the article and further details are available within the backmatter of the website version of this article.

1. Lutein as One of the Important Carotenoids

Carotenoids are a class of fat-soluble pigments that include carotenes (β-carotene, lycopene) and xanthophylls (astaxanthin, lutein, fucoxanthin). Carotenes are hydrocarbon carotenoids, while xanthophylls are the oxygenated versions of carotenes. Humans cannot synthesize carotenoids and must obtain them through their diet. These compounds are mainly produced by plants, fungi, and microorganisms, with a total of 1204 natural carotenoids identified [1]. They primarily absorb light at wavelengths of 400–550 nm. In photosynthetic organisms, carotenoids play a crucial role in protecting photosynthetic organisms against photodamage and supporting their oxygenic photosynthesis [2].

Carotenoids have many health benefits, as listed in Table 1. Astaxanthin is the strongest antioxidant among carotenoids, with significantly stronger antioxidant activity and free radical inhibitory activity than vitamin E, β-carotene, and lutein [3]. The synthetic version is used in aquaculture to give fish (i.e., salmon) and crustaceans their pinkish color. β-Carotene is well-known as a precursor to vitamin A and is also used as natural orange-yellow color in the food industry [4]. Lutein is widely used as an antioxidant, food coloring agent, and nutritional supplement in cosmetics, food, health products, and medicine [5]. Other carotenoids, including zeaxanthin and fucoxanthin, also offer health benefits (Table 1).

Table 1. Health benefits of six carotenoids confirmed by human studies, their natural sources, and recommended dose (modified from Gong and Bassi, 2016 and Ren et al., 2021) [6,7].

Carotenoid	Health Benefits	Natural Sources	Recommended Dose
Astaxanthin	Strong anti-oxidant property Anti-inflammatory effects Anti-cancer Cardiovascular health	Shrimp; Salmon; Crabs; Microalgae (*Haematococcus pluvialis*) Phaffia rhodozyma	4–12 mg/day
β-Carotene	Prevent night blindness Anti-oxidant property Prevents liver fibrosis	Pumpkin; Mango; Carrots; Microalgae (*Dunaliella salina*)	600 µg RE */day
Lutein	Prevents cataract and age-related macular degeneration Anti-oxidant property Anti-cancer Prevents cardiovascular diseases	Marigold flower; Yolk; Broccoli; Microalgae; Orange-yellow fruits; Leafy green vegetables	6 mg/day
Zeaxanthin	Anti-cancer Anti-inflammatory Anti-allergy Against UV, skin redness	Marigold flower; Maize; Orange peppers; Microalgae; Scallions	2 mg/day
Fucoxanthin	Anti-obesity Anti-oxidant property	Macroalgae; Microalgae	/

* RE, retinol equivalent.

The global carotenoid market was valued at approximately USD 1.8 billion in 2021 and is projected to reach USD 2 billion by 2026 [8]. The astaxanthin market is projected to surpass USD 800 million by 2026. The β-carotene market is expected to be worth USD 620 million by 2026, with a CAGR of 3.8% from 2018 to 2026 [9]. The lutein market is projected to reach USD 358 million by the end of 2024 [8]. The global fucoxanthin market reached a valuation of USD 95 million in 2020 and is projected to attain a value of USD 113.5 million by 2026 [10]. Other carotenoids like zeaxanthin, lycopene, and canthaxanthin are also gaining support and increasing market interest due to their health benefits. The market shares by application are dominated by animal feed (34.8%), followed by food and beverages (26.1%), dietary supplements (23.5%), pharmaceuticals (9.2%), and cosmetics (6.5%) [8]. The global carotenoid market is poised for substantial growth, driven by increasing demand across various applications and the rising awareness of their health benefits.

In the evolving market of carotenoids, there is a significant contrast between natural and synthetic sources, each with its own cost implications, safety considerations, and market share dynamics. Carotenoids produced by chemical synthesis can be up to three times cheaper than those derived from natural sources. Consequently, natural carotenoids only capture 10–20% of the market share due to high production costs [5]. However, this ratio is changing as concerns about safety and environmental impact grow on synthetic carotenoids and the higher health benefits of natural carotenoids become clear [11]. Synthetic carotenoids are predominantly used in applications such as animal feed and as colorants, whereas natural carotenoids are preferred for use in medicine and food supplements [11]. Another major reason for the preference towards natural carotenoids is their significant differences in forms and bioactivity compared to synthetic sources, as natural carotenoids are usually complex mixtures of various isomers, whereas synthetic carotenoids typically consist of a single form [11]. The market value of synthetic carotenoids is relatively low, ranging from USD 250 to USD 2000 per kilogram, whereas natural carotenoids from plant sources range from USD 350 to USD 7500 per kilogram [12]. This price difference reflects the higher consumer demand and perceived benefits of natural carotenoids over their synthetic counterparts.

2. Microalgae as Producers of Lutein and Other Carotenoids

Microalgae are emerging as significant producers of valuable carotenoids, with biosynthesis occurring in their chloroplasts. Microalgae are a diverse group of eukaryotic aquatic microorganisms that can thrive in fresh, brackish, or marine water, with over 200,000 species classified into various phylogenetic groups [13]. Depending on the species, microalgae can be cultivated photoautotrophically, using light as an energy source to assimilate CO_2 for biomass production, or heterotrophically, utilizing an organic carbon source for energy. Some species can grow mixotrophically, using both light and organic carbon sources. Photoautotrophic microalgae can be cultivated in open raceway ponds (ORPs) or closed photobioreactors (PBRs) [13,14]. These systems can be placed outdoors using natural sunlight, indoors with artificial lighting, or in greenhouses with natural sunlight. ORP systems are characterized by lower investment and operational costs but are more susceptible to biological contamination compared to closed PBRs [15]. In contrast, PBRs offer easier control over cultivation parameters and higher biomass productivity. Heterotrophic microalgae are cultivated in fermenters (closed systems) where an organic carbon source is added. These fermenters can achieve higher cell densities (25–125 g DW/L) compared to PBRs (0.5 g DW/L) [13]. Heterotrophic cultivation in well-controlled bioreactors is gaining commercial attention for pigment production due to its ability to overcome challenges related to CO_2 and light supply, as well as contamination and land requirements in open autotrophic systems [16]. Microalgae growth can be managed in various modes: batch, fed-batch, semi-continuous, and continuous modes [13]. In batch mode, microalgae are grown with all nutrients provided at the start, continuing until nutrients deplete. It is simple to operate, which also reduces the risk of contamination, but it generally results in lower biomass production. In fed-batch mode, nutrients are periodically added during cultivation, extending the exponential growth phase and increasing biomass production. However, this method requires more nutrient monitoring and has more risk of contamination compared to batch operation. In semi-continuous mode, a portion of the culture is periodically harvested at a certain time point, and fresh medium is added to maintain growth. This method maintains high biomass productivity but needs periodic intervention, which increases the labor intensity and the risk of contamination. In continuous mode, part of the culture is continuously harvested while new medium is added to the system. This method offers stable biomass productivity and requires low labor cost, but it is more difficult to control and more prone to contamination compared to the other operation modes.

Carotenoids produced by microalgae can be categorized into primary and secondary types, each serving distinct functions. Primary carotenoids, such as β-carotene, lutein, fucoxanthin, lycopene, and violaxanthin, are essential for photosynthesis, playing a crucial role in light harvesting and protecting chlorophyll from photodamage. Secondary carotenoids, such as astaxanthin and canthaxanthin, are not directly involved in photosynthesis but can act as antioxidants, protecting the cells from damage caused by stressors such as high light intensity, nutrient deprivation, high salinity, or oxidative stress [17]. In response to these stress factors, secondary carotenoids can be overproduced in microalgae. Several primary carotenoids, such as β-carotene, also accumulate under stress conditions and act as secondary metabolites [18]. Despite the ability to cultivate microalgae for its high-value carotenoids with health benefits, only astaxanthin and β-carotene are produced on a commercial scale by *Haematococcus pluvialis* [19,20] and *Dunaliella salina* [21], respectively.

In contrast to astaxanthin and β-carotene, lutein production from microalgae is not yet commercially established. Currently, natural lutein is primarily produced from marigold flowers [5]. However, several microalgal species offer promising prospects for the production of this carotenoid. There is high interest in natural lutein because of its potential health benefits (Table 1). However, high culture costs and low carotenoid yields are bottlenecks for the commercialization of lutein from microalgae. Lin et al. (2015) estimated that microalgae need to have a lutein content exceeding 1% of their dry weight (DW) to be considered commercially viable for production [22]. Unlike secondary carotenoids, which can be overproduced under stress conditions, the overproduction of primary carotenoids like lutein is much more challenging due to its connection to the growth performance of microalgae. Cultivation conditions can be optimized to enhance the biomass productivity (g DW/L/day) and lutein content (mg/g DW) in microalgae. However, the optimal parameters for maximizing the lutein content can reduce the biomass productivity, thereby lowering the overall lutein productivity (mg/L/day) [23]. Therefore, a balanced approach is necessary to optimize both lutein content and biomass productivity for maximum overall lutein production. Metabolic engineering, which enables the targeted upregulation of specific carotenoid biosynthesis pathways, might offer a solution to enhance lutein productivity [5]. This review exclusively focuses on enhancing lutein content and productivity through strain improvement via random mutagenesis and metabolic engineering, alongside cultivation strategies such as fed-batch, batch, and semi-continuous methods. Unlike reviews that emphasize cultivation parameters for increasing carotenoid content broadly, our review specifically concentrates on lutein and offers a comparative analysis across various microalgal species.

Downstream processing (harvesting, cell disruption, extraction, and purification) of microalgal biomass for lutein production are also crucial steps determining the production costs but will not be covered here, as they were recently extensively reviewed by Gong and Bassi (2016) [6] and Zohra et al. (2022) [24].

3. Random Mutagenesis to Increase the Lutein Production

Random mutagenesis is a versatile and powerful technique employed in genetic research and biotechnology to induce genetic variation and enhance desirable traits in organisms [25]. This approach involves the deliberate induction of mutations across an organism's genome without targeting specific genes, thus generating a diverse pool of genetic variants [26]. The method stands in contrast to new genomic techniques (NGTs), which involve specific, intentional changes to particular genes. The process of random mutagenesis typically involves exposing microalgal cultures to physical or chemical mutagens [25,26]. Physical mutagens such as UV radiation or gamma rays create DNA damage that can lead to mutations during DNA repair processes [25]. Chemical mutagens, like ethyl methanesulfonate (EMS) and N-methyl-N'-nitro-N-nitrosoguanidine (MNNG), introduce changes by chemically altering nucleotides, causing mispairing and subsequent mutations during replication [27]. Random mutagenesis does not require any previous knowledge about the genetics and metabolism of the target organism and the development of molecular tools, which can be time-consuming and expensive [28]. This advantage is particularly significant for organisms with limited or inaccessible genomic information and can be used for the assignment of phenotypes (traits/characteristics) to a certain gene/genotype and the expansion of the current understanding of the biology and metabolism of several microalgae species [26].

One of the significant applications of random mutagenesis is in the field of microalgal biotechnology, particularly for the enhancement of valuable compounds like lutein (Table 2). Mixotrophic cultivation of a *Chlorella sorokiniana* MB-1-M12 mutant, created by a 1 h MNNG treatment, resulted in a higher lutein content and lutein productivity (7.52 mg/g DW; 3.63 mg/L/day) compared to the wild-type strain *C. sorokiniana* MB-1 (5.86 mg/g DW;

2.56 mg/L/day) [29]. Ren et al. (2022) also treated microalga *Chromochloris zofingiensis* with MNNG, resulting in the mutant *Cz-pkg* that could reach a lutein content of 6.28 mg/g DW with a lutein productivity of 10.57 mg/L/day, which was 2.5- and 8.5-fold higher, respectively, than that of the wild-type [30]. Characterization of the mutant *Cz-pkg* with single-nucleotide polymorphism (SNP) analysis to pinpoint the exact mutation(s) revealed a T to A substitution in the cGMP-dependent protein kinase (PGK), leading to a premature stop codon (UAG) [30]. As a result, the photosystem of the *Cz-pgk* mutant can work effectively for uptake of both inorganic (CO_2) and organic carbon sources (glucose) under mixotrophic conditions, thereby enhancing its capacity for lutein production [30].

Following mutagenesis, the challenge lies in screening and identifying mutants with the desired trait—in this case, increased lutein content. Random mutagenesis introduces mutations randomly, leading to many neutral or deleterious changes that necessitate extensive screening to identify beneficial mutants. This screening process can be labor-intensive, time-consuming, and resource-intensive [26]. Additionally, the randomness of the mutations can cause unintended effects, with changes occurring in genes unrelated to the targeted trait, resulting in unforeseen alterations in other phenotypic characteristics. Effective screening methods for lutein include the use of norflurazon and nicotine. Norflurazon, an herbicide that inhibits phytoene desaturase in the carotenoid biosynthesis pathway, disrupts the production of downstream carotenoids, including lutein [31]. Nicotine, a specific inhibitor of lycopene β-cyclase, affects lutein biosynthesis by interfering with the conversion of lycopene to β-carotene [32]. Screening for mutants that tolerate and thrive in the presence of norflurazon and/or nicotine allows for the identification of strains with modifications in the carotenoid biosynthesis pathway, potentially leading to an increased lutein content. Cordero et al. (2011) used these screening methods to identify mutants with a higher lutein content [32]. *C. zofingiensis* was chemically mutagenized with MNNG and spread onto media supplemented with norflurazon or nicotine. From all the mutants, two mutants resistant to norflurazon showed a 53–55% increase in lutein content relative to the wild-type, reaching values of 7.0 mg/g DW. One mutant resistant to nicotine exhibited a 1.4-fold higher lutein content than that of the wild-type, reaching 6.4 mg/g DW [32].

A complementary approach to identifying mutants with higher lutein content involves a color-based colony screening approach. This technique leverages the visual differences in pigmentation that result from variations in carotenoid content among different mutants. For example, mutants with a higher lutein content will appear more intensively yellow. Huang et al. (2018) used this screening approach to select for a higher zeaxanthin, lutein, and β-carotene content in *C. zofingiensis* cells treated with the chemical mutagen MNNG and grown on media containing glucose [33]. The selected mutant, *CZ-bkt1*, with a premature stop codon in β-carotene ketolase, accumulated zeaxanthin up to 0.34 mg/g DW, lutein up to 7.12 mg/g DW, and β-carotene up to 1.51 mg/g DW [33]. Stress conditions known to induce carotenoid biosynthesis were used to further increase the carotenoid content in the mutant. High-light radiation (460 μmol/m^2/s) and nitrogen deficiency could increase the zeaxanthin content up to 7.00 mg/g DW, lutein up to 13.81 mg/g DW, and β-carotene up to 7.18 mg/g DW when induced by high-light irradiation and nitrogen deficiency [33].

4. Metabolic Engineering for Enhanced Production of Lutein

In contrast to random mutagenesis, new genomic techniques (NGTs) can introduce very specific, intentional changes to particular genes. Metabolic engineering using NGTs is a powerful approach for steering cell metabolism by modifying specific pathway enzymes or regulatory proteins. It utilizes a toolbox that includes gene knock-out/knock-in techniques, as well as gene repression and overexpression applications. Gene knock-out techniques, such as clustered regulatory interspaced short palindromic repeats/Cas (CRISPR/Cas) [34], zinc-finger nucleases (ZFNs) [35], and transcription activator-like effector nucleases (TALENs) [36], result in the modification of specific genes to abolish undesired pathways [37]. Knock-in techniques integrate new genetic material to introduce novel functionalities. Gene repression restricts the expression of certain genes to redirect metabolic flux and includes RNA interference (RNAi), a technique used to degrade specific mRNA molecules, preventing their translation into proteins [38]. CRISPR interference (CRISPRi), another technique under gene repression, interferes with the transcription machinery [39]. Gene overexpression techniques, achieved through promotor engineering or copy number amplification, can generate a many-fold overexpression of a specific gene that is responsible for a rate-determining step. Metabolic engineering techniques are key approaches to improve microalgal biomass productivity and suitability on an industrial scale in terms of high-value compound production (such as lutein), growth rate, mode of nutrition, and synergism between lutein accumulation and growth [5].

Metabolic engineering relies on a good knowledge of the biosynthetic pathway and involved genes. The carotenoid biosynthetic pathway in microalgae is intricate and involves several enzymatic steps, each of which can serve as a potential target for metabolic engineering. Precursors of the carotenoid biosynthesis pathway can be retrieved from the mevalonate (MVA) pathway and/or the methyl-D-ertythritol phosphate pathway (MEP). However, in green microalgae, the MVA pathway has been lost, leaving the MEP pathway as the sole pathway for isopentenyl phyrophosphate (IPP) synthesis. The carotenoid biosynthetic pathway is well described in Figure 1 [39,40].

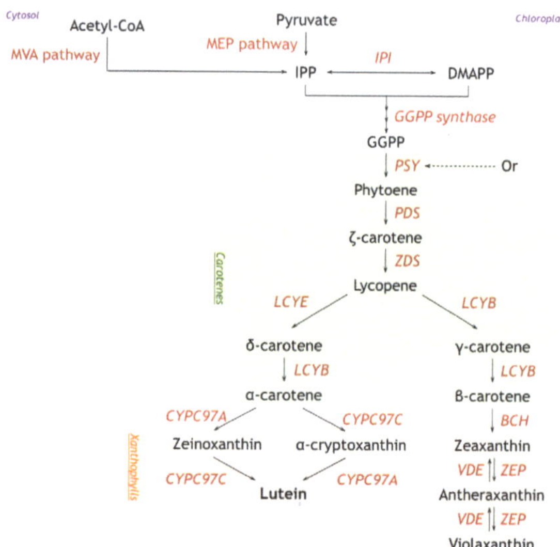

Figure 1. Schematic overview of the carotenoid biosynthesis pathway. The carotenoid biosynthesis starts with the synthesis of isopentyl pyrophosphate (IPP) through either the mevalonate (MVA) pathway or the methyl-D-erythritol phosphate (MEP) pathway located in the chloroplast [40]. IPP isomerizes to dimethylallyl diphosphate (DMAPP) and together they produce the immediate precursor of the carotenoid synthesis, geranylgeranyl pyrophosphate (GGPP). The condensation of two GGPP molecules leads to the formation of a colorless carotene, phytoene, by enzyme phytoene synthase (PSY). This PSY-catalyzed reaction is known to be the most important rate-limiting step in the carotenoid biosynthetic pathway [41]. Phytoene is converted to lycopene via a multi-step desaturation reaction catalyzed by phytoene desaturase (PDS) and ζ-carotene desaturase (ZDS) [42]. The cyclization of lycopene is the branching point of the carotenoid synthesis pathway into an α-branch and β-branch. In the α-branch, lycopene-ε-cyclase (LCYE) and lycopene β-cyclase (LCYB) catalyze lycopene to α-carotene, and then cytochrome P450 β-hydroxylase (CYPC97A) and cytochrome P450 ε-hydroxylase (CYPC97C) convert α-carotene to lutein [40]. In the β-branch, lycopene β-cyclase (LCYB) catalyzes lycopene to β-carotene, which is further hydroxylated by β-carotene hydroxylase (BCH) to form zeaxanthin [7]. Zeaxanthin epoxidase (ZEP) then converts zeaxanthin to violaxanthin. Violaxanthin can be converted to neoxanthin or enter the xanthophyll cycle, where it can be reversibly converted back to zeaxanthin via antheraxanthin to balance light harvesting and photoprotection.

Regulatory mechanisms in this pathway include transcriptional regulation, where the genes encoding carotenoid biosynthetic enzymes are regulated by (1) environmental factors such as light and nutrient availability, (2) feedback inhibition, where end products inhibit upstream enzymes, and (3) post-transcriptional regulation, which influences mRNA stability and translation efficiency [43]. One important example of post-transcriptional regulation is by the Orange (Or) gene, which is known to affect carotenoid accumulation and is involved in stabilizing the PSY gene, ensuring that PSY remains active and functional within the cell [44]. Different strategies can be used to increase the content of carotenoids, either directly by overexpression of endogenous enzymes involved in the carotenoid biosynthesis pathway, or indirectly by inhibition of competition pathways and thus redistributing the metabolic flow towards certain pigments [5].

Applications of metabolic engineering in microalgae for overproduction of carotenoids, and lutein in particular, are listed in Table 2. To increase the overall carotenoid production, the PSY gene and Or gene are good candidates for metabolic engineering, as they catalyze a rate-limiting step in the carotenoid biosynthesis pathway. Overexpression of the Or gene in *Chlamydomonas reinhardtii* using a strong dual-promotor system led to a lutein production from 0.69 mg/L to 1.04 mg/L and a β-carotene production from 0.18 mg/L to 0.24 mg/L, respectively [45]. Functional analysis of Or genes in various orange-flesh melon fruits identified a single-nucleotide polymorphism known as "golden SNP", which converts a highly conserved arginine to histidine and is responsible for an elevated carotenoid accumulation [44]. Site-directed mutagenesis was used by Yazdani et al. (2021) to introduce this "golden SNP" in the Or gene in *C. reinhardtii* and overexpression resulted in a 4-fold increase in α-carotene, 3.1-fold increase in lutein, 3.2-fold increase in β-carotene, and 3.1-fold increase in violaxanthin [46]. Overexpression of the PSY gene itself can also result in a higher overall carotenoid content, as shown by Velmurugan et al. (2023) [47]. The overexpression of the endogenous PSY gene in *D. salina* led to a substantial increase in lutein and β-carotene content, up to 5.4-fold and 7.2-fold compared to the wild-type [47].

Considering the metabolic branches originating from lycopene, overexpression of LCYB or LCYE could affect the productivity of carotenoids in each branch pathway. The overexpression of LCYE could redirect the lycopene flux towards the α-branch, with an increase in lutein production as a result. Tokunuaga et al. (2021) overexpressed the LCYE gene in *C. reinhardtii* using a dual-promotor system, resulting in a significant increase in lutein concentrations (2.3-, 2.0-, and 2.6-fold, respectively) compared to the wild-type [48]. Lou et al. (2021) used the LCYE gene from *C. vulgaris*, which is rich in lutein, for the heterologous expression in *C. reinhardtii*, where it showed an enhanced lutein content from 0.583 mg/g DW to 1.357 mg/g DW [49]. Both studies indicate that the conversion of lycopene to α-carotene can be increased by homologous or heterologous expression of the LCYE gene.

Table 2. Application of random mutagenesis and metabolic engineering in microalgae to increase carotenoid content.

Source Species	Host Strain	Target Gene	Technique	Carotenoid	Reference
C. sorokiniana	/	/	Random mutagenesis (MNNG treatment)	Increased lutein content up to 7.25 mg/g DW with a productivity of 2.56 mg/L/day.	[29]
C. zofingiensis	/	/	Random mutagenesis (MNNG treatment)	Increased lutein content up to 6.25 mg/g DW with a productivity of 10.57 mg/L/day	[30]
C. sorokiniana	/	/	Random mutagenesis (MNNG treatment)	Increased lutein content up to 7.0 mg/g DW and 6.4 mg/L/day	[32]
C. zofingiensis	/	/	Random mutagenesis (MNNG treatment)	Increased zeaxanthin (up to 7.0 mg/g DW), lutein (up to 13.81 mg/g DW) and β-carotene (7.18 mg/g DW).	[33]

Table 2. *Cont.*

Source Species	Host Strain	Target Gene	Technique	Carotenoid	Reference
C. reinhardtii	Endogenous	Or	Overexpression Or gene using dual-promotor system	Lutein production increase from 0.69 mg/L to 1.04 mg/L and from 0.18 mg/L to 0.24 mg/L	[45]
C. reinhardtiii	Endogenous	Or	Overexpression Or gene	Increased α-carotene (1.9-fold higher), lutein (2-fold higher), β-carotene (2.1-fold higher) and violaxanthin (2.1-fold higher) content compared to WT.	[46]
C. reinhardtii	Endogenous	Or	Overexpression Or gene with single amino acid substitution using site-directed mutagenesis	Increased α-carotene (4-fold higher), lutein (3.1-fold higher), β-carotene (3.2-fold higher) and violaxanthin (3.1-fold higher) content compared to WT.	[46]
C. reinhardtii	*Brassica oleracea*	Or	Heterologous expression Or gene.	Increased lutein (1.5-fold higher: 112.4 pg/cell to 73.0 pg/cell lutein (WT)) and astaxanthin content (2-fold higher: 0.41 pg/cell to 0.2 pg/cell (WT))	[50]
C. reinhardtii	*Mesorhizobium loti* and *Sulfuri-hydrogenibium yellowstonense*	CA	Heterologous expression of CA gene	Increased lutein concentration from 4.41 mg/L (WT) to 8.89 mg/L (CA from *Ml*) and 7.07 mg/L (CA from *SY*).	[51]
D. salina	Endogenous	PSY	Overexpression of PSY gene.	Increased lutein (7.6-fold higher) and β-carotene (5.4-fold higher) content compared to WT.	[47]
D. salina	*H. pluvialis*	PSY	Heterologous expression of PSY gene.	Increased lutein (7.2-fold higher) and β-carotene (2.4-fold higher) conten compared to WT.	[47]
C. reinhardtii	*D. salina*	PSY	Heterologous expression LCYE gene.	Increased lutein (2.6-fold higher) content compared to WT.	[32]
Scenedesmus	/	PSY	Expression of synthetic PSY gene.	Increased β-carotene content from 10.8 mg/g (WT) cell to 30 mg/g cell.	[28]
C. reinhardtii	Endogenous	LCYE	Overexpression of LCYE gene.	Increased lutein (at least 2-fold higher) content.	[48]
C. reinhardtii	*C. vulgaris*	LCYE	Heterologous of LCYE gene.	Increased lutein content (2.3-fold higher) compared to WT.	[49]
H. pluvialis	Endogenous	PDS	Overexpression of PDS gene with single amino acid substitution using site-directed mutagenesis.	Increased lutein (1.5 µg/g DW to 1.9 µg/g DW), zeaxanthin (142 µg/g DW to 214 µg/g DW), β-carotene (532 µg/g DW to 728 µg/g DW) and astaxanthin content compared to WT.	[52]
C. zofingiensis	Endogenous	PDS	Overexpression of PDS gene with single amino acid substitution using site-directed mutagenesis.	Increased total carotenoid content with 32.1% and astaxanthin with 54.1%.	[53]
C. reinhardtii	/	LCYE	Knock-out of the LCYE gene using CRISPR/Cas (NHEJ)	Increased zeaxantin content (up to 60%).	[54]
C. reinhardtii	/	LCYE	Knock-out of the LCYE gene using CRISPR/Cas (HDR)	Increased zeaxantin (0.31 mg/L (WT) to 0.59 mg/L), antheraxanthin (0.28 mg/L (WT) to 0.63 mg/L) and violaxanthin (1.3 mg/L (WT) to 2.3 mg/L) content.	[55]

According to the studies mentioned, Or, PSY, and ε-LCY are the key targets for metabolic engineering to increase the lutein content. PSY regulates the overall flow towards carotenoid synthesis, while ε-LCY directs the flow specifically towards the α-carotene branch [56]. Therefore, overexpression of the PSY gene influences the total carotenoid content, while overexpression of the ε-LCY gene only positively affects the lutein content. Aside from the genes involved in the carotenoid biosynthetic pathway, other genes involved in microalgae metabolism might be good targets for metabolic engineering. Lin et al. (2022) showed that the heterologous expression of carbonic anhydrases (CAs) in *C. reinhardtii* with the aim to increase photosynthetic capability and carbon capture could increase biomass from lutein production from 4.41 mg/L to 8.89 and 7.07 mg/L [51].

5. Cultivation Strategies to Increase Microalgal Lutein Production

Many microalgal species, including *Chlorella* sp., *Chlamydomonas* sp., *Desmosdesmus* sp., and *Scenedesmus* sp., produce lutein [57]. An important step before implementing a cultivation strategy is understanding the effect of cultivation parameters on lutein production of microalgae, which is reviewed by Hu et al. (2018) [16], Liu et al. (2021) [8], and Suparmaniam et al. (2024) [23]. Under photoautotrophic or mixotrophic conditions, high light intensity (often > 750 $\mu mol/m^2/s$) induces carotenogenesis, leading to higher lutein content in microalgae [23]. Furthermore, adequate CO_2 supply and the removal of the built-up O2 are essential for biomass growth, especially in photoautotrophic conditions. Under heterotrophic conditions, glucose is the optimal carbon source, and urea is the most effective nitrogen source for lutein production, outperforming carbon sources, such as glycerol and acetate, and nitrogen sources, such as nitrate and ammonium [58–61]. Under mixotrophic cultivation, acetate is the preferred carbon source, while nitrate and urea are the preferred nitrogen sources for lutein production. Importantly, high initial concentrations of carbon sources (often > 50–100 g/L) such as glucose and acetate can inhibit growth and reduce lutein production due to substrate inhibition [16]. Aeration also strongly affects lutein productivity since the heterotrophic metabolism of microalgae requires oxygen for growth [60]. Other than supplying sufficient oxygen for biomass production, oxygen levels do not influence the lutein content [16]. Nitrogen availability is critical for biomass productivity and lutein production, as nitrogen depletion leads to lutein degradation [57]. For certain microalgal species such as *Chlamydomonas* and *Scenedesmus* species, it has been reported that the lutein content remains high until the onset of nitrogen depletion [61,62]. Finally, each microalga has a specific optimal pH and temperature range for growth. Generally, increasing temperature up to a stress limit enhances lutein productivity, while temperatures beyond this threshold severely affect cellular growth and survival. On the contrary, cooler temperatures decrease the nutrient uptake rate and slow down both microalgal growth and lutein productivity [57]. However, Léon-Vaz et al. (2023) demonstrated that certain microalgal species produce higher lutein content at elevated light intensity (500 $\mu mol/m^2/s$) and 10 °C, compared to lower light intensity (100 $\mu mol/m^2/s$) and 20 °C [63]. Consequently, the optimal temperature for lutein production varies by species and must be determined experimentally.

In the following part of this review, innovative cultivation strategies for various microalgae species will be discussed, focusing on lutein productivity (mg/L/day) and lutein content (mg/g DW). The strategies resulting in the highest lutein production are shown in Tables 3 and 4.

5.1. Cultivation Strategies for Chlorella Species

Mixotrophic and heterotrophic conditions have been explored for lutein production in various *Chlorella* species, including *C. sorokiniana*, *C. protothecoides* (now *Auxenochlorella protothecoides*), and *C. vulgaris*. Researchers place particular emphasis on cultivation strategies to maximize lutein production in strains of *C. sorokiniana*. *C. sorokiniana* FZU60 is an excellent candidate for large-scale lutein production due to its rapid growth and high lutein content, reaching 8.29–11.22 mg/g DW under photoautotrophic and mixotrophic conditions, and 2.33–4.42 mg/g DW under heterotrophic conditions. Based on the reviewed articles (Table 3), the optimal conditions for cultivating *C. sorokiniana* FZU60 include a temperature range of 30–33 °C, light intensity of 150–750 μmol/m^2/s depending on culture density, and an optimal pH of 7.5. Suitable media are BG-11 and Mann and Meyer's. Depending on the culture density, the aeration requirements vary around 0.02–0.2 vvm with 2–2.5% CO$_2$ for photoautotrophic or mixotrophic growth. For optimal heterotrophic growth, the dissolved oxygen (DO) is crucial and should be controlled at 20–50% air saturation using aeration or agitation (stirring).

Two-stage strategies, starting from fed-batch mixotrophic conditions and transitioning to photoautotrophic conditions, achieve higher lutein productivities for *C. sorokiniana* FZU60 compared to single-stage batch and fed-batch modes under mixotrophic conditions. Xie et al. (2020) used a two-stage strategy to maximize lutein productivity of *C. sorokiniana* FZU60 [64]. Initially, this strain was cultivated under fed-batch mixotrophic conditions for 3 days under white light 350 μmol/m^2/s, maintaining 1 g/L sodium acetate (NaAc) by adding a 400 g/L NaAc solution when the DO reached 7 mg/L. The culture was then shifted to photoautotrophic conditions for 4 days using the same light conditions, with concentrated BG11 medium added when 90% of the nitrate was consumed. This strategy achieved a higher lutein productivity (4.67 mg/L/day) compared to a batch mixotrophic operation (3.59 mg/L/day), while achieving a lutein content of 9.51 mg/g. Furthermore, Xie et al. (2019) discovered that a two-stage strategy to produce lutein in *C. sorokiniana* FZU60 can be operated in semi-continuous mode [65]. Initially, *C. sorokiniana* FZU60 was cultivated under fed-batch mixotrophic conditions in BG-11 medium with 1 g/L NaAc pulses every 12 h for 1.5 days under white light (150 μmol/m^2/s). Subsequently, 92.5% of the culture was shifted to photoautotrophic conditions with the same light intensity and 100-fold BG11 medium for another 1.5 days. The other 7.5% was used as the seed for a new cycle in the first stage. This strategy enhanced the average lutein productivity to 11.57 mg/L/day, compared to 4.78 mg/L/day in batch culture and 4.20 mg/L/day in fed-batch culture, while reaching a lutein content of on average 9.57 mg/g DW after each cycle.

The nutrient feeding dosage and light intensity also significantly impact the lutein productivity. Ma et al. (2020) compared different fed-batch strategies for *C. sorokiniana* FZU60 in BG-11 medium under mixotrophic conditions (750 μmol/m^2/s) [66]. They found that feeding a constant NaAc concentration of 1 g/L yielded higher lutein productivity (8.04 mg/L/day) compared to the gradient feeding of NaAc (6.11 mg/L/day). Additionally, the authors explored the effect of the light intensity of the second stage. Higher lutein productivity and lutein content were achieved with 750 μmol/m^2/s (8.25 mg/L/day; 11.22 mg/g DW) compared to 150 μmol/m^2/s (5.62 mg/L/day; 9.85 mg/g DW).

Xie et al. (2022) investigated different nutrient feeding strategies for *C. sorokiniana* FZU60 under heterotrophic conditions in a 5 L fermenter [59]. Importantly, the DO was controlled at 20% by adjusting the stirring speed. When DO levels increased, new media with glucose were fed into the fermenter. Enhancements in lutein productivity and lutein concentration in the fermenter were observed using a 3-fold concentration of Mann and Myer's medium with urea (82.50 mg/L/day; 415.93 mg/L after 6 days) compared to 1-fold (53.93 mg/L/day; 273.86 mg/L after 6 days) and 6-fold (50.37 mg/L/day; 254.90 mg/L after 6 days) concentrations. This observation was explained by the more stable substrate addition and operational parameters (stirring speed and DO) compared to the 1-fold, and less substrate inhibition compared to the 6-fold. Notably, the lutein content of *C. sorokiniana* FZU60 under fed-batch heterotrophic growth was only 2.57 mg/g DW.

Table 3. Cultivation strategies to increase the lutein production in *Chlorella* sp.

Microalgae	Cultivation Mode	Reactor Volume (L)	Strategy *	Lutein Content (mg/g DW)	Lutein Production (mg/L)	Lutein Productivity (mg/L/day)	Reference
C. sorokiniana FZU60	Two-stage semi-continuous (5 cycles)	1	1st Fed-batch Mixo (BG-11 with 1 g/L NaAc every 12 h and 150 μmol/m^2/s) After 1.5 days, 92.5% medium replacement which is transferred to 2nd stage. 7.5% to new cycle 1st stage. 2nd Batch Photo (150 μmol/m^2/s)	9.57 (Day 3, average of the 5 cycles)	17.35 (Day 3, average of the 5 cycles)	11.57 (average of the 5 cycles)	[65]
C. sorokiniana FZU60	Two-stage	50	1st Fed-batch Mixo (acetate and 350 μmol/m^2/s) 2nd Fed-batch Photo (BG11 and 350 μmol/m^2/s)	9.51 (Day 7)	33.55 (Day 7)	4.67 (average over 7 days)	[64]
C. sorokiniana FZU60	Fed-batch	1	Mixo (acetate and 750 μmol/m^2/s)	8.29 (Day 7)	32.16 (Day 4)	8.04 (average over 4 days)	[66]
C. sorokiniana FZU60	Two-stage	1	1st Fed-batch Mixo (acetate and 750 μmol/m^2/s) 2nd Batch Photo (750 μmol/m^2/s)	11.22 (Day 8)	65.96 (Day 8)	8.25 (average over 8 days)	[66]
C. sorokiniana FZU60	Fed-batch	5	Hetero (Mann and Myer's with glucose and urea)	2.57 (Day 6)	415.93 (Day 6)	82.50 (average)	[59]
C. sorokiniana MB-1-M12	Semi-continuous	1	Batch Mixo (acetate and 150 μmol/m^2/s) After glucose depletion, 75% medium replacement	4.98 (Day 7 in 2nd cycle)	11.95 (Day 7 in 2nd cycle)	6.61 (2nd cycle average)	[67]
C. sorokiniana MB-1-M12	Batch	1	Batch Photo (150 μmol/m^2/s)	6.01 (Day 4)	16.40 (Day 5)	3.56 (average)	[68]
C. sorokiniana MB-1-M12	Batch	1	Batch Mixo (acetate and 150 μmol/m^2/s)	7.00 (Day 5)	18.04 (Day 5)	5.15 (average)	[68]
C. sorokiniana MB-1-M12	Batch	1	Batch Hetero (glucose)	2.31 (Day 7)	7.71 (Day 4)	1.88 (average)	[68]
C. sorokiniana MB-1-M12	Two-stage	1	1st Batch Photo (150 μmol/m^2/s) 2nd Batch Hetero (glucose)	4.75 (Day 10)	24.97 (Day 10) 20.5 (after day 6)	1.75 (average)	[68]
C. sorokiniana MB-1-M12	Two-stage	1	1st Batch Hetero (glucose) 2nd Batch Photo (150 μmol/m^2/s)	6.52 (Day 6)	34.62 (Day 9)	2.86 (average)	[68]
C. sorokiniana MB-1-M12	Two-stage	1	1st Batch Mixo (acetate and 150 μmol/m^2/s) 2nd Batch Hetero (glucose)	3.50 (Day 10)	19.07 (Day 10) 17.5 (after day 7)	1.3 (average)	[68]
C. sorokiniana MB-1-M12	Two-stage	1	1st Batch Hetero (glucose) 2nd Batch Mixo (acetate and 150 μmol/m^2/s)	6.17 (Day 10)	33.64 (Day 10)	3.42 (average)	[68]

Table 3. *Cont.*

Microalgae	Cultivation Mode	Reactor Volume (L)	Strategy *	Lutein Content (mg/g DW)	Lutein Production (mg/L)	Lutein Productivity (mg/L/day)	Reference
C. sorokiniana MB-1-M12	Fed-batch	1	Fed-batch Hetero (glucose)	3.40 (Day 11)	39.50 (Day 11)	3.24 (average)	[69]
C. sorokiniana MB-1-M12	Two-stage semi-continuous (3 cycles)	1	1st Fed-batch Hetero (glucose and urea) After highest biomass accumulation, 75% medium replacement which is transferred to 2nd stage. 25% to new cycle 1st stage. 2nd Batch Mixo (acetate and 150 μmol/m^2/s)	6.77 (1ste cycle; day 11) 6.61 (2nd cycle, day 17) 6.53 (3th cycle, day 23)	76.00 (1ste cycle; day 11) 80.88 (2nd cycle, day 17) 81.77 (3th cycle, day 23)	1st stage ($\pm$ 6.17 average) 2nd stage ($\pm$2.86 average)	[69]
C. sorokiniana MB-1-M12	Two-stage semi-continuous (3 cycles)	5	1st Fed-batch Hetero (glucose and urea) After highest biomass accumulation, 75% medium replacement which is transferred to 2nd stage. 25% to new cycle 1st stage. 2nd Batch Mixo (acetate and 150 μmol/m^2/s)	8.19 (1ste cycle; day 14) 8.09 (2nd cycle, day 18) 8.71 (3th cycle, day 21)	181.11 (1ste cycle; day 14) 153.60 (2nd cycle, day 18) 169.17 (3th cycle, day 21)	1st stage ($\pm$20.02 average) 2nd stage ($\pm$5.55 average)	[69]
Chlorella protothecoides CS-41	Two-stage	30	1st Fed-batch Hetero (glucose and urea) After 10 days temperature shifted from 28 °C to 32 °C 2nd Batch (nutrient limited phase)	5.35 (Day 14) 3.8 (after 10 days)	209.08 (Day 14) 200 (after 10 days)	19.18 (average)	[70]
Chlorella minutissima MCC-27	Batch	2	Batch Photo (Constant 260 μmol/m s)	6.37 (Day 5)	22.1 (Day 5)	4.32 (average)	[71]
Chlorella minutissima MCC-27	Batch	2	Batch Photo (linear increase from 75 μmol/m s to 260 μmol/m s)	8.24 (Day 5)	26.75 (Day 5)	5.35 (average)	[71]
Chlorella vulgaris	Fed-batch	5	Hetero (glucose and urea)	5.32 (Day 5)	252.75 (Day 5)	67.4 (average)	[72]

* Photoautotrophic growth (photo), mixotrophic growth (mixo), and heterotrophic growth (hetero).

C. sorokiniana MB-1-M12 is another promising strain for lutein production, developed through random mutagenesis with MNNG [31]. This lutein-rich mutant has lutein content ranging from 4.98 to 8.71 mg/g DW under phototrophic and mixotrophic conditions, and from 2.31 to 4.9 mg/g DW under heterotrophic conditions. While optimal parameters for pH, light intensity, and aeration are similar to previously described *C. sorokiniana* FZU60, the optimal temperature range for maximizing lutein production in *C. sorokiniana* MB-1-M12 is slightly lower, at 27–28 °C. Chen et al. (2019) showed that the medium replacement ratios used in semi-continuous mode also impact the lutein productivity of *C. sorokiniana* MB-1-M12 under mixotrophic conditions (150 $\mu mol/m^2/s$ with NaAc). They evaluated three medium replacement ratios (25%, 50%, 75%) over six repeated cycles, replacing the medium when carbon was depleted [67]. Semi-continuous cultivation with 75% medium replacement (6.61 mg/L/day) resulted in higher lutein productivity than batch (3.43 mg/L/day) and other replacement ratios (50%: 3.79 mg/L/day; 25%: 2.76 mg/L/day).

Chen et al. (2021) evaluated four different two-stage strategies for *C. sorokiniana* MB-1-M12 in BG-11 medium over 10 days: TSAH—initiated as autotrophic and switched to heterotrophic on day 3, TSHA—initiated as heterotrophic and switched to autotrophic on day 4, TSMH—initiated as mixotrophic and switched to heterotrophic on day 5, and TSHM—initiated as heterotrophic and switched to mixotrophic on day 4 [68]. The strategies that started with heterotrophic conditions obtained higher maximum lutein concentration and lutein content compared to the strategies initiated with photoautotrophic and mixotrophic conditions and ending with heterotrophic conditions (Table 3). In addition, Chen et al. (2022) further improved the TSHM strategy by integrating fed-batch and semi-continuous operational strategies for the *C. sorokiniana* MB-1-M12 [69]. During the heterotrophic phase, glucose and urea were fed into the reactor to maintain glucose concentrations between 2.0 and 7.5 g/L and ensure sufficient nitrogen. When the highest biomass was reached in the reactor, the fed-batch heterotrophic mode was switched to mixotrophic by transferring 75% of the medium to the mixotrophic mode, while the remaining 25% continued in a new fed-batch heterotrophic phase. This strategy was applied for three cycles in a 1 L reactor, achieving lutein production (average of 79.55 mg/L) and lutein content (average of 6.44 mg/g DW). In addition, this strategy was successfully scaled to a 5 L reactor, achieving lutein production (average of 167.96 mg/L) and lutein content (average of 8.33 mg/g DW).

Finally, Shi et al. (2002) investigated a strategy to enhance lutein productivity in *Chlorella protothecoides* CS-41 (*Auxenochlorella protothecoides*) under heterotrophic conditions in a 30 L fermenter [70]. The strategy involved initiating a fed-batch culture with glucose (40 g/L) and urea (7 g/L) for 10 days at 28 °C. Subsequently, the culture was exposed to nutrient-limited conditions at 32 °C for 84 h (equivalent to 3.5 days). After 10 days, the lutein concentration in the reactor reached 200 mg/L, with the strain exhibiting a lutein content of 3.8 mg/g DW. Following the period of nutrient limitation of 3.5 days, the lutein content increased to 5.35 mg/g DW. However, the lutein concentration only slightly rose to a maximum of 209.08 mg/L due to a decline in microalgal biomass in the reactor.

5.2. Cultivation Strategies for Other Microalgal Species

Several species of *Scenedesmus* exhibit the capacity to produce lutein. These green algae thrive in freshwater habitats and can resist high light intensities (1625–1900 $\mu mol/m^2/s$). The optimal lutein production occurs at temperatures of 30–40 °C and the preferred pH range for cultivation is between 6 and 8, with Mann and Myers medium commonly employed for lutein production. Aeration levels of 0.2–0.5 vvm, coupled with CO_2 supplementation ranging from 2.5 to 10%, enhance growth. Under photoautotrophic conditions, lutein content is 4.8–5.5 mg/g DW, around 2.55 mg/g DW under mixotrophic conditions, and around 1.49 mg/g DW under heterotrophic conditions, which is generally lower compared to *Chlorella* species (Tables 3 and 4). Ho et al. (2014) selected *S. obliquus* FSP-3 from the six *S. obliquus* strains in photoautotrophic conditions for its higher lutein productivity capacities [61]. Different light-related strategies were evaluated for this strain in batch, including various types of fluorescent lamps (e.g., TL5, T8, and helix lamps) and light intensities (from 30 to 600 µmol/m^2/s). The full white light spectrum (410–610 nm) is more favorable for lutein production than monochromatic green (480–580 nm), blue (435–515 nm), and red (600–690 nm) LED light sources. The optimal lutein productivity (4.08 mg/L/day) was obtained when using a white TL5 fluorescent lamp at a light intensity of 300 µmol/m^2/s. Florez-Miranda et al. (2017) tested a two-stage strategy to increase the lutein productivity of *S. incrassatulus* CLHE-Si01 [73]. The heterotrophic stage was performed in batch mode using glucose as a carbon source. Once glucose was consumed, the cultures were transferred to a photoautotrophic stage (230 µm/m^2/s). After 24 h of photoinduction, the lutein productivity reached 3.10 mg/L/day. Chen et al. (2019) found that a fed-batch strategy under mixotrophic conditions (150 µmol/m^2/s, 12 h/12 h) continuously feeding Detmer's medium with 20 g/L glucose led to a lutein productivity of 4.96 mg/L/day for *S. obliquus* CWL-1 [74].

The freshwater microalga *Desmodesmus* also has the capacity to accumulate lutein in its cells. Xie et al. (2013) identified *Desmodesmus* sp. F51 as the best strain for lutein production [75]. Various growing media were tested, and Modified Bristol's medium was selected as optimal for lutein productivity (3.05 mg/L/day) under phototrophic conditions (150 µmol/m^2/s). To further enhance the lutein production, a fed-batch strategy was employed using different nitrate concentrations (1.1, 2.2, 4.4, 8.8, and 17.6 mM). The highest lutein productivity (3.56 mg/L/day) and content (5.05 mg/g DW) were achieved with 2.2 mM nitrate pulse-feeding, with minimal differences observed between other concentrations. Interestingly, Ahmed et al. (2019) discovered that the synergistic effect of the plant hormones salicylic acid and succinic acid can enhance nitrate assimilation and increase lutein production in *Desmodesmus* sp. [76]. Supplementing with 100 µM salicylic acid and 2.5 mM succinic acid under phototrophic conditions in batch culture achieved a maximal lutein content of 7.01 mg/g DW and lutein productivity of 5.11 mg/L/day. Additionally, a fed-batch strategy involving nitrate, succinic acid, and salicylic acid further enhanced the lutein content and lutein productivity, reaching 7.50 mg/g DW and 5.78 mg/L/day, respectively.

Chlamydomonas is a green microalga with high light resistance and the capacity to produce lutein in fresh and salt water. Similar to *Scenedesmus* and *Desmodesmus*, fewer cultivation strategies have been explored to increase lutein production in this genus. Different light and temperature strategies were evaluated for *Chlamydomonas* sp. [77]. However, the reported lutein productivities and lutein content (2.52–4.24 mg/g DW) are generally lower compared to those of *Chlorella* (Table 3). However, as discussed earlier, its well-understood genetics make it an ideal candidate for genetic manipulation, allowing us to potentially enhance its lutein production.

Based on the literature reviewed, *Chlorella* sp. appears to be superior to *Scenedesmus* sp., *Chlamydomonas* sp., and *Desmodesmus* sp. for lutein production. The most effective cultivation strategy, yielding the highest lutein productivity (82.50 mg/L/day), involves cultivating *C. sorokiniana* FZU60 under heterotrophic conditions in fed-batch mode with 3-fold concentrated Mann and Myer's medium supplemented with glucose and urea, while controlling dissolved oxygen (DO) in the fermenter [59]. Over 6 days, this approach achieves a lutein concentration of 415.93 mg/L due to high biomass concentration in the fermenter. Controlling operational parameters such as pH, ensuring effective gas supply (ambient air and CO_2), and maintaining dissolved oxygen (DO) levels in the fermenter are crucial for achieving higher biomass production and consequently enhancing lutein productivity. However, under these heterotrophic conditions, the lutein content is only 2.57 mg/g DW. To enhance the lutein content in *Chlorella*, two-stage strategies can be employed. For instance, using fed-batch heterotrophic conditions followed by mixotrophic conditions for *C. sorokiniana* MB-1-M12, as proposed by Chen et al. (2022) [69], results in a lutein content of 8.19 mg/g DW after 14 days of cultivation. Conversely, the fed-batch mixotrophy followed by mixotrophic conditions for *C. sorokiniana* FZU60, as suggested by Ma et al. (2020) [66], achieves a lutein content of 11.22 mg/g DW after 8 days of cultivation. These two-stage strategies were demonstrated to run in semi-continuous mode, optimally with a medium replacement ratio of 75% or higher.

Table 4. Cultivation strategies to increase the lutein production in other microalgae.

Microalgae	Cultivation Mode	Reactor Volume (L)	Strategy *	Lutein Content (mg/g DW)	Lutein Production (mg/L)	Lutein Productivity (mg/L/day)	Reference
Scenedesmus almeriensis	Batch	2	Photo (1625 μE/m^2/s)	5.5	/	4.77	[78]
Scenedesmus almeriensis	continuous mode (dilution rate 0.3 L/day)	2	Photo (1625 μE/m^2/s)	5.4	/	3.8	[79]
Scenedesmus obliquus FSP-3	Batch	1	Photo (white TL5 fluorescent 300 μmol/m^2/s)	4.80 (Day 5)	20.5 (Day 5)	4.08 (average)	[61]
Scenedesmus incrassatulus CLHE-Si01	Two-stage	6	1st Batch Hetero (glucose) After glucose was consumed 2nd Batch Photo (150 μmol/m^2/s)	1.49 (Day 7)	/	3.10 (average)	[73]
Scenedesmus obliquus CWL-1	Fed-batch	7	Mixo (glucose and 150 μmol/m^2/s 12 h/12 h)	2.55 (Day 9)	27.3 (Day 9)	4.96 (Day 5)	[74]
Chlamydomonas sp. JSC4	Batch	1	Photo (625 μmol/m^2/s)	3.82	/	5.08	[62]
Desmodesmus sp. F51	Fed-batch	1	Photo (nitrate and 150 μmol/m^2/s)	5.05 (Day 6)	16.5 (Day 6)	3.56 (Day 6)	[75]
Desmodesmus sp.	Fed-batch	1	Photo (nitrate, succinic acid and salicylic acid)	7.5 (Day 4)	18.9 (Day 6)	5.78	[76]
Coccomyxa onubensis	Batch		Photo (100 mM NaCl)	6.7 (Day 3)		1.63	[80]

* Photoautotrophic growth (photo), mixotrophic growth (mixo), and heterotrophic growth (hetero).

6. Future Directions and Challenges

6.1. Conclusion and Future Directions in Metabolic Engineering for Microalgal Lutein Production

Random mutagenesis and metabolic engineering of microalgae for enhanced lutein production presents a promising avenue for sustainable and efficient lutein biosynthesis, as both techniques were able to increase the lutein content in microalgae (Table 2). On the one hand, random mutagenesis involves the alteration of genetic material in a non-specific manner, leading to the discovery of beneficial mutations that can increase lutein production. On the other hand, metabolic engineering allows for the precise modification of specific genes known to be involved in the lutein biosynthetic pathway, offering a targeted approach to enhancing production.

Comparing studies to determine the most efficient different random mutagenesis and metabolic engineering technique is challenging because lutein content is often reported in different units, such as mg/g DW, pg/cell, or not reported at all. Overall, random mutagenesis resulted in a higher lutein content compared to metabolic engineering (Table 2), largely because it has been applied to microalgal species that already exhibit a high natural lutein production. On the other hand, metabolic engineering has predominantly focused on *C. reinhardtii*, primarily because its genome is one of the first to be fully sequenced, and it has well-established transformation protocols.

In this review, the highest found lutein content (13.81 mg/g DW) was from *C. zofingiensis* mutant, obtained via random mutagenesis, exposed to stress conditions (nitrogen deficiency and high light irradiation of 460 μmol/m^2/s) [33]. However, this was achieved at a very small scale (50–100 mL), and further evaluation of lutein productivity is necessary to compare it with reported values from cultivation strategies (Tables 3 and 4). Exploring various cultivation strategies could be beneficial and intriguing avenues for further research, as demonstrated by Chen et al. in 2017 with their *C. sorokiniana* MB-1-M12 mutant generated through random mutagenesis [29].

Interestingly, some genes modified through random mutagenesis, such as the cGMP-dependent protein kinase, have shown significant effects on lutein biosynthesis [30]. The identification of such genes is valuable because it highlights specific genetic targets that can be further studied and precisely modified through metabolic engineering. With the increasing availability of genomic knowledge and techniques, metabolic engineering research can shift to microalgal species, such as *Chlorella* sp. and *Scenedesmus* sp., that naturally produce higher levels of lutein.

In the context of utilizing microalgae biomass with enhanced lutein content, consideration must be given to two key regulatory aspects: novel food regulation and genetically modified organism (GMO) legislation. Some dried microalgal biomass with high lutein content can be consumed directly or after cell disruption to enhance bioavailability, serving as food or a food supplement. In the EU, this is allowed for several species such as *A. protothecoides*, *C. sorokiniana*, *C. vulgaris*, and *Parachlorella kessleri*, among others like *Scenedesmus vacuolatus*. Other species, such as *C. reinhardtii*, are in the pipeline of novel food regulation, while *Desmodesmus* species and other *Scenedesmus* species are currently not allowed as food.

Using random mutagenesis on permitted microalgal species with a history of safe consumption before 1997, such as *Chlorella*, poses no issues for regulatory approval. However, microalgae approved under the novel food regulation would require a new application under this regulation if random mutagenesis is used. Furthermore, the use of GMO could be controversial and face stricter regulations requiring careful evaluation and approval. Due to the relatively strict regulatory framework for GMOs in the EU, most commercial applications of gene editing technologies, including transgenic microalgae, have occurred outside the EU.

6.2. Comparing the Optimal Cultivation Strategies for Lutein Production

As seen from Tables 3 and 4, employing fed-batch heterotrophic cultivation in a controlled fermenter results in the highest lutein productivity. Two-stage strategies, transitioning from heterotrophic to mixotrophic or photoautotrophic cultivation conditions, often result in slower lutein production and lower biomass density in the second stage. This contributes to lower final lutein concentrations in the PBR compared to those achieved with fed-batch heterotrophic conditions. Nevertheless, the lutein content is higher in these two-way strategies, which might be important for further downstream processing. Microalgal biomass with higher lutein content may facilitate greater extraction efficiencies during downstream processing, owing to a higher concentration gradient that enhances the diffusion of lutein from the cell interior to the extraction solvent. Consequently, achieving high lutein productivity from microalgae initially cultivated under heterotrophic conditions, which typically have lower lutein content, may require increased resources such as chemicals, energy, and water. However, this hypothesis warrants further investigation, as no comparative studies were found in the literature.

Energy consumption for artificial lighting under mixotrophic and phototrophic conditions can significantly increase the production cost of microalgae, which is not the case for heterotrophic cultivation. Perez-López et al. (2014) found that replacing artificial lighting with sunlight reduces both environmental impact and cost, though it can also lower biomass productivity [81]. Interestingly, Dineshkumar et al. (2016) tested various light strategies (constant light intensity, and linearly and exponentially increasing light intensity) for *Chlorella minutissima* [71]. They found that a linear light strategy not only increased lutein productivity and lutein content (5.35 mg/L/day; 8.24 mg/g DW) compared to constant illumination at 260 μmol/m^2/s (4.32 mg/L/day; 6.37 mg/g DW), but also reduced light energy consumption by 32%. Apart from the energy costs associated with lighting, mixotrophic and phototrophic conditions can reduce the ecological footprint by lowering CO_2 emissions, thanks to their carbon sequestration capabilities, which is not the case for heterotrophic cultivation.

Scalability of the used strategy is also important for commercialization and reducing production costs of microalgae. Additionally, a single-stage process might be easier to operate compared to a two-way strategy. Jeon et al. (2014) demonstrated that fed-batch heterotrophic cultivation for lutein production is scalable to a commercial level [72]. They observed that lutein concentration remained consistent for *C. vulgaris* when scaling up from a 5 L batch (252.75 mg/L) to fed-batches in a 25 m^3 system (260.55 mg/L) and further to fed-batches in a 240 m^3 system (263.13 mg/L). Scaling the two-way strategy to commercial levels has not been attempted. As seen in Tables 3 and 4, most research on microalgal lutein production is predominantly confined to laboratory conditions with volumes often limited to 1 L bottles, PBR, or small-scale fermenters. Scaling the two-way strategy could present several challenges. For example, if the culture becomes dense under heterotrophic conditions in the first stage, it will need to be diluted for efficient light penetration when transferred to the second stage under mixotrophic or photoautotrophic conditions. Diluting the culture in closed systems would require larger PBRs, which would significantly increase the capital expenditure (CAPEX). Suparmaniam et al. (2024) hypothesize that two-way strategies can be upscaled by first cultivating the microalgae heterotrophically in fed-batch mode until the early exponential phase [23]. In the second stage, the microalgal cultures would be transferred to an open raceway pond (ORP) system operated under mixotrophic conditions. However, using ORPs under mixotrophic conditions might be challenging due to the increased risk of contamination because of the presence of a carbon source, resulting in culture crashes. Operating in photoautotrophic conditions for the second stage, ensuring the carbon source is fully utilized before changing conditions, might be a possible solution to minimize contamination risks in ORPs.

Under heterotrophic and mixotrophic conditions, both microalgae and bacteria also compete for organic carbon, with bacteria often outnumbering microalgae due to their shorter doubling times. As a result, keeping anexic conditions through the scaling process will be important to avoid culture crashes or contamination issues, which could raise safety concerns for human consumption. Selecting lutein-producing species that can resist extreme environments, such as low-pH and high-salt conditions, might offer a solution for contamination during cultivation. These resilient species could enhance the stability and efficiency of large-scale lutein production systems. For example, Bermejo et al. (2018) demonstrated that the acidophilic eukaryotic microalga *Coccomyxa onubensis*, which can endure moderate salt stress and low pH (pH 2.5), can accumulate lutein under phototrophic conditions (140 μmol/m^2/s) with the addition of 100 mM NaCl, achieving up to 6.7 mg/g DW [80]. However, this species grows much slower compared to *Chlorella* species (Table 3), resulting in lower lutein productivity (1.63 mg/L/day).

6.3. Comparing Microalgae and Marigold for Lutein Production

6.3.1. Advantages of Microalgal Production for Lutein Compared to Marigold

Natural lutein is currently produced commercially from marigold flowers, specifically *Tagetes patula* or *Tagetes erecta*. However, several advantages have led to growing interest in using microalgae for lutein production (Figure 2). The lutein content in the dry petal powders of *T. erecta* and *T. patula* ranges from 0.829 to 27.946 g/kg and 0.597 to 12.31 g/kg, respectively, depending on environmental conditions, growth stages, and genetic variation [22]. As a result, the lutein content in marigold petals can be higher than in dried microalgae (Tables 3 and 4). However, microalgae have a growth rate that is 5–10 times faster than that of higher plants, which can significantly increase their lutein productivity [22]. Moreover, microalgae can be cultivated year-round, unlike seasonal marigold flowers. Li et al. (2015) compared the lutein yield, assuming an available lutein content of 20 g/kg in dry petal powder and 5 g/kg in dry microalgal powder [22]. Because of the higher growth rates and year-round production of microalgal biomass, the annual lutein production rate can reach 350–750 kg/hectare, whereas for marigolds, it is approximately 120 kg/hectare. The marigold production is also labor-intensive and predominantly located in economically upcoming countries such as China, India, and some African nations, which are locations that are prone to climate change (extreme temperature, drought, and heavy rainfall), which could affect the marigold production. Li et al. (2015) also stated that a lutein content of at least 10 g/kg (1% DW) in microalgae is deemed essential for commercial viability [22]. As discussed in previous sections, specific *Chlorella* strains can achieve lutein content ranging from 4 to 11 mg/g DW, depending on the cultivation conditions and strategies used, indicating that this criterion can be met with optimized cultivation methods. Furthermore, advancements in random mutagenesis and metabolic engineering offer potential for surpassing these values significantly, as shown for the *C. zofingiensis* mutant (13.81 mg/g DW) described by Huang et al. (2018) [33].

Figure 2. The advantages of using microalgae as a source for lutein production.

Additionally, microalgal cultivation requires no arable land and 2–10 times less water compared to marigold flowers. While microalgae require more nitrogen and phosphorus—1.5 and 2 times more, respectively—they need 3.5 times less potassium than marigold flowers [22]. Using food waste hydrolysate and other side streams can serve as sustainable and cost-effective nutrient sources, further enhancing the sustainability of microalgal lutein production. For instance, Wang et al. (2020) tested *Chlorella* sp. GY-H4 mixotrophic cultivation using food waste hydrolysate supplemented with 20 g/L glucose in semi-continuous mode, achieving a lutein productivity of 10.5 mg/L/day and a lutein content of 8.9 mg/g DW [82].

Additionally, ricotta cheese whey and cane molasses have shown potential as culture media for growing *Chlorella* species [83,84]. Tran et al. (2014) explored cost reduction by recycling the cultivation medium; however, this approach was shown to decrease both biomass and lutein content [85]. It is essential to note that legislation will require side streams used for microalgal lutein production to meet food-grade standards to ensure safety and compliance with health regulations.

6.3.2. Disadvantages of Microalgal Production for Lutein Compared to Marigold

The primary drawback of lutein production from microalgae is its energy-intensive process, particularly in the downstream processing (harvesting, cell disruption, extraction, and purification) of the biomass [86]. To remove the larger volume of water in microalgal culture, extracting microalgal suspension requires more energy than for marigold flowers [22]. Conventional techniques like centrifugation consume significant energy, whereas sedimentation is time-consuming and carries a risk of lutein degradation. Drying requires similar amounts of energy for both marigold flowers and microalgae. Cell disruption is often necessary to increase carotenoid yields but is more energy-intensive than crushing/powdering the marigold flowers due to microalgae's smaller cell size (3–10 μm), which reduces disruption efficiency [6]. Due to this, the energy needed for microalgal cell disruption ranges from 33 to 530 MJ/kg, about 1000 times higher than the energy required for crushing marigold flowers (800 kJ/kg) [22]. The composition, thickness, and size of microalgae cell walls dictate the energy demand for disruption. Lutein-rich *Chlorella* species, known for their strong cell walls, typically require cell disruption. In contrast, species such as *C. reinhardtii*, with more fragile cell walls, do not require cell disruption [87].

Despite microalgae requiring less water during cultivation compared to marigold flowers, the extraction phase for microalgae demands more solvents, water, and energy, attributed to stronger lutein bonds within the microalgal biomass and its small size [22]. Microwave-assisted extraction and supercritical fluid extraction are explored as potential alternatives for traditional solvent extraction methods due to their thermal stability and high efficiency, despite higher operational costs and energy demands [15]. Furthermore, an intense purification step is needed to eliminate water, chlorophyll, and other compounds bound to the free lutein in the microalgal cells. Advanced methods like chromatography offer high purification efficiency but are costly and challenging to scale up [15]. In contrast, marigold flower extracts contain primarily lutein and zeaxanthin esters, simplifying the extraction and purification processes. Typically, solvent extraction with n-hexane is employed to extract oleo-resin from milled dry flower petals, occasionally aided by KOH to release free lutein. For microalgae, solvent extraction, purification, and residual water removal for lutein purification are more laborious, with excess solvent requirements [24]. Balancing effectiveness with affordability remains a challenge in optimizing microalgal lutein extraction and purification processes.

Lutein extracted from marigolds can be marketed as an additive (E161b) in the European Union. Furthermore, it is granted GRAS status by the United States Food and Drug Administration for use as a health supplement promoting eye health [22]. In contrast, commercialization of purified lutein extracted from microalgal species would require regulatory approval. Alternatively, microalgae species rich in lutein can be used as a whole, bypassing the need for extraction and purification steps. However, consumer acceptance of this approach may be hindered by the distinct flavor and color characteristics of microalgae [88]. While marigolds are still the main source, research is ongoing to improve the efficiency and cost-effectiveness of microalgae for lutein production. In the future, we might see a shift towards microalgae as the preferred method.

7. Conclusions

The review highlights the potential and challenges of enhancing lutein production in microalgae through random mutagenesis and metabolic engineering. Both techniques have shown promise, with random mutagenesis achieving higher lutein content due to its application to naturally high-producing species, while metabolic engineering offers precision in modifying specific genes. The highest lutein content was observed in a *C. zofingiensis* mutant under stress conditions, yet scalability and productivity comparisons require further research. Regulatory considerations must be taken into account when using certain microalgae species, especially when employing genetic modification techniques.

Optimal cultivation strategies suggest fed-batch heterotrophic cultivation in controlled fermenters for the highest lutein productivity, though two-stage strategies, involving a transition from heterotrophic to mixotrophic or photoautotrophic conditions, show potential for higher lutein content, particularly in *Chlorella* species. The scalability of these cultivation strategies remains a challenge, with heterotrophic conditions being easier to scale than mixotrophic or photoautotrophic conditions. The review compares microalgal and marigold lutein production, highlighting microalgae's faster growth rates and year-round cultivation advantages. Despite the current challenge of energy-intensive downstream processing in microalgae, advancements in extraction and purification technologies hold promise for overcoming these hurdles.

In conclusion, microalgae offer promising avenues for sustainable lutein production, particularly with advancements in genetic techniques for obtaining high-producing lutein species. Testing these species with the proposed cultivation strategies and scaling up the cultivation process are crucial for commercial viability. With ongoing research focused on optimizing cultivation and processing methods, microalgae have the potential to surpass marigolds as the preferred source of lutein in the future.

Author Contributions: B.C., E.V., K.V.L., L.N. and J.R.: Conceptualization, methodology, software, formal analysis, investigation, resources, data curation, writing—original draft preparation, writing—review and editing, supervision and funding acquisition. All authors have read and agreed to the published version of the manuscript.

Funding: This publication is based upon work from COST action CA18238 (Ocean4Biotech), supported by COST program 2019–2023 (European Cooperation in Science and Technology) and upon the European Union (Horizon Europe) project GeneBEcon, Grant Agreement N° 101061015. EV is supported by Research Foundation Flanders (FWO-SB research project N° 1SHGX24N).

Conflicts of Interest: L.N. is employed by ScotBio. The remaining authors declare that the research was conducted in the absence of any commercial or financial relationships that could be construed as a potential conflict of interest.

References

1. Yabuzaki, J. Carotenoids Database: Structures, Chemical Fingerprints and Distribution among Organisms. *Database* **2017**, *2017*, bax004. [CrossRef] [PubMed]
2. Sun, T.; Li, L. Toward the 'Golden' Era: The Status in Uncovering the Regulatory Control of Carotenoid Accumulation in Plants. *Plant Sci.* **2020**, *290*, 110331. [CrossRef] [PubMed]
3. Donoso, A.; González-Durán, J.; Muñoz, A.A.; González, P.A.; Agurto-Muñoz, C. Therapeutic Uses of Natural Astaxanthin: An Evidence-Based Review Focused on Human Clinical Trials. *Pharmacol. Res.* **2021**, *166*, 105479. [CrossRef] [PubMed]
4. Duan, X.; Xie, C.; Hill, D.R.A.; Barrow, C.J.; Dunshea, F.R.; Martin, G.J.O.; Suleria, H.A.R. Bioaccessibility, Bioavailability and Bioactivities of Carotenoids in Microalgae: A Review. *Food Rev. Int.* **2024**, *40*, 230–259. [CrossRef]
5. Cao, K.; Cui, Y.; Sun, F.; Zhang, H.; Fan, J.; Ge, B.; Cao, Y.; Wang, X.; Zhu, X.; Wei, Z.; et al. Metabolic Engineering and Synthetic Biology Strategies for Producing High-Value Natural Pigments in Microalgae. *Biotechnol. Adv.* **2023**, *68*, 108236. [CrossRef] [PubMed]
6. Gong, M.; Bassi, A. Carotenoids from Microalgae: A Review of Recent Developments. *Biotechnol. Adv.* **2016**, *34*, 1396–1412. [CrossRef] [PubMed]
7. Ren, Y.; Sun, H.; Deng, J.; Huang, J.; Chen, F. Carotenoid Production from Microalgae: Biosynthesis, Salinity Responses and Novel Biotechnologies. *Mar. Drugs* **2021**, *19*, 713. [CrossRef]
8. Liu, C.; Hu, B.; Cheng, Y.; Guo, Y.; Yao, W.; Qian, H. Carotenoids from Fungi and Microalgae: A Review on Their Recent Production, Extraction, and Developments. *Bioresour. Technol.* **2021**, *337*, 125398. [CrossRef]
9. Silva, S.C.; Ferreira, I.C.F.R.; Dias, M.M.; Barreiro, M.F. Microalgae-Derived Pigments: A 10-Year Bibliometric Review and Industry and Market Trend Analysis. *Molecules* **2020**, *25*, 3406. [CrossRef]
10. Sun, H.; Wang, J.; Li, Y.; Yang, S.; Di, D.; Tu, Y.; Liu, J. Synthetic Biology in Microalgae towards Fucoxanthin Production for Pharmacy and Nutraceuticals. *Biochem. Pharmacol.* **2024**, *220*, 115958. [CrossRef]

11. Saini, R.K.; Keum, Y.-S. Microbial Platforms to Produce Commercially Vital Carotenoids at Industrial Scale: An Updated Review of Critical Issues. *J. Ind. Microbiol. Biotechnol.* **2019**, *46*, 657–674. [CrossRef]
12. Ram, S.; Mitra, M.; Shah, F.; Tirkey, S.R.; Mishra, S. Bacteria as an Alternate Biofactory for Carotenoid Production: A Review of Its Applications, Opportunities and Challenges. *J. Funct. Foods* **2020**, *67*, 103867. [CrossRef]
13. Elissen, H.J.; van der Wiede, R.Y. *Productie van Zoutwateralgen Voor Toepassingen in Food (En Feed) Deelrapport II: Werkpakketten 1&2 van Project Foodgrade Productie van Zoutwateralgen: Deelrapport II: Batchexperimenten Met Zoute Reststromen En Verschillende Mariene Algensoorten*; ACRRES: Lelystad, The Netherlands, 2018.
14. Carvalho, J.C.M.; Matsudo, M.C.; Bezerra, R.P.; Ferreira-Camargo, L.S.; Sato, S. Microalgae Bioreactors. In *Algal Biorefineries: Volume 1: Cultivation of Cells and Products*; Bajpai, R., Prokop, A., Zappi, M., Eds.; Springer: Dordrecht, The Netherlands, 2014; pp. 83–126. ISBN 978-94-007-7494-0.
15. Suparmaniam, U.; Lam, M.K.; Uemura, Y.; Lim, J.W.; Lee, K.T.; Shuit, S.H. Insights into the Microalgae Cultivation Technology and Harvesting Process for Biofuel Production: A Review. *Renew. Sustain. Energy Rev.* **2019**, *115*, 109361. [CrossRef]
16. Hu, J.; Nagarajan, D.; Zhang, Q.; Chang, J.; Lee, D. Heterotrophic Cultivation of Microalgae for Pigment Production: A Review. *Biotechnol. Adv.* **2018**, *36*, 54–67. [CrossRef]
17. Ren, Y.; Deng, J.; Huang, J.; Wu, Z.; Yi, L.; Bi, Y.; Chen, F. Using Green Alga Haematococcus Pluvialis for Astaxanthin and Lipid Co-Production: Advances and Outlook. *Bioresour. Technol.* **2021**, *340*, 125736. [CrossRef]
18. Shi, T.Q.; Wang, L.R.; Zhang, Z.X.; Sun, X.M.; Huang, H. Stresses as First-Line Tools for Enhancing Lipid and Carotenoid Production in Microalgae. *Front. Bioeng. Biotechnol.* **2020**, *8*, 610. [CrossRef] [PubMed]
19. Jannel, S.; Caro, Y.; Bermudes, M.; Petit, T. Novel Insights into the Biotechnological Production of Haematococcus Pluvialis-Derived Astaxanthin: Advances and Key Challenges to Allow Its Industrial Use as Novel Food Ingredient. *J. Mar. Sci. Eng.* **2020**, *8*, 789. [CrossRef]
20. Li, J.; Zhu, D.; Niu, J.; Shen, S.; Wang, G. An Economic Assessment of Astaxanthin Production by Large Scale Cultivation of Haematococcus Pluvialis. *Biotechnol. Adv.* **2011**, *29*, 568–574. [CrossRef]
21. Monte, J.; Ribeiro, C.; Parreira, C.; Costa, L.; Brive, L.; Casal, S.; Brazinha, C.; Crespo, J.G. Biorefinery of Dunaliella Salina: Sustainable Recovery of Carotenoids, Polar Lipids and Glycerol. *Bioresour. Technol.* **2020**, *297*, 122509. [CrossRef] [PubMed]
22. Lin, J.-H.; Lee, D.-J.; Chang, J.-S. Lutein Production from Biomass: Marigold Flowers versus Microalgae. *Bioresour. Technol.* **2015**, *184*, 421–428. [CrossRef]
23. Suparmaniam, U.; Kee, M.; Wei, J.; Shi, I.; Lai, B.; Chin, F.; Hoong, S.; Lim, S.; Ling, Y.; Loo, P. Abiotic Stress as a Dynamic Strategy for Enhancing High Value Phytochemicals in Microalgae: Critical Insights, Challenges and Future Prospects. *Biotechnol. Adv.* **2024**, *70*, 108280. [CrossRef] [PubMed]
24. Zohra, F.; Medjekal, S. Microalgal Carotenoids: A Promising Alternative to Synthetic Dyes. *Algal Res.* **2022**, *66*, 102823. [CrossRef]
25. Bleisch, R.; Freitag, L.; Ihadjadene, Y.; Sprenger, U.; Steingröwer, J.; Walther, T.; Krujatz, F. Strain Development in Microalgal Biotechnology—Random Mutagenesis Techniques. *Life* **2022**, *12*, 961. [CrossRef] [PubMed]
26. Trovão, M.; Schüler, L.M.; Machado, A.; Bombo, G.; Navalho, S.; Barros, A.; Pereira, H.; Silva, J.; Freitas, F.; Varela, J. Random Mutagenesis as a Promising Tool for Microalgal Strain Improvement towards Industrial Production. *Mar. Drugs* **2022**, *20*, 440. [CrossRef] [PubMed]
27. Dasan, Y.K.; Lam, M.K.; Chai, Y.H.; Lim, J.W.; Ho, Y.C.; Tan, I.S.; Lau, S.Y.; Show, P.L.; Lee, K.T. Unlocking the Potential of Microalgae Bio-Factories for Carbon Dioxide Mitigation: A Comprehensive Exploration of Recent Advances, Key Challenges, and Energy-Economic Insights. *Bioresour. Technol.* **2023**, *380*, 129094. [CrossRef] [PubMed]
28. Fathy, W.A.; Techen, N.; Elsayed, K.N.M.; Essawy, E.A.; Tawfik, E.; Abdelhameed, M.S.; Hammouda, O.; Ross, S.A. Insights into Random Mutagenesis Techniques to Enhance Biomolecule Production in Microalgae: Implications for Economically Viable Bioprocesses. *Int. Aquat. Res.* **2023**, *15*, 85–102.
29. Chen, J.-H.; Chen, C.-Y.; Chang, J.-S. Lutein Production with Wild-Type and Mutant Strains of Chlorella Sorokiniana MB-1 under Mixotrophic Growth. *J. Taiwan Inst. Chem. Eng.* **2017**, *79*, 66–73. [CrossRef]
30. Ren, Y.; Deng, J.; Lin, Y.; Huang, J.; Chen, F. Developing a Chromochloris Zofingiensis Mutant for Enhanced Production of Lutein under CO_2 Aeration. *Mar. Drugs* **2022**, *20*, 194. [CrossRef] [PubMed]
31. Breitenbach, J.; Zhu, C.; Sandmann, G. Bleaching Herbicide Norflurazon Inhibits Phytoene Desaturase by Competition with Cofactors. *J. Agric. Food Chem.* **2001**, *49*, 5270–5272. [CrossRef]
32. Cordero, B.F.; Obraztsova, I.; Couso, I.; Leon, R.; Vargas, M.A.; Rodriguez, H. Enhancement of Lutein Production in Chlorella Sorokiniana (Chorophyta) by Improvement of Culture Conditions and Random Mutagenesis. *Mar. Drugs* **2011**, *9*, 1607–1624. [CrossRef]
33. Huang, W.; Lin, Y.; He, M.; Gong, Y.; Huang, J. Induced High-Yield Production of Zeaxanthin, Lutein and B-carotene by a Mutant of *Chlorella zofingiensis*. *J. Agric. Food Chem.* **2018**, *31*, 891–897. [CrossRef] [PubMed]
34. Ran, F.A.; Hsu, P.D.; Wright, J.; Agarwala, V.; Scott, D.A.; Zhang, F. Genome Engineering Using the CRISPR-Cas9 System. *Nat. Protoc.* **2013**, *8*, 2281–2308. [CrossRef] [PubMed]
35. Carroll, D. Genome Engineering with Zinc-Finger Nucleases. *Genetics* **2011**, *188*, 773–782. [CrossRef] [PubMed]
36. Sun, N.; Zhao, H. Transcription Activator-like Effector Nucleases (TALENs): A Highly Efficient and Versatile Tool for Genome Editing. *Biotechnol. Bioeng.* **2013**, *110*, 1811–1821. [CrossRef] [PubMed]

37. Castro, N.G.; Bjelic, J.; Malhotra, G.; Huang, C.; Alsaffar, S.H. Comparison of the Feasibility, Efficiency, and Safety of Genome Editing Technologies. *Int. J. Mol. Sci.* **2021**, *22*, 10355. [CrossRef]
38. Boettcher, M.; McManus, M.T. Choosing the Right Tool for the Job: RNAi, TALEN, or CRISPR. *Mol. Cell* **2015**, *58*, 575–585. [CrossRef]
39. Ghavami, S.; Pani, A. CRISPR Interference and Its Applications. In *Progress in Molecular Biology and Translational Science*; Elsevier: Amsterdam, The Netherlands, 2021; pp. 123–140.
40. Velmurugan, A.; Muthukaliannan, G.K. Genetic Manipulation for Carotenoid Production in Microalgae an Overview. *Curr. Res. Biotechnol.* **2022**, *4*, 221–228. [CrossRef]
41. Sirohi, P.; Verma, H.; Singh, S.K.; Singh, V.K.; Pandey, J.; Khusharia, S.; Kumar, D.; Kaushalendra; Teotia, P.; Kumar, A. Microalgal Carotenoids: Therapeutic Application and Latest Approaches to Enhance the Production. *Curr. Issues Mol. Biol.* **2022**, *44*, 6257–6279. [CrossRef]
42. Liang, M.-H.; Li, X.-Y. Involvment of Transcription Factors and Regulatory Proteins in the Regulation of Carotenoid Accumulation in Plants and Algae. *J. Agric. Food Chem.* **2023**, *71*, 18660–18673. [CrossRef]
43. Tamaki, S.; Mochida, K.; Suzuki, K. Diverse Biosynthetic Pathways and Protective Functions against Environmental Stress of Antioxidants in Microalgae. *Plants* **2021**, *10*, 1250. [CrossRef]
44. Chayut, N.; Yuan, H.; Ohali, S.; Meir, A.; Sa'ar, U.; Tzuri, G.; Zheng, Y.; Mazourek, M.; Gepstein, S.; Zhou, X.; et al. Distinct Mechanisms of the ORANGE Protein in Controlling Carotenoid Flux. *Plant Physiol.* **2017**, *173*, 376–389. [CrossRef]
45. Morikawa, T.; Uraguchi, Y.; Sawayama, S. Overexpression of DnaJ-Like Chaperone Enhances Carotenoid Synthesis in Chlamydomonas Reinhardtii. *Appl. Biochem. Biotechnol.* **2018**, *184*, 80–91. [CrossRef]
46. Yazdani, M.; Croen, M.G.; Fish, T.L.; Thannhauser, T.W.; Ahner, B.A. Overexpression of Native ORANGE (OR) and OR Mutant Protein in Chlamydomonas Reinhardtii Enhances Carotenoid and ABA Accumulation and Increases Resistance to Abiotic Stress. *Metab. Eng.* **2021**, *68*, 94–105. [CrossRef]
47. Velmurugan, A.; Muthukaliannan, G.K. Homologous and Heterologous Expression of Phytoene Synthase Gene in Marine Microalgae Dunaliella Salina and Its Potential as Aquaculture Feed. *Aquac. Int.* **2023**, *31*, 3125–3144. [CrossRef]
48. Tokunaga, S.; Morimoto, D.; Koyama, T.; Sawayama, S. Enhanced Lutein Production in Chlamydomonas Reinhardtii by Overexpression of the Lycopene Epsilon Cyclase Gene. *Appl. Biochem. Biotechnol.* **2021**, *193*, 1967–1978. [CrossRef]
49. Lou, S.; Lin, X.; Liu, C.; Anwar, M.; Li, H.; Hu, Z. Molecular Cloning and Functional Characterization of CvLCYE, a Key Enzyme in Lutein Synthesis Pathway in Chlorella Vulgaris. *Algal Res.* **2021**, *55*, 102246. [CrossRef]
50. Kumari, S.; Vira, C.; Lali, A.M.; Prakash, G. Heterologous Expression of a Mutant Orange Gene from Brassica Oleracea Increases Carotenoids and Induces Phenotypic Changes in the Microalga Chlamydomonas Reinhardtii. *Algal Res.* **2020**, *47*, 101871.
51. Lin, J.Y.; Effendi, S.S.W.; Ng, I.S. Enhanced Carbon Capture and Utilization (CCU) Using Heterologous Carbonic Anhydrase in Chlamydomonas Reinhardtii for Lutein and Lipid Production. *Bioresour. Technol.* **2022**, *351*, 127009. [CrossRef]
52. Steinbrenner, J.; Sandmann, G. Transformation of the Green Alga Haematococcus Pluvialis with a Phytoene Desaturase for Accelerated Astaxanthin Biosynthesis. *Appl. Environ. Microbiol.* **2006**, *72*, 7477–7484. [CrossRef]
53. Liu, J.; Sun, Z.; Gerken, H.; Huang, J.; Jiang, Y.; Chen, F. Genetic Engineering of the Green Alga Chlorella Zofingiensis: A Modified Norflurazon-Resistant Phytoene Desaturase Gene as a Dominant Selectable Marker. *Appl. Microbiol. Biotechnol.* **2014**, *98*, 5069–5079. [CrossRef]
54. Song, I.; Kim, J.; Baek, K.; Choi, Y.; Shin, B.; Jin, E. The Generation of Metabolic Changes for the Production of High-Purity Zeaxanthin Mediated by CRISPR-Cas9 in Chlamydomonas Reinhardtii. *Microb. Cell Fact.* **2020**, *19*, 220. [CrossRef]
55. Kneip, J.S.; Kniepkamp, N.; Jang, J.; Mortaro, M.G.; Jin, E.; Kruse, O.; Baier, T. CRISPR/Cas9-Mediated Knockout of the Lycopene ε-Cyclase for Efficient Astaxanthin Production in the Green Microalga Chlamydomonas Reinhardtii. *Plants* **2024**, *13*, 1393. [CrossRef]
56. Fu, Y.; Wang, Y.; Yi, L.; Liu, J.; Yang, S.; Liu, B.; Chen, F.; Sun, H. Lutein Production from Microalgae: A Review. *Bioresour. Technol.* **2023**, *376*, 128875. [CrossRef]
57. Zheng, H.; Wang, Y.; Li, S.; Nagarajan, D.; Varjani, S.; Lee, D.; Chang, J. Recent Advances in Lutein Production from Microalgae. *Renew. Sustain. Energy Rev.* **2022**, *153*, 111795. [CrossRef]
58. Shi, X.-M.; Zhang, X.-W.; Chen, F. Heterotrophic Production of Biomass and Lutein by Chlorella Protothecoides on Various Nitrogen Sources. *Enzyme Microb. Technol.* **2000**, *27*, 312–318. [CrossRef]
59. Xie, Y.; Zhang, Z.; Ma, R.; Liu, X.; Miao, M.; Ho, S.-H.; Chen, J.; Leong, Y.K.; Chang, J.-S. High-Cell-Density Heterotrophic Cultivation of Microalga Chlorella Sorokiniana FZU60 for Achieving Ultra-High Lutein Production Efficiency. *Bioresour. Technol.* **2022**, *365*, 128130. [CrossRef]
60. Qu, L.; Ren, L.-J.; Huang, H. Scale-up of Docosahexaenoic Acid Production in Fed-Batch Fermentation by Schizochytrium Sp. Based on Volumetric Oxygen-Transfer Coefficient. *Biochem. Eng. J.* **2013**, *77*, 82–87. [CrossRef]
61. Ho, S.-H.; Chan, M.-C.; Liu, C.-C.; Chen, C.-Y.; Lee, W.-L.; Lee, D.-J.; Chang, J.-S. Enhancing Lutein Productivity of an Indigenous Microalga Scenedesmus Obliquus FSP-3 Using Light-Related Strategies. *Bioresour. Technol.* **2014**, *152*, 275–282. [CrossRef]
62. Ma, R.; Zhao, X.; Xie, Y.; Ho, S.-H.; Chen, J. Enhancing Lutein Productivity of Chlamydomonas Sp. via High-Intensity Light Exposure with Corresponding Carotenogenic Genes Expression Profiles. *Bioresour. Technol.* **2019**, *275*, 416–420. [CrossRef]
63. León-Vaz, A.; León, R.; Vigara, J.; Funk, C. Exploring Nordic Microalgae as a Potential Novel Source of Antioxidant and Bioactive Compounds. *New Biotechnol.* **2023**, *73*, 1–8. [CrossRef]

64. Xie, Y.; Li, J.; Ho, S.-H.; Ma, R.; Shi, X.; Liu, L.; Chen, J. Pilot-Scale Cultivation of Chlorella Sorokiniana FZU60 with a Mixotrophy/Photoautotrophy Two-Stage Strategy for Efficient Lutein Production. *Bioresour. Technol.* **2020**, *314*, 123767. [CrossRef]

65. Xie, Y.; Li, J.; Ma, R.; Ho, S.-H.; Shi, X.; Liu, L.; Chen, J. Bioprocess Operation Strategies with Mixotrophy/Photoinduction to Enhance Lutein Production of Microalga Chlorella Sorokiniana FZU60. *Bioresour. Technol.* **2019**, *290*, 121798. [CrossRef] [PubMed]

66. Ma, R.; Zhang, Z.; Ho, S.-H.; Ruan, C.; Li, J.; Xie, Y.; Shi, X.; Liu, L.; Chen, J. Two-Stage Bioprocess for Hyper-Production of Lutein from Microalga Chlorella Sorokiniana FZU60: Effects of Temperature, Light Intensity, and Operation Strategies. *Algal Res.* **2020**, *52*, 102119. [CrossRef]

67. Chen, J.-H.; Chen, C.-Y.; Hasunuma, T.; Kondo, A.; Chang, C.-H.; Ng, I.-S.; Chang, J.-S. Enhancing Lutein Production with Mixotrophic Cultivation of Chlorella Sorokiniana MB-1-M12 Using Different Bioprocess Operation Strategies. *Bioresour. Technol.* **2019**, *278*, 17–25. [CrossRef] [PubMed]

68. Chen, J.-H.; Kato, Y.; Matsuda, M.; Chen, C.-Y.; Nagarajan, D.; Hasunuma, T.; Kondo, A.; Chang, J.-S. Lutein Production with Chlorella Sorokiniana MB-1-M12 Using Novel Two-Stage Cultivation Strategies—Metabolic Analysis and Process Improvement. *Bioresour. Technol.* **2021**, *334*, 125200. [CrossRef] [PubMed]

69. Chen, J.-H.; Nagarajan, D.; Huang, Y.; Zhu, X.; Liao, Q.; Chang, J.-S. A Novel and Effective Two-Stage Cultivation Strategy for Enhanced Lutein Production with Chlorella Sorokiniana. *Biochem. Eng. J.* **2022**, *188*, 108688. [CrossRef]

70. Shi, X.-M.; Chen, F. High-Yield Production of Lutein by the Green Microalga Chlorella Protothecoidesin Heterotrophic Fed-Batch Culture. *Biotechnol. Prog.* **2002**, *18*, 723–727. [CrossRef]

71. Dineshkumar, R.; Subramanian, G.; Dash, S.K.; Sen, R. Development of an Optimal Light-Feeding Strategy Coupled with Semi-Continuous Reactor Operation for Simultaneous Improvement of Microalgal Photosynthetic Efficiency, Lutein Production and CO_2 Sequestration. *Biochem. Eng. J.* **2016**, *113*, 47–56. [CrossRef]

72. Jeon, J.Y.; Kwon, J.-S.; Kang, S.T.; Kim, B.-R.; Jung, Y.; Han, J.G.; Park, J.H.; Hwang, J.K. Optimization of Culture Media for Large-Scale Lutein Production by Heterotrophic Chlorella Vulgaris. *Biotechnol. Prog.* **2014**, *30*, 736–743. [CrossRef]

73. Flórez-Miranda, L.; Cañizares-Villanueva, R.O.; Melchy-Antonio, O.; Martínez-Jerónimo, F.; Ortíz, C.M.F. Two Stage Heterotrophy/Photoinduction Culture of Scenedesmus Incrassatulus: Potential for Lutein Production. *J. Biotechnol.* **2017**, *262*, 67–74. [CrossRef]

74. Chen, W.-C.; Hsu, Y.-C.; Chang, J.-S.; Ho, S.-H.; Wang, L.-F.; Wei, Y.-H. Enhancing Production of Lutein by a Mixotrophic Cultivation System Using Microalga Scenedesmus Obliquus CWL-1. *Bioresour. Technol.* **2019**, *291*, 121891. [CrossRef]

75. Xie, Y.; Ho, S.-H.; Chen, C.-N.N.; Chen, C.-Y.; Ng, I.-S.; Jing, K.-J.; Chang, J.-S.; Lu, Y. Phototrophic Cultivation of a Thermo-Tolerant Desmodesmus Sp. for Lutein Production: Effects of Nitrate Concentration, Light Intensity and Fed-Batch Operation. *Bioresour. Technol.* **2013**, *144*, 435–444. [CrossRef]

76. Ahmed, N.R.; Manirafasha, E.; Pan, X.; Chen, B.-Y.; Lu, Y.; Jing, K. Exploring Biostimulation of Plant Hormones and Nitrate Supplement to Effectively Enhance Biomass Growth and Lutein Production with Thermo-Tolerant Desmodesmus Sp. F51. *Bioresour. Technol.* **2019**, *291*, 121883. [CrossRef]

77. Ma, R.; Zhao, X.; Ho, S.-H.; Shi, X.; Liu, L.; Xie, Y.; Chen, J.; Lu, Y. Co-Production of Lutein and Fatty Acid in Microalga Chlamydomonas Sp. JSC4 in Response to Different Temperatures with Gene Expression Profiles. *Algal Res.* **2020**, *47*, 101821. [CrossRef]

78. Sánchez, J.F.; Fernández-Sevilla, J.M.; Acién, F.G.; Cerón, M.C.; Pérez-Parra, J.; Molina-Grima, E. Biomass and Lutein Productivity of Scenedesmus Almeriensis: Influence of Irradiance, Dilution Rate and Temperature. *Appl. Microbiol. Biotechnol.* **2008**, *79*, 719–729. [CrossRef]

79. Sánchez, J.F.; Fernández, J.M.; Acién, F.G.; Rueda, A.; Pérez-Parra, J.; Molina, E. Influence of Culture Conditions on the Productivity and Lutein Content of the New Strain Scenedesmus Almeriensis. *Process Biochem.* **2008**, *43*, 398–405. [CrossRef]

80. Bermejo, E.; Ruiz-Domínguez, M.C.; Cuaresma, M.; Vaquero, I.; Ramos-Merchante, A.; Vega, J.M.; Vílchez, C.; Garbayo, I. Production of Lutein, and Polyunsaturated Fatty Acids by the Acidophilic Eukaryotic Microalga Coccomyxa Onubensis under Abiotic Stress by Salt or Ultraviolet Light. *J. Biosci. Bioeng.* **2018**, *125*, 669–675. [CrossRef]

81. Pérez-López, P.; González-García, S.; Jeffryes, C.; Agathos, S.; McHugh, E.; Walsh, D.; Murray, P.; Moane, S.; Feijoo, G.; Moreira, M. Life Cycle Assessment of the Production of the Red Antioxidant Carotenoid Astaxanthin by Microalgae: From Lab to Pilot Scale. *J. Clean. Prod.* **2014**, *64*, 332–344. [CrossRef]

82. Wang, X.; Zhang, M.-M.; Sun, Z.; Liu, S.-F.; Qin, Z.-H.; Mou, J.-H.; Zhou, Z.-G.; Lin, C.S.K. Sustainable Lipid and Lutein Production from Chlorella Mixotrophic Fermentation by Food Waste Hydrolysate. *J. Hazard. Mater.* **2020**, *400*, 123258. [CrossRef] [PubMed]

83. Ribeiro, J.E.S.; Martini, M.; Altomonte, I.; Salari, F.; Nardoni, S.; Sorce, C.; da Silva, F.L.H.; Andreucci, A. Production of Chlorella Protothecoides Biomass, Chlorophyll and Carotenoids Using the Dairy Industry by-Product Scotta as a Substrate. *Biocatal. Agric. Biotechnol.* **2017**, *11*, 207–213. [CrossRef]

84. Liu, J.; Sun, Z.; Zhong, Y.; Gerken, H.; Huang, J.; Chen, F. Utilization of Cane Molasses towards Cost-Saving Astaxanthin Production by a Chlorella Zofingiensis Mutant. *J. Appl. Phycol.* **2013**, *25*, 1447–1456. [CrossRef]

85. Tran, D.; Doan, N.; Louime, C.; Giordano, M.; Portilla, S. Growth, Antioxidant Capacity and Total Carotene of Dunaliella Salina DCCBC15 in a Low Cost Enriched Natural Seawater Medium. *World J. Microbiol. Biotechnol.* **2014**, *30*, 317–322. [CrossRef]

86. Barba, F.J.; Grimi, N.; Vorobiev, E. New Approaches for the Use of Non-Conventional Cell Disruption Technologies to Extract Potential Food Additives and Nutraceuticals from Microalgae. *Food Eng. Rev.* **2015**, *7*, 45–62. [CrossRef]

87. Gille, A.; Trautmann, A.; Posten, C.; Briviba, K. Bioaccessibility of Carotenoids from Chlorella Vulgaris and Chlamydomonas Reinhardtii. *Int. J. Food Sci. Nutr.* **2016**, *67*, 507–513. [CrossRef]

88. Coleman, B.; Van Poucke, C.; Dewitte, B.; Ruttens, A.; Moerdijk-Poortvliet, T.; Latsos, C.; De Reu, K.; Blommaert, L.; Duquenne, B.; Timmermans, K.; et al. The Potential of Microalgae as Flavoring Agent for Plant-Based Seafood Alternatives. *Futur. Foods* **2022**, *5*, 100139. [CrossRef]

MDPI AG
Grosspeteranlage 5
4052 Basel
Switzerland
Tel.: +41 61 683 77 34

Marine Drugs Editorial Office
E-mail: marinedrugs@mdpi.com
www.mdpi.com/journal/marinedrugs

9 783725 835539